Engineering Materials

This series provides topical information on innovative, structural and functional materials and composites with applications in optical, electrical, mechanical, civil, aeronautical, medical, bio- and nano-engineering. The individual volumes are complete, comprehensive monographs covering the structure, properties, manufacturing process and applications of these materials. This multidisciplinary series is devoted to professionals, students and all those interested in the latest developments in the Materials Science field, that look for a carefully selected collection of high quality review articles on their respective field of expertise.

Indexed at Compendex (2021) and Scopus (2022)

Pankaj Kumar · Vandna Sharma · Vinayak Adimule
Editors

Cholesteric Liquid Crystals

Dielectric and Optical Applications

Springer

Editors
Pankaj Kumar
Centre for Liquid Crystal Research
Chitkara University Institute of Engineering and Technology
Chitkara University
Patiala, Punjab, India

Vinayak Adimule
Department of Chemistry
Angadi Institute of Technology and Management
Belagavi, Karnataka, India

Vandna Sharma
Centre for Liquid Crystal Research
Chitkara University Institute of Engineering and Technology
Chitkara University
Patiala, Punjab, India

Jožef Stefan Institute
Ljubljana, Slovenia

ISSN 1612-1317 ISSN 1868-1212 (electronic)
Engineering Materials
ISBN 978-981-96-3820-8 ISBN 978-981-96-3821-5 (eBook)
https://doi.org/10.1007/978-981-96-3821-5

This Springer imprint is published by the registered company Springer Nature Singapore Pte Ltd.
The registered company address is: 152 Beach Road, #21-01/04 Gateway East, Singapore 189721, Singapore

We dedicate this book to our mentor, guide renowned liquid crystal scientist Dr. Pankaj Kumar who have worked along a side us and contributed so much to our current understanding of the liquid crystal problems and deepen the avenues for routes of advanced innovations. Dr. Pankaj Kumar showered to us prodigious love, affection, support, encouragement constantly and taught us so much in all the interdisciplinary and multidisciplinary areas of liquid crystals. To him we owe profound admiration, encouragement and gratitude.

Preface

The research and applications in the area of liquid crystals (LCs) which represent a fascinating blend of physics, chemistry, and material science, flourishing in depth of opportunities for scientific and industrial advancements. Among the different types of LCs, cholesteric liquid crystals (CLCs) possess their unique helical structures, remarkable optical and mechanical properties. The industrial applications of CLCs find great significance for disseminating cutting-edge information in disciplines such as display technology, sensors, actuators, fluorescence imaging, advanced photonics, and biomedical devices. The book is a detailed guide composed of 13 systematically arranged chapters, aimed at promoting the transformation of scientific and technological applications of CLCs. The book covers a wide range of CLCs' fundamental properties, and cutting-edge advancements in the field of interdisciplinary, multidisciplinary applications. Each chapter explores specific aspects of CLCs, from their molecular structure, physical properties, and related industrial, academic, technological significance. The book provides detailed coverage of broad spectrum of audiences, including research scholars, academicians, and industrialists who seek a deeper understanding of the CLC materials.

The book begins with an introductory chapter "Introduction to Cholesteric Liquid Crystals, Preparation and Device Architectures" outlining the fundamental properties of CLCs and emulsions using chiral and achiral dopants, the architecture of CLCs, as well as other applications such as optical anti-counterfeiting and modifiable optical signals. The chapter insights into fluorescence, key characteristics of CLCs, polymer stabilization, temperature effects, etc. Chapters "Recent Advances in Optical Storage and Display Applications of Cholesteric Liquid Crystals" and "Recent Advances in the Optical Properties of Cholesteric Liquid Crystals" give a comprehensive overview of the role of CLCs in revolutionizing optical technologies such as optical storage and display applications and advancements in optical properties. Chapter "Role of Chiral Dopants, Luminescence Applications of Cholesteric Liquid Crystals" aims to explore the role of chiral dopants in modulating the luminescence behavior of CLCs, with a focus on their applications in displays, sensing, biological imaging, and photonic devices. Chapter "The Chemistry of Synthesis of Cholesteric Liquid Crystals: Their Self-Assembly and Colour Interferences" includes the overview on chemistry involved in creating CLCs, emphasizing the molecular architecture, the function of chiral dopants, and the variables affecting the self-assembly of CLCs. Chapter "Recent Advances in the Smart Window Applications of Cholesteric Liquid Crystals" explores the CLCs advancements in smart windows, effect of heat, solar radiation, incident light, applied electric field, and temperature to make them energy efficient and durable. Chapter "Impact on the Dynamic Responses of Nanoparticles Embedded Cholesteric Liquid Crystal Composites" focuses on the effect of introducing various organic and inorganic nanomaterials with different size, shape, and composition on the mechanism of interaction in CLC mixtures, and their advancements in various applications such as photonics, smart windows, and materials science. Chapter "Tunable Photoluminescence of Nanoparticles Doped Cholesteric Liquid Crystals Composites for Photonic Applications" describes the recent studies carried out on photoluminescence (PL) properties by doping nanoparticles (NPs) to CLC mixtures. Further, the chapter discusses enhancement in intensity and quantum yield, as well as reductions in full width at half maximum, Tuning of PL intensity led to new applications in high-performance luminescent displays, photonic devices, and sensors. In ninth chapter "Electro-Optical and Dielectric Properties of Nanomaterial Dispersed Cholesteric Liquid Crystal Composites", selective reflection-dependent and related applications of CLCs such as CLC photo-switches for display applications, biomedical imaging, and optical communication have been discussed in detail. The chapter also provides information about role of introducing nanomaterials, their detailed properties such as dielectric and electro-optical have been summarized. Chapter "Bistable Cholesteric Liquid Crystals: Electro-optic Applications" introduces the fabrication and working principle of bistable CLCs, their advanced applications, controlling architecture of bistability, and their applications in electro-optical devices. Chapter "Effect of Nanomaterial's Dispersion on the Dynamic Responses of Cholesteric Liquid Crystals" discusses the effect of introducing nanomaterials in CLCs and their advanced applications in the field of dielectric, optical displays, sensors, smart materials, luminescent displays, UV modulators, dye lasers, UV detectors etc. It also covers briefly the performance of nanomaterials dispersed in CLCs and their analysis based on type, size, shape, concentration, and mixing methods. Chapter "Recent Advances on Polymer-Stabilized Cholesteric Liquid Crystal: Development and Applications" describes the role of the helical structure of CLCs, advancements in optical properties such as selective light reflection, circular dichroism, and rotational etc. Further, the chapter also focuses on polymer-stabilized CLCs and their electro-optical responses on thin films, particles, fiber, and droplets containing broadening bandwidth, red and blue tuning, switching the reflection notch, and the thermal response. Chapter "Enhancement in Optical and Dielectric Properties of Quantum Dots Doped Cholesteric Liquid Crystal" discusses the role of quantum dots (QDs) doping in the CLCs, and related applications such as uses of CdSe QDs, PbS QDs, Mn-doped ZnS QDs, CdTe/$CsPbBr_3$, $CuInS_2$ QDs for tuneable photonic devices, biosensors, color filters, and smart windows.

Finally, we would like to acknowledge all the authors for their finest contributions, heartfelt thanks and highest respect to Dr. Pankaj Kumar, Pro-Vice Chancellor Chitkara University, Chitkara University Research and Innovation Network

(CURIN), Rajpura, Punjab, India for guidance, support, encouragement and selfless dedication. We would also like to express sincere thanks to our reviewers, for providing timely and scientific feedback. We would also thank people from Springer Nature, Singapore for their constant support, and timely suggestions. We hope that the book delivers its purpose as the arena for presenting and exchanging ideas of research in the field of advanced dielectric and optical applications of CLCs.

Patiala, India Pankaj Kumar
Patiala, India Vandna Sharma
Belagavi, India Vinayak Adimule

Acknowledgements

I Dr. Vinayak Adimule and Dr. Vandna Sharma extend our heartfelt thanks to Dr. Pankaj Kumar, Pro-Vice Chancellor, Chitkara University, Patiala, Punjab, India for relentless efforts, professional guidance, unending encouragement, and support were instrumental in bringing this book. We would also like to acknowledge the exceptional team at Springer Nature, Singapore for their timely support, encouragement for publishing this book.

Special thanks to the members of my research team, led by Dr. Pankaj Kumar, Dr. Vandna Sharma, Dr. Rangappa Keri, Dr. Santosh Nandi, whose dedication and hard work pursued gathering accurate information have enriched the contents of this book immensely.

We express our profound appreciation to participating premier institutions such as Chitkara University, Punjab, India; Condensed Matter Physics, Jozef Stefan Institute, Ljubljana, Slovenia; Dr. B. R. Ambedkar National Institute of Technology, Jalandhar, Punjab, India; Amity University, Uttar Pradesh, Noida, GNA University, Phagwara, Punjab, India; Guru Nanak College Budhlada, Mansa, Punjab, India; KLE Technological University, Hubli, India, whose unwavering support, understanding and timely support were indispensable for bringing this book.

Lastly, we thank the readers and supporters who have taken keen interest in our work. Your participation, feedback motivates us to explore and writing continuously towards unending field of liquid crystals.

Contents

Editors and Contributors

About the Editors

Dr. Pankaj Kumar has more than 18 years of research experience in the field of liquid crystal (LC) and related technology. Dr. Kumar has worked in DRDO and CSIR (Government of India) sponsored projects as JRF and SRF. From December 2009 to June 2012, also worked as a Post Doctoral Fellow (PDF) at Chonbuk National University, South Korea and a recipient of the MERCK International award at IMID-2011 for the best research of the year in LCD technology. Dr. Kumar is also a recipient of various national best paper awards for the outstanding contribution to LC research and has hands-on experience of LC cell fabrication, alignment techniques, image analysis, dielectric spectroscopy, conoscopy and expertise in use of sophisticated instruments like polarizing microscope, temperature controller and UV-Vis. Currently, Dr. Kumar's group has experimentally investigated the Electro-optical and anchoring responses of LCs doped with nanoscale particles. POM, conoscopy, FESEM and AFM results confirm that by dispersing a controllable concentration of nano scale particles, we can achieve aggregation free highly uniform alignment of LC in the cell without using any alignment layer. Currently, investigator is engaged to study the basic and applied aspects of Nematic/FLC mixtures, normal/reverse mode PDLCs and cholesteric LC light shutter etc. Recently, Dr. Kumar and his group have granted two patents and filed three patents in LC technology. Recently, he was also invited as visiting scientist for collaborative research work at Jozef Stefan Institute, Ljubljana, Slovenia. He has published more than 100 research articles and two books as well as submitted several book chapters in Scopus, Q-ratted journals and Conference proceedings with high impact factor. Dr. Kumar has attended and presented papers in National and International Conferences as a keynote speaker and invited speaker. He is a life member of various national and international societies: Indian Science Congress Association (ISCA), Materials Research Society of India (MRSI), Indian Liquid Crystal Society (ILCS), Institute of Smart Structures and System (ISSS), as well as Former member of many research societies: Korean Information Display Society (KIDS), Society for Information Display (SID), Punjab Academy of Science, India. Presently, he is working as Professor and Pro Vice-Chancellor (Research), Centre for Liquid Crystal Research, Chitkara University Institute of Engineering and Technology, Chitkara University, Rajpura, Patiala-140401, Punjab, India. Ten students along with a DST INSPIRE fellow and one SERB-SRF are awarded Ph.D. degree under the supervision of Dr. Kumar and Six Ph.D. students are currently working under his supervision.

Dr. Vandna Sharma has more than nine years of research experience in the field of liquid crystals and related areas. Dr. Sharma worked as *Post Doctoral Fellow as a Scientific Associate* in the prestigious European Research Council (ERC) Project at the Department of Soft Matter Physics, *Jozef Stefan Institute, Ljubljana, Slovenia (Europe).* During Her Ph.D., Dr. Sharma also has worked as a *DST INSPIRE Fellow* and successfully completed *DST INSPIRE* Fellowship project. Dr. Sharma is also a recipient of various research awards *Young Scientist Award-2019* by Punjab Academy of Sciences, *Dewan Jawahar Lal Nayar Memorial Prize-2015* by Indian Liquid Crystal Society, Several *Best Poster Paper Awards*, International Travel Support (ITS-2018) by DST-SERB to attend 27th International Liquid Crystal Conference (ILCC-2018), Kyoto, Japan. Dr. Sharma has hands on experience of liquid crystal cell fabrication, alignment techniques, image analysis, dielectric spectroscopy and expertise in use of sophisticated instruments like polarizing optical microscopy, temperature controller, impedance analyzer, UV-Visible spectrophotometer. As well as, Dr. Sharma has excellence in handling of the instruments like Upright and Inverted *POM, DSC, UV-VIS Spectrophotometer, Nanoscribe for micro 3D printings, Laser Tweezer, AFM, LCR Meter, Microinjector and Micromanipulator, Digital Storage Oscilloscope, Q-switch nanosecond Laser, STED Optical Setup, He-Ne Laser, Spin Coating Unit*, and lots of other small measuring instruments like UV curing chamber, Frequency generators, Vacuum oven, Lock in Amplifier etc. Currently, we are engaged to study the basic and applied aspects of nematic mixtures, vertically aligned liquid crystals, cholesteric liquid crystal light shutters, normal and reverse mode polymer dispersed liquid crystals, nanomaterials-liquid crystals, blue phases, polymer-stabilized blue phase liquid crystals for display applications and preparation of liquid crystal optical micro-resonators for photonic devices. She is a life member of various national research societies: Indian Science Congress Association (ISCA), Indian Liquid Crystal Society (ILCS) and Math Tech Thinking Foundation (MTTF) as well as former member of American Chemical Society (ACS). She has published more than 35 research articles, and submitted several book chapters in Scopus, Q-rated journals and conference proceedings with high impact factor. Recently, Dr. Sharma was also invited as a visiting scientist for collaborative research work at Jozef Stefan Institute, Ljubljana, Slovenia.

Presently, Dr. Sharma is working as Assistant Professor, Centre for Liquid Crystal Research, Chitkara University Institute of Engineering and Technology, Chitkara University, Rajpura, Patiala-140401, Punjab, India. Four Ph.D. scholars are currently working under her supervision.

Dr. Vinayak Adimule has 15 years of research experience as senior Scientist, and associate research scientist in R&D organizations of TATA (Advinus Therapeutics Ltd.), Astra Zeneca India, Trans Chem Ltd. Expert in the area of Medicinal Chemistry, Material Chemistry, Material electronics, Photocatalysis, Polymer nanocomposites, Nanoelectronics, Actuators, and Sensors. Dr. Adimule was Overseas Research Fellow of Nanyang Global Research Center, Bangkok, Malaysia. He has published more than 150 research articles, and 15 books/several book chapters in Scopus, Q-ratted journals with high impact factor, attended and presented papers at National and International Conferences as Keynote speaker and distinguished speaker. He has published books in Wiley, Trans-Tech, etc., Chaired/Keynote speaker/Organizing secretory, Editor for International and National conferences, Editorial Board Member, and Life Member of many international societies/research institutions. He is editor of the special issues published in Springer, Elsevier, and Wiley. Dr. Adimule is a recognized research guide at Visvesvarayya Technological University, Belagavi, Karnataka, India.

Presently, he is working as Professor and Dean of Research at Angadi Institute of Technology and Management, Belagavi, Karnataka, India. Scientific Advisor for various government and research organizations, Industries, and R&D organizations. He is supervising four research scholars for Ph.D. Research interests include Nano Electronics, Liquid Crystals, Liquid crystal Nanostructures, Optoelectronics, Material Chemistry, Sensors and Actuators, Bio Nanomaterials, and Medicinal Chemistry.

Contributors

Vinayak Adimule Department of Chemistry, Angadi Institute of Technology and Management (AITM), Belagavi, Karnataka, India

Gangadhar Bagihalli Engineering Chemistry, KLE Institute of Technology, Airport Road, Gokul, Hubballi, Karnataka, India

Raghavendra P. Bakale Department of Chemistry, Jain College of Engineering, Belagavi, Karnataka, India

Praveen Barmavatu Department of Mechanical Engineering, Faculty of Engineering, Universidad Tecnológica Metropolitana, Santiago, Chile

Riccardo Castagna CNR, Institute of Heritage Science, Sesto Fiorentino, FI, Italy;
URT-CNR, Universit'a Di Camerino (UNICAM), Camerino, MC, Italy

Garima Chauhan Institute of Foreign Trade and Management, Muradnagar, India

Madhavaprasad Dasari Department of Physics, GITAM Institute of Science, GITAM (Deemed to be University), Visakhapatnam, India

Ridhima Gahrotra Department of Applied Sciences, Chitkara University Institute of Engineering and Technology, Chitkara University, Punjab, India

Arvind K. Gathania Department of Physics and Photonic Science, National Institute of Technology, Hamirpur, Himachal Pradesh, India

Bojappa Ginni VTU Recognized Research Centre in Chemistry, KLE Institute of Technology, Hubballi, Karnataka, India

Savita Hanaji Department of Engineering Science and Humanities, KLE Technological University, Dr. M. S. Sheshgiri Campus, Udyambag, Belagavi, Karnataka, India

Sushant Kakati Department of Engineering Physics, KLE Institute of Technology, Airport Road, Gokul, Hubballi, Karnataka, India

Manpreet Kaur Dr. B R Ambedkar National Institute of Technology, Jalandhar, Punjab, India

Ramneek Kaur P. G. Department of Physics, Mata Gujri College, Fatehgarh Sahib, India

Rangappa Keri Centre for Nano and Material Sciences (CNMS), Jain (Deemed-to-be University), Jain Global Campus, Ramanagaram, Kanakapura, Bangalore, Karnataka, India

Khushboo Dr. B R Ambedkar National Institute of Technology, Jalandhar, Punjab, India

Ajay Kumar Department of Physics, Dharma Samaj College (Raja Mahendra Pratap Singh State University, Aligarh), Aligarh, Uttar Pradesh, India

Pankaj Kumar Centre for Liquid Crystal Research, Chitkara University Institute of Engineering and Technology, Chitkara University, Punjab, India

Rishi Kumar Department of Physics, Guru Nanak College Budhlada, Mansa, Punjab, India

Sandeep Kumar Raman Research Institute, Sadashivanagar, Bengaluru, India; Department of Chemistry, Nitte Meenakshi Institute of Technology, Yelahanka, Bengaluru, India

Praveen Malik Dr. B R Ambedkar National Institute of Technology, Jalandhar, Punjab, India

S. N. Mathad Department of Engineering Physics, KLE Institute of Technology, Gokul, Hubballi, Karnataka, India

Santosh Nandi Department of Chemistry, KLE Technological University, Dr. M. S. Sheshgiri Campus, Udyambag, Belagavi, Karnataka, India

S. S. Nesargi Chemistry Section, Department of Chemistry, KLE Technological University, Dr. M. S. Sheshgiri Campus, Udyambag, Belagavi, Karnataka, India

Pooja Centre for Liquid Crystal Research, Chitkara University Institute of Engineering and Technology, Chitkara University, Punjab, India

Jai Prakash Department of Physics, Aligarh Muslim University, Aligarh, India

P. Priscilla Department of Applied Physics, Amity Institute of Applied Sciences, AmityUniversity Uttar Pradesh, Noida, India

Neeraj Puri School of Natural Sciences, GNA University, Phagwara, Punjab, India

K. K. Raina M. S. Ramaiah University of Applied Sciences, Bangalore, India

Amid Ranjkesh Department of Condensed Matter Physics, Jožef Stefan Institute, Ljubljana, Slovenia

Sheela A. Sangam Department of Chemistry, K.L.S Gogte Institute of Technology, Belagavi, Karnataka, India

Harsh Sharma Dr. B R Ambedkar National Institute of Technology, Jalandhar, Punjab, India

Ronak Kumar Sharma Dr. B R Ambedkar National Institute of Technology, Jalandhar, Punjab, India

Vandna Sharma Centre for Liquid Crystal Research, Chitkara University Institute of Engineering and Technology, Chitkara University, Punjab, India

Gautam Singh Department of Applied Physics, Amity Institute of Applied Sciences, AmityUniversity Uttar Pradesh, Noida, India

Supreet Department of Physics, Amity School of Applied Sciences, Amity University Haryana, Gurgaon, India

Mahendran Vellaichamy Department of Condensed Matter Physics, Jožef Stefan Institute, Ljubljana, Slovenia

Abbreviations, Symbols and Meaning

2D	Two-dimensional
5PCH	4-trans-pentyl cyclohexyl cyanobenzene
AC	Alternating current
AR	Augmented reality
ATO	Antimony doped tin oxide
AuNPs	Gold nanoparticles
BP	Blue Phases
c	Concentration of the chiral components
CCD	Charge-coupled device
CD	Circular dichroism
CdSe	Cadmium selenide
CLC	Cholesteric liquid crystal
CLCEs	Chiral liquid crystal elastomers
CLCP	Cholesteric liquid crystal polymer
CNTs	Carbon nanotubes
COC	Cholesteryl oleyl carbonate
CP	Circularly polarized
CPL	Circularly polarized luminescence
CQDs	Carbon quantum dots
CR	Contrast ratio
CuS	Copper sulfide
DC	Direct current
DFT	Density functional theory
DHM	Digital holographic microscopy
DS	Dynamic scattering
DSC	Differential scanning calorimetry
EC	Electrochromic
EHD	Electro-hydrodynamic
EISA	Evaporation-induced self-assembly
ERLC	Electrical-responsive liquid crystal
ET	Energy transfer

f-AgNPs	Functionalized silver nanoparticles
FC	Focal conic
FNPs	Ferroelectric nanoparticles
FRET	Förster resonance energy transfer
FTIR	Fourier transform infrared spectroscopy
FWHM	Full width at half maxima
GQDs	Graphene quantum dots
GRIN	Gradient-index
HABA	(2-(4-hydroxyphenylazo)benzoic) acid
HMDs	Head-mounted displays
HOMO	Highest occupied molecular orbital
HPC	Hydroxypropyl cellulose
H-state	Homeotropic state
HTP	Helical twisting power
InP/ZnS	Indium phosphide/zinc sulfide
IR	Infrared
ITO	Indium tin oxide
LC	Liquid crystal
LCDs	Liquid crystal displays
LCMs	Liquid crystal monomers
L-CP	Left-circularly polarized
LCPNs	Liquid crystal polymer networks
LEDs	Light emitting diodes
LH	Left handed
LUMO	Lowest occupied molecular orbital
LWGP	Light waveguide plate
n_e	Extra-ordinary refractive index
NiNBs	Nickel nano bowls
NLC	Nematic liquid crystal
n_o	Ordinary refractive index
NPs	Nanoparticles
NYCL	4′-(1,3-dimethyl-3-chloro)propoxy-4-cyanobiphenyl
OC	Optical components
ODA	Octadecylamine
P	Pitch
PAL	Photo alignment layer
PB	Pancharatnam–berry
PBG	Photonic band gap
PCMs	Phase change materials
PD&SLC	Polymer dispersed and polymer stabilized liquid crystal
PDLC	Polymer dispersed liquid crystal
PIPS	Polymerization-induced phase separation
PL	Photoluminescence
PMMA	Poly methyl methacrylate
POM	Polarized optical microscope

POSS	Polyhedral oligomeric silsesquioxane
PRLC	Photo-responsive liquid crystal
PSCLCs	Polymer-stabilized cholesteric liquid crystals
PSCT	Polymer-stabilized cholesteric texture
PSLC	Polymer-stabilized liquid crystal
PSNLCs	Polymer-stabilized nematic liquid crystals
P-state	Planar state
PVA	Poly(vinyl alcohol)
QDs	Quantum dots
R	Reflection
R-CP	Right-circularly polarized
RET	Resonance energy transfer
R-H	Relative-humidity
RH	Right handed
S	Order parameter
SPR	Surface plasmon resonance
SR	Stabilization rate
SWCNTs	Single-walled carbon nanotubes
T	Transmission
TC	Thermochromic
TDDFT	Time-dependent density functional theory
TEM	Transmission electron microscopy
TiO_2	Titanium oxide
TLTC	Thermally light-transmittance controlled
T_{max}	Maximum transmittance
T_{min}	Minimum transmittance
T_{opaque}	Transmittance in opaque state
TOPO	Trioctylphosphine oxide
$T_{transparent}$	Transmittance in transparent state
UV	Ultraviolet
VR	Virtual reality
V_{th}	Threshold voltage
XRD	X-ray diffraction
ZnO	Zinc oxide
α	Distribution parameter coefficient
Δn	Birefringence
$\Delta\varepsilon'$	Dielectric anisotropy
$\Delta\lambda$	Reflection band
ε	Dielectric anisotropy
ε'	Real part of permittivity
ε''	Imaginary part of permittivity
ε_∞	High frequency dielectric constant
ε_s	Low frequency dielectric constant
θ	Angle between the director and the molecular axis
λ	Bragg's reflection wavelength

σ_{ac}	AC conductivity
τ_o	Average relaxation time
ω	Angular frequency

Introduction to Cholesteric Liquid Crystals, Preparation and Device Architectures

Vinayak Adimule, Rangappa Keri, and Santosh Nandi

Abstract Cholesteric liquid crystals (CLC) have recently attracted enormous interest in the development of soft materials due to their unique ability to change their refractive index, tuning of optical density states, selective and specific diffraction wavelengths, formation of supramolecular helical architecture, self-organized pitch gradient with nanostructures, non-uniform distribution of pitch gradient etc. present chapter focuses on the physical and optical properties of CLC with broad refection characteristics, formation of CLC with chiral and achiral dopants, their florescence characteristics, broad range of optical applications, wide range of device architecture used in the modulation of optical active intelligent materials etc. The chapter also illustrates the formation of CLC emulsions their basic device architecture, illustration of the optical microscopy, and in the applications of optical anti-counterfeiting, modifiable optical signals, response to the external circuit etc. The chapter also covers the synthesis of CLC, their basic physical and chemical characteristics, fluorescent nature, polymer stabilized CLC, flexibility in the molecular design, temperature responsive CLC, their basic optical device architecture. The chapter also enlighten briefly about the CLC integrated system used in the various optical and electrical applications.

Keywords Basic principles · Fabrication · Chiral molecules · Fluorescence · Device architecture · Optical applications · Optical counterfeiting

---

V. Adimule
Department of Chemistry, Angadi Institute of Technology and Management (AITM), Savagaon Road, Belagavi, Karnataka 590009, India

R. Keri (✉)
Centre for Nano and Material Sciences (CNMS), Jain (Deemed-to-be University), Jain Global Campus, Ramanagaram, Kanakapura, Bangalore, Karnataka 562112, India
e-mail: sk.rangappa@jainuniversity.ac.in

S. Nandi (✉)
Department of Chemistry, KLE Technological University, Dr. M. S. Sheshgiri Campus, Udyambhag, Belagavi, Karnataka 590008, India
e-mail: ssnandi@klescet.ac.in

P. Kumar et al. (eds.), *Cholesteric Liquid Crystals*, Engineering Materials,
https://doi.org/10.1007/978-981-96-3821-5_1

1 Introduction

Cholesteric liquid crystals received wide interest in recent decades due to their ability to form supramolecular helical self-organized architecture having various optical and electrical properties. The first discovery of CLC was made with the benzoate compounds [1, 2]. In contrast with the nematic liquid crystal (NLC) phase which are purely oriented towards the elongated order, CLC exhibit helical structure with twist axis perpendicular to the local directions. The molecular chirality thus exists in CLC as layered structures and the orientational plane perpendicular to the axis rotates with constant angle. They are characterized by pitch (p) which is the distance along the helical axis with orientation of 360°. The handedness of the helical structure of CLC can be determined from the twist sense measurements and strongly dependent on the configuration of the chiral dopants in CLC. The CLC can be prepared by two general methods. One is synthesis of asymmetric structures by the introduction of chiral carbon atoms [3, 4]. In this case P of the CLC inherently dependent on the nature of the chiral groups in chiral dopants and the other one is addition of chiral optically active compounds on NLC phase provided miscibility conditions are satisfied. In this case P of the CLC dependent on the content of the chiral dopants. Due to unique helical structures of CLC they exhibit distinct properties such as selective reflection [5–8], circular dichroism [9], optical rotation [10], electro-optical properties [11] etc. The CLC therefore can find applications in the field of optical data storage [12], polarization independent devices [13], polarizer free optical reflective devices [14], pressure and temperature sensors [15, 16],

In recent years various temperature responsive devices were developed using CLC for the application of smart windows, energy saving actuators, smart labels, temperature sensors etc. the polymer dispersed liquid crystal (PDLC), polymer stabilized liquid crystal (PSLC) and CLC were used as small mesogens for the construction of optically tuneable devices. The thermochromic nature of the CLC and their optical anti-profiting properties demonstrate the potential tunability of the CLC with the optical and optoelectronic applications of CLC [17–20]. The typical helical form of CLC as shown in Fig. 1. The reflection of the circularly polarized light by the CLC displayed in Fig. 2.

2 Selective Light Reflection and Mechanism of Broad Band Reflection

Under normal irradiation of white light on the CLC show varieties of colours and the color varies with the incident angle. The selectivity of the reflection depends on Braggs law. Periodic change in the helical structure of the CLC change their optical reflectivity upon the incident light along the direction of the wavelength. The band width directly relates to the average of ordinary and extraordinary wavelengths. The band width is independent of the Δn and p. The smaller the value of Δn and p, the

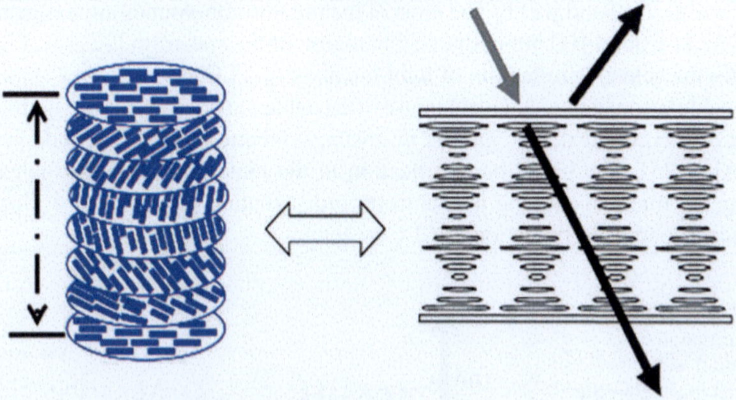

Fig. 1 Typical Helical form of CLC (Reproduced with permission from reference [21])

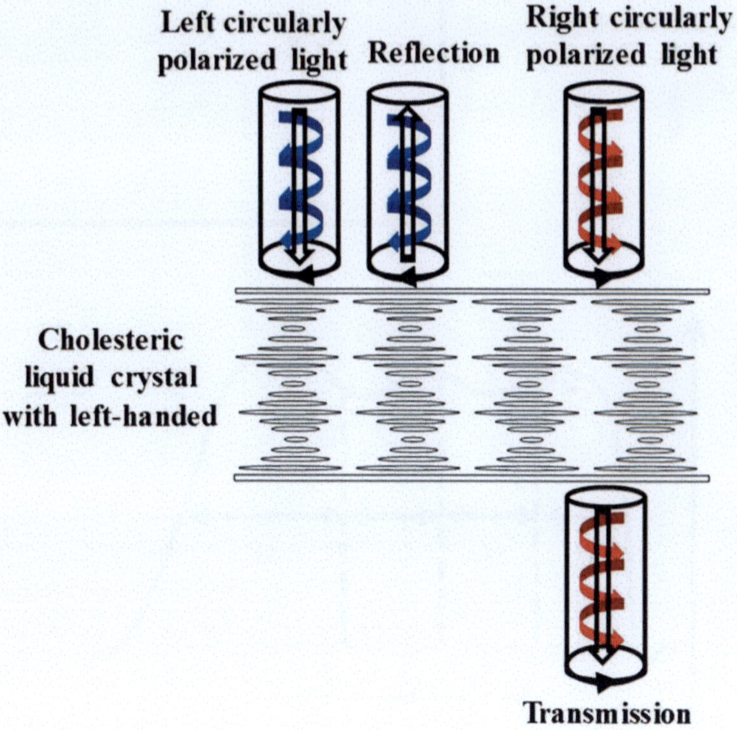

Fig. 2 Illustration of reflection of circularly polarized light of CLC (Reproduced with permission from reference [21])

smaller will be the band width. The most of the organic com*p*ounds are n is generally 0.3 and $\Delta\lambda$ is around 100 nm in the visible region of the spectrum [22–24]. Figure 3 illustrates the selective reflection of light intensity of CLC.

The modulation of the $\Delta\lambda$ with increase in the value of Δn will become significantly difficult due to colour defect, greater viscosity, synthetic complexity, and synthetic complexity [25]. The broad band reflection in the materials can be visualized by selecting the materials having non-uniform *p* distribution [26]. Figure 4 illustrates the typical broad band reflection in CLC systems.

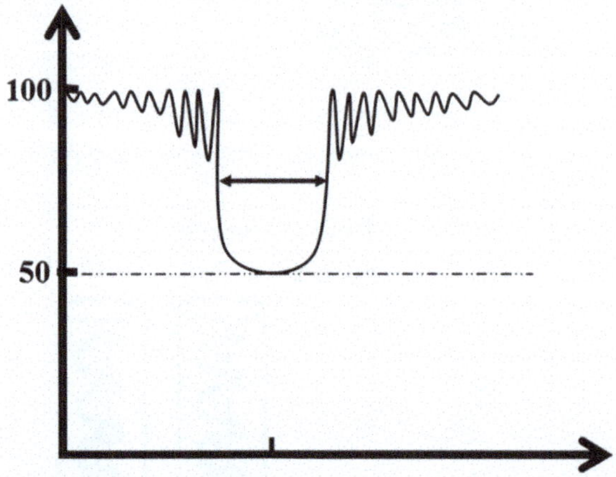

Fig. 3 Typical illustration of light reflection in CLC (Reproduced with permission from reference [21])

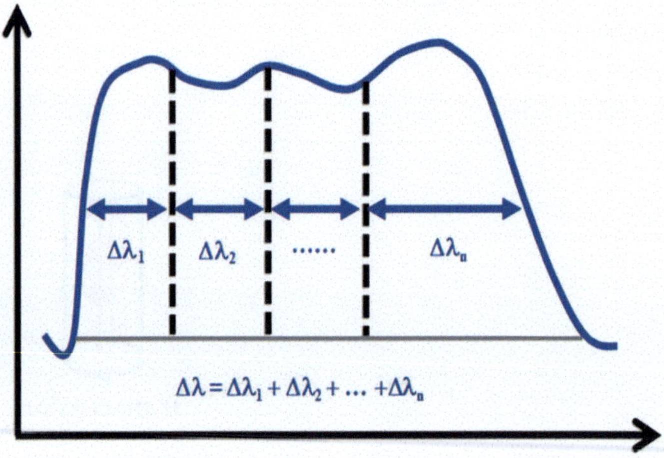

Fig. 4 Typical illustration of the broadening of bandwidth of CLC (Reproduced with permission form reference [21])

2.1 *Methods to Modify Broad Band Reflection and Thermal Diffusion Induced Broad Band Reflection*

The demand in the application related to colour display proposes the selective light reflection band width in CLC with narrow p value. The attempts to increase the broad band width of the reflection by non-uniform distribution of the p value. The phenomenon of mutual diffusion of the interface resulted from the stacking the CLC layers with the addition of Li-ion. The stacked right-handed CLC reflect selectively red, blue, green radiations [27–30].

By the incorporation of CLC layers and non-uniform distribution of the P, Wu et al. synthesized broad band circularly polarized light in the wavelength from 400 to 736 nm [31] by the thermal dispersion of CLC polymer. Transmission spectra of CLC and the polymer films shown in Fig. 5. Mitov et al. synthesized thin film having band width of 300 nm covering the entire range of visible region of the spectrum [32]. The results showed that appropriate p broadening depends on the heat treatment time and the processing time. Figure 6 illustrates the mechanism of thermal diffusion of CLC with p gradient [33].

3 Typical Preparation of Cholesteric Liquid Crystal (CLC) and Their Uniqueness

Calamitic liquid crystals (LC) commonly consist of rigid mesogen and flexible spacers. The core (rigid mesogens) consists of aromatic/hexagonal ring systems. The core of the CLC liked with the soft tails with crystal-like structure and liquid like rheology as shown in Fig. 7. The introduction of achiral NLC phase the chiral dopants converts them into CLC [34]. The CLC is similar to chiral NLC phase with slight difference in the rotational off sets of anisotropic mesogen units. The introduction of chiral dopants brings helical rotations either in left or right-handed. This rotation depends on the stereochemistry of the chiral dopants. For the unpolarized light maximum of 50% will be reflected. The simplified Bragg's equation can be defined as in Eq. (1) [35],

$$\lambda = p \times \mathrm{n}_{\mathrm{avg}} \tag{1}$$

where, λ—Bragg's reflection wavelength, p—is defined as the distance over the full (2π) rotations of the CLC.

Helical pitch of the CLC is determined from the helical twisting power (HTP). The HTP can be defined as the measure of the compelling planes of the CLC to the depth of the layer.

The pitch (p) can be given by the Eq. (2) [36],

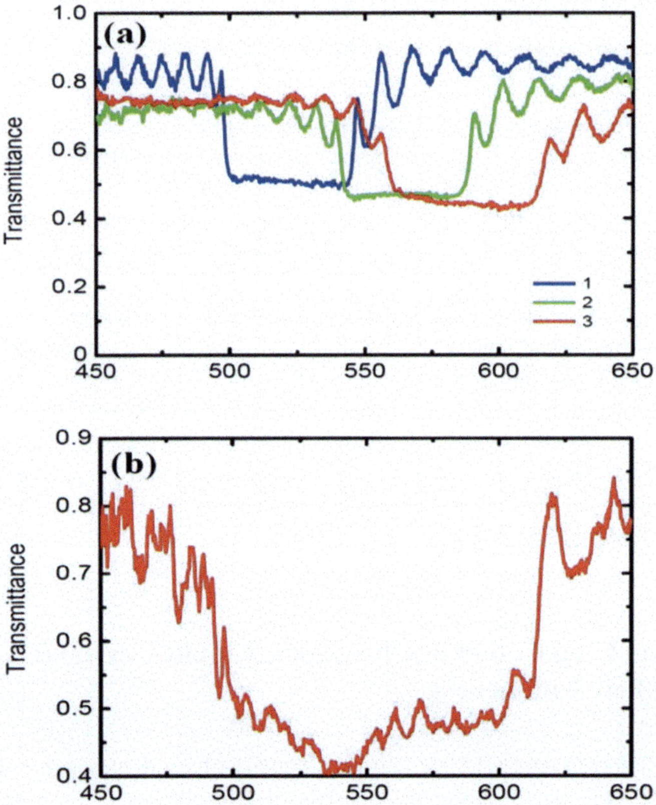

Fig. 5 Transmission spectra of **a** three individual CLC and **b** stacked CLC polymer films (Reproduced with permission form reference [21])

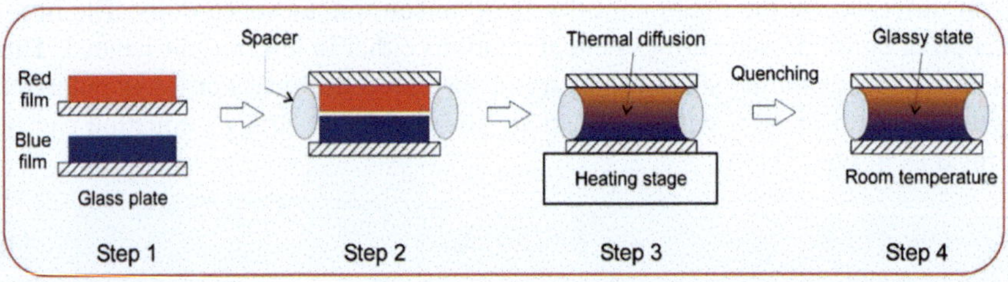

Fig. 6 Typical illustration of the thermal diffusion with P gradient (Reproduced with permission form reference [21])

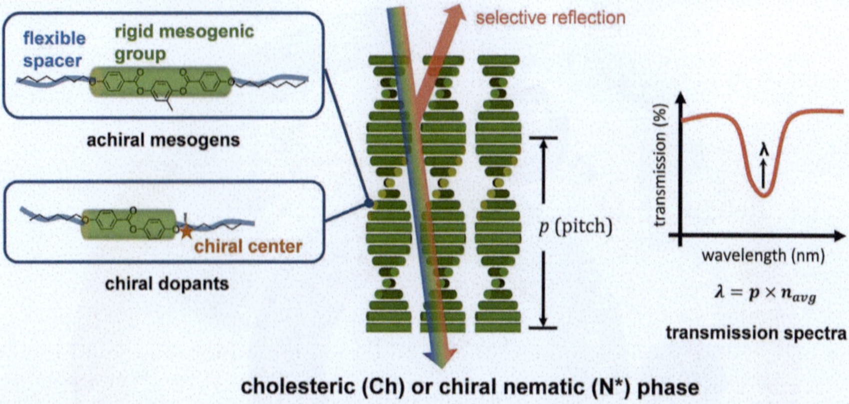

Fig. 7 Schematic illustration of CLC, pitch, introduction of achiral/chiral dopants and their transmission spectra (Reproduced with permission from reference [38]). Open access Creative Common Attributions 4.0

$$p = \frac{1}{\mathrm{HTP} \times \mathrm{c} \times \mathrm{cee}} \tag{2}$$

where, c—concentration of the chiral components in CLC and ee—enantiomeric excess (commonly taken as 1). The positive sign indicates handedness induced by the HTP towards right side and negative sign indicates handedness induced by the HTP is on left side. The ordinary (n_0) and extraordinary reflective indices and the width of the reflection band (Δλ) related by the Eq. (3) [37].

$$\Delta\lambda = p \times \Delta \mathrm{n} = p \times (\mathrm{n_e} - \mathrm{n_0}) \tag{3}$$

The CLC undergo self-assembled structures and the colours of the CLC can be achieved without the phase separation of the materials. Due to the liquid like behavior of the CLC, can be easily visualized by layered structures and makes it more efficient in fabrication of various devices such as cells, coatings, 3D objects, films etc.

3.1 Typical Fabrication of CLC Emulsions and Their Device Architecture

There is increasing interest towards the incorporation of CLC in the interface of liquid phase structures having spherical confinements. Especially when CLC dispersed in water and in presence of surfactants. Such systems give rise to innovative way of formation of CLC intricate structures and high responsive soft materials. They find robust applications in the field of sensing technology, soft robotics, photonics, etc.

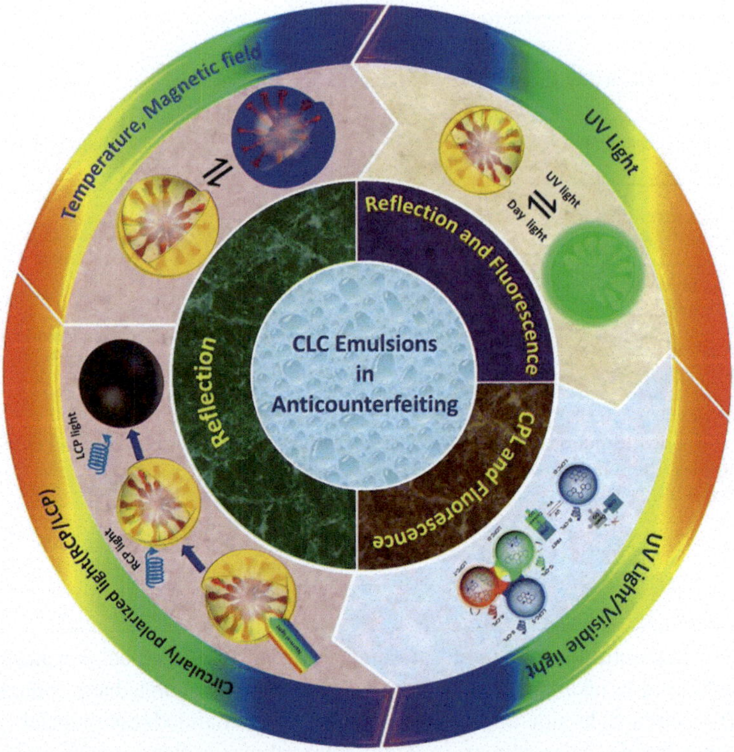

Fig. 8 Schematic illustration of optical anticounterfeiting using CLC emulsions (Reproduced with permission form reference [54]). Open access Creative Common Attributions 4.0

[39–41]. Many literatures illustrate the application and mechanism of anticounterfeiting [42–44] using chromic materials, luminescent and photoluminescent materials. The CLC emulsion systems were also reported for the photonic anticounterfeiting [45]. Such emulsions of CLC generate colour encryptions and find diversified photonic applications. The anticounterfeiting involves the process of encrypting and decryption of the sensitive information. Figure 8 illustrates the comprehensive visual representation of the CLC anticounterfeiting technologies, there future advancements and application towards luminescent, information security devices.

3.2 *Structural Arrangement of CLC*

When a chiral molecule is introduced into the achiral NLC, cholesteric mesogens are created having distinct physical properties. The chirality to the NLC involves the formation of liquid crystal photonic materials having specific Braggs reflections. The formation of helical structure ensures the brags reflectors to align along with the reflective index with the incident light. CLC behave as photonic nanostructures due to their periodicity of the detector directions. The modification in the *p* results in structural variation of CLC which makes different colours [46, 47]. Figure 9 illustrates the formation of CLC from various types of LC, their structural orientations, director orientation towards the incident light. Using CLC it is possible to design the structure having multimode reflecting colours, fluorescence, circularly polarized luminescence (CPL) for obtaining high level of anticounterfeiting properties. Using CPL, CLC can be used to fabricate and form various CPL-CLC systems. They are mainly of two types. First is optical rotation and the second is selective reflection of CPL [14, 14].

During the preparation of CLC, commonly binapthaly derivatives and cholesterol derivatives used as chiral dopants [48]. Various strategies are used to derive CLC, such as grafting chiral/fluorescent materials, doping fluorescent /luminescent molecules to NLC etc. The doping fluorescent/luminescent molecules to NLC, which can combine chirality and fluorescence is the unique way to construct emulsive LC molecules [51, 52]. This kind of fluorescent/luminescent chiral molecules offers several advantages to the CLC systems. Most often stimuli responsive molecules, especially light responsive molecules offer greater advantages. Molecules of the class spiropyran, azobenzene, diary ethylene was commonly used. Over a period of time,

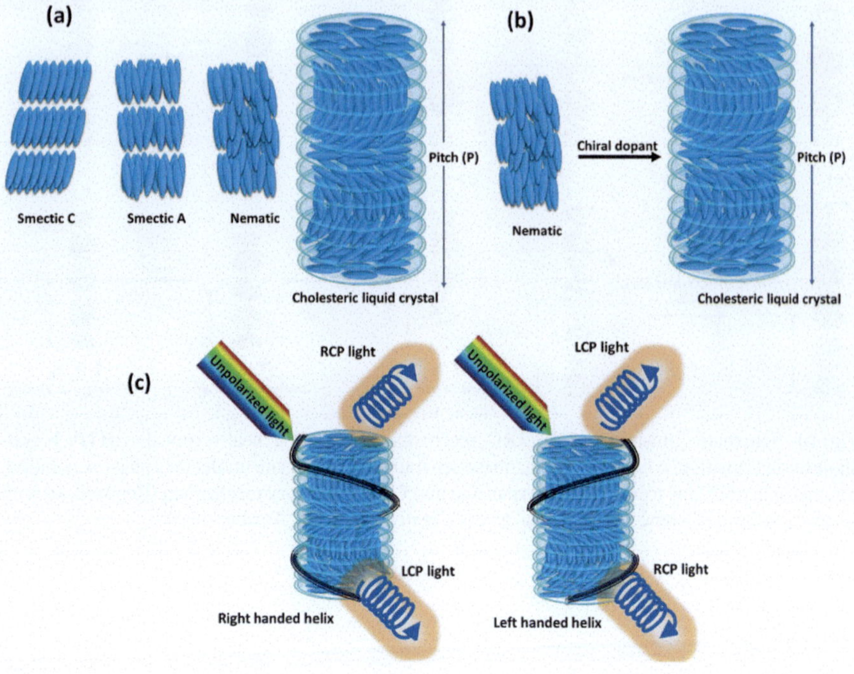

Fig. 9 **a** Schematic illustration of the arrangement of CLC molecules. **b** Helical arrangement of CLC after doping chiral molecules. **c** illustration of director orientation in right/left-handed in helical structure of CLC (Reproduced with permission form reference [54]). Open access Creative Common Attributions 4.0

the stimuli responsive materials dynamically change their colour due to external heat, light radiation, mechanical stress and electromagnetic radiation etc. Wide varieties of orientational structures were created with the CLC. Varieties of smart materials such as chemical and biological sensors, programmable laser, reflective colour coatings etc. using CLC droplets [53]. Thus, formed structures were elucidated by CLC helical pitch, size and shape of the CLC droplets, elastic constants, size ration of the droplets etc. (Fig. 10).

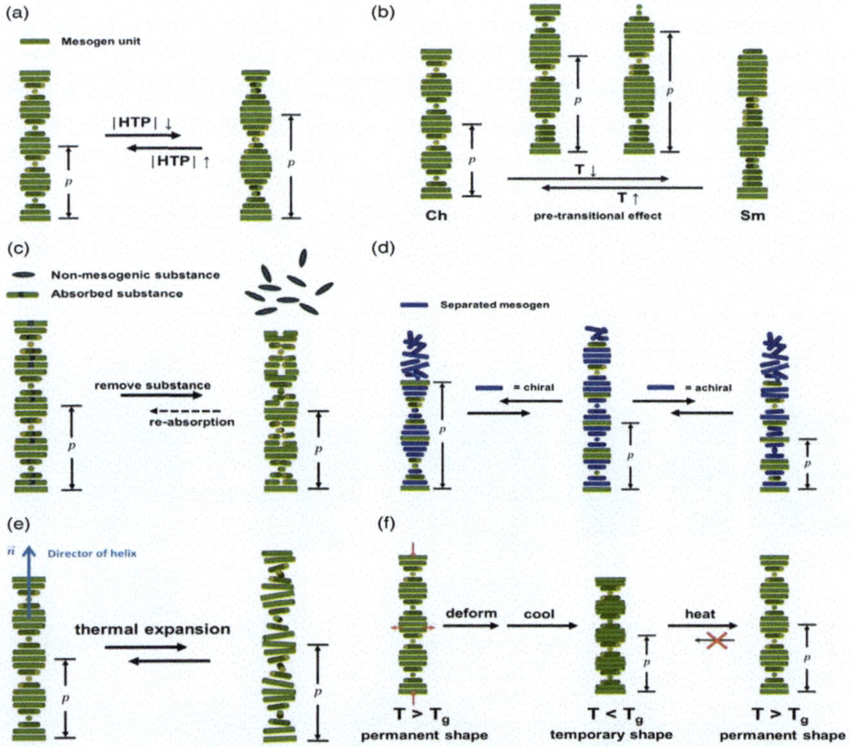

Fig. 10 Schematic illustration of TRRC mechanism in CLC **a** heat-responsive HTP, **b** Sm-cholesteric transition, **c** non mesogenic phase separation, **d** mesogenic molecules phase separation, **e** increase in pitch due to anisotropic expansion and **f** shape memory mechanism (Reproduced with permission form reference [38]). Open access Creative Common Attributions 4.0

4 Temperature Responsive CLC Systems and Their Preparation/Alignment

A number of processes are involved in the preparation of CLC systems using chiral dopants. The change in the colour and reflective indices with the change in the temperature can be achieved via various mechanisms. The pitch (*p*) will be the determining factor for the change in the reflective index and colour of the CLC systems [54].

4.1 Isotrpization

The temperature responsive reflective colours were induced to NLC molecules during nematic to isotropic transitions. During this change, during this process, the structural colour disappears. Isotrpization will be present in almost all CLC heated system. The temperature4 responses will be initiated during reversible isotropization [55].

4.2 Chiral Dopants HTP (Heat Responsive) During Preparation of CLC

The preparation of CLC requires chiral dopant molecules having chiral centre. Individual chiral molecules will have different HTP values and HTP values of chiral dopants are temperature dependent. The winding/unwinding of the helices in the CLC causes change in the *p* value during heating/cooling process. The process is independent of the thickness of the sample but dependent on the P of the thickness of the layers in CLC [56].

4.3 Transitions from Smectic to CLC

Smectic form is an ordered structures and some CLC have smectic structure at lower temperature. Exceptional reflections achieved when P of the unwinding helices increases to infinite length with decrease in temperature. This temperature range is relatively small and the transition from smectic to CLC is a gradual growth of smectic over CLC at increasing *p* value [57]. With decrease in temperature, reflection is redshifted and is attributed to the pre transitional helix unwinding.

4.4 Phase Separation

The incorporated substances like water, chiral dopants, and other volatile components from the CLC system alters the *p* value and decrease in *p* value is due to the removal of water and other volatile components from CLC. However, removal of the chiral dopant increases the *p* value depending upon whether the dopant is reabsorbed by the CLC. The electron rich, high oxygen concentration and defect centres present in the inorganic oxides nanomaterials [58, 59] also can be used to separate the phases present in the CLC systems. The components can be removed by varieties of processes such as precipitation, evaporation, phase separation etc. [60].

4.5 Thermal Expansion

For Planar CLC polymer, anisotropic expansion results along the direction of the helical axis. During this process P increases and spectra is redshifted. The anisotropic expansion also depends on the decoration of nanostructured materials on the surface of the CLC [61–63].

5 Architecture of CLC Deswelling Mechanism (Polymer Dispersed LC (PDLC) Fabrication)

Emulsification method employed for the dispersion of CLC into PDLC system using aqueous solutions. Due to the immiscibility of the CLC in aqueous solution, CLC introduced in the form of a droplets. After coating process on the surface of the substrate, it is kept for drying. The CLC on the surface of the substrate converted into spherical form due to deswelling mechanism resulting from the evaporation of the water molecules. The compression of the CLC occurs on the surface of the substrate and planar helical orientation results from the polymer. Figure 11 illustrate the deswelling mechanism of the fabrication of the CLC into PDLC system. The reflective CLC coatings formed with different sized droplets of CLC and depend on the rate of mixing of emulsification [64]. Poly vinyl alcohol commonly used for the preparation of CLC droplets during emulsifications. Additional methods developed and incorporated for the PDLC system comprising of CLC. Emulsion based method commonly incorporated for the PDLC CLC system. The transparency of the PDLC system retarded as a result of increase in the size of the droplets during the emulsification of PDLC-CLC system.

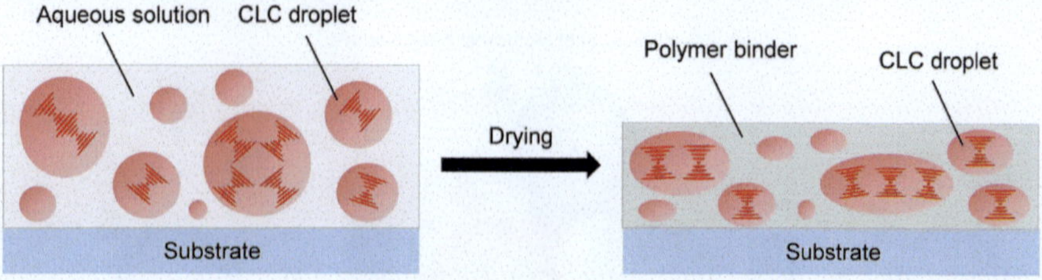

Fig. 11 Schematic illustration of the deswelling mechanism of CLC alignment during drying process (Reproduced with permission form reference [38]). Open access Creative Common Attributions 4.0

5.1 *Fluorescence Applications of CLC, Typical Construction and Their Device Architecture*

The luminescent application of the CLC comprises of a number of luminescent dissymmetry factors [65]. This property is attributed to the helical structures of the CLC and the nature of the CLC. The recent advancements include by making use of the thermosensitive CLC and creating laser array by manipulating the wavelength. Based on the polarization and wavelength of the encrypted radiation of the laser array, the information can be stored. Figure 12 illustrates the Förster resonance energy transfer (FRET) driven information storage system in CLC based on the circularly polarized light (CPL) as well as the fluorescence, response time and the encryption of the laser radiation. Multidimensional devices were developed with selective CPL having multi responsive system, chirality, intensity, and wavelength for information encoding process [66]. Several nanostructured materials [67–70] can be used for the development of colours, change in the *p* and HTP values of CLC. The excellent processing performance and fluorescence properties of phototunable CLC is highly desirable. Using solids Li et al. developed phototunable CLC capsules through interfacial polarization and adjustable CPL luminescence materials. The such materials were characterized by enhance fluorescence, full colour emission, time responsive characteristics, and CPL for effective usage in the information encoding [71].

A six pairs of LCPC films were utilized to create high-security bar codes from initial 1D blue fluorescent codes to 4D full-color CPL codes with increasing irradiation time. The schematic process is illustrated in the encryption and decryption of 4D information (Fig. 12).

5.2 *Fabrication of Photonic Capsule by Dye with CLC (Microencapsulation of CLC in PDLC Systems)*

Microencapsulation of CLC with uniform distribution of angularly polarized results into the development of thermos responsive devices. The materials are incorporated

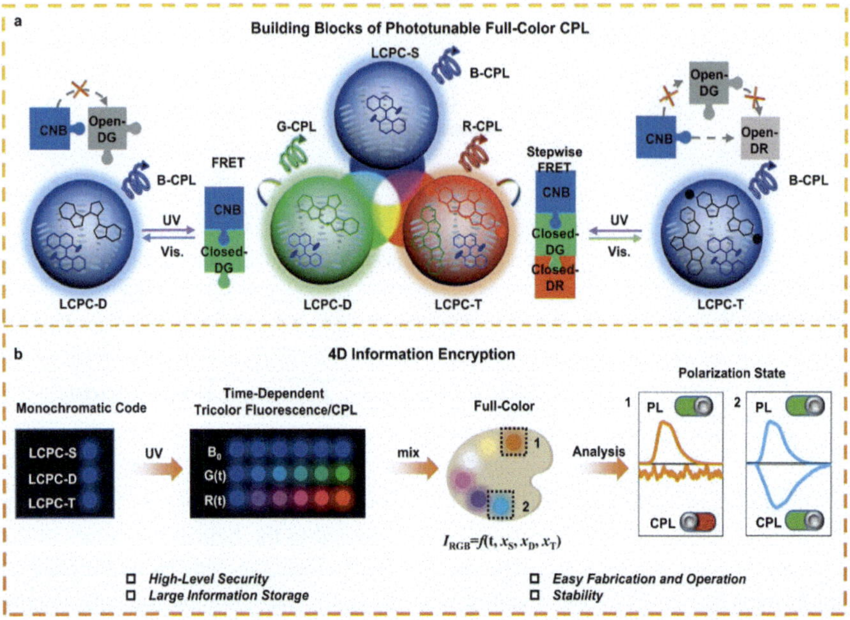

Fig. 12 **a** Light-driven FRET process involved in CLC building blocks for phototunable CPL, **b** 4D design of fluorescence encryption consisting of fluorescence, time response and CPL characteristics (Reproduced with permission form reference [54]). Open access Creative Common Attributions 4.0

into the capsule made off silicone methacrylate (photocured). Figure 13 illustrates the typical fabrication of the photocured CLC with the alignment laser. The planar anchoring conditions were achieved by using polyvinyl alcohol and remnant glycerol from the emulsion solution along the helical axis extending to radical direction [72]. The CLC are preserved which shows red to green colour shift at nearer to room temperature. Fluorescence dye was used in CLC systems which facilitates the omni axial lasing and they find applications in the field of image processing.

6 Architecture of Fluorescent Additives to CLC and Mechanism of Anticounterfeiting Labels

CLC with one colour are not the best to fulfil the demands to met the actual applications. CLC embedded with other materials such as various nanostructures, dyes, dispersion of surfactants etc. can escalate the applications and improves the complexity of the CLC environment for luminesce and fluorescence applications. With vibrant colours CLC usage increases, especially in fluorescence which utilizes

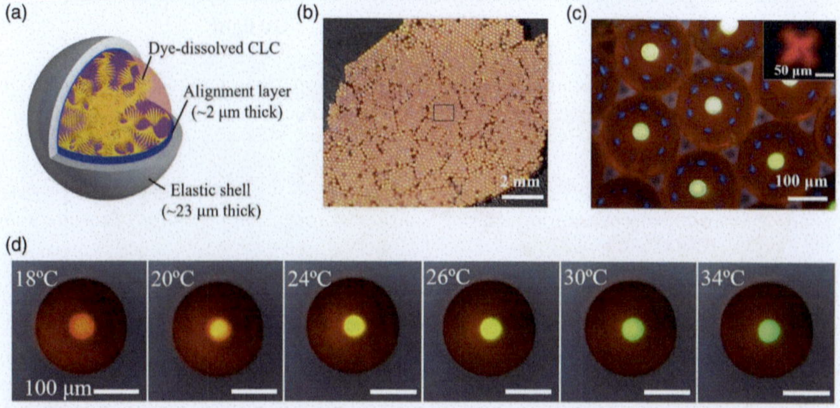

Fig. 13 **a** Schematic illustration of the fabrication of photonic capsule composed of CLC with dye and their alignment layer. **b–c** Optical microscopy image of spherical shaped CLC capsules. **d** TRRC of CLC microcapsules observed in optical microscopy (Reproduced with permission form reference [38]). Open access Creative Common Attributions 4.0

anticounterfeiting technologies. Figure 14 illustrates the schematic fabrication, architecture for anticounterfeiting technology adopted for the additives in CLC. The fluorescent invisible patterns are encoded and disclosed when on usage. The technology provides inherent advantages such as in information encryption and in anticounterfeiting. Various nanomaterials can be used to disperse over the CLC to increase the fluorescence into the LC molecules is one of the promising technologies. The LC thus converted into fluorescence cholesteric liquid crystals have dual fluorescence properties embodied with structural colours. Such system act as luminous dyes and encrypt information. The CLC such systems exhibit colours in white light and fluorescence under UV light. The technology developed by Qin et al. [73] for-patterning geminate ink using flow focusing glass capillary technology. This method involves usage of DC5 (cyno substituted oligo (p-phenylenevinylene) derivatives as luminescent material.

7 Conclusions

In the present book chapter, the preparation of CLC their physiochemical properties were illustrated in details. The chapter covers the broader area of the physical parameters such as pitch, optical reflections. Fluorescence, the theory of widening the bandwidth, reflectance etc. luminescence, temperature dependency of CLC, role of chiral dopants, and importance of formation of inherent helical nature. The chapter also covers the various fabrication procedures employed in the CLC preparations, role of chiral dopants, deswelling mechanism, size of the droplets during PDLC fabrication, etc. This chapter also focus on the fabrication and architecture of CLC

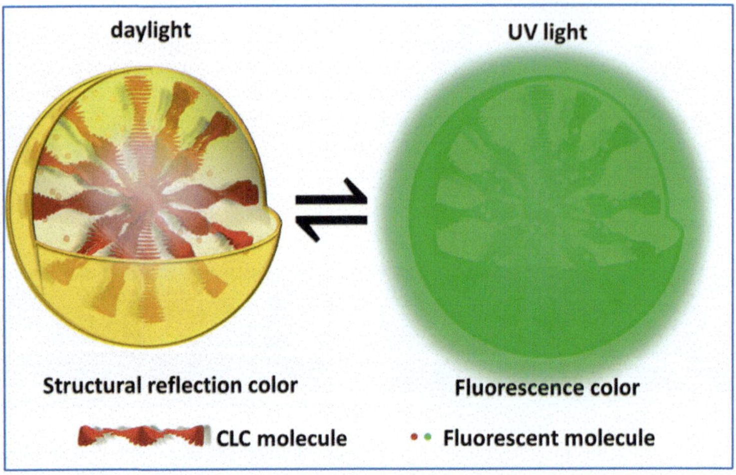

Fig. 14 Schematic illustration of the mechanism of anticounterfeiting labels obtained by adding fluorescent additives in to CLC (Reproduced with permission form reference [54]). Open access Creative Common Attributions 4.0

systems for fluorescent applications, anticounterfeiting technology developed for information encryption and fabrication of photonic technology with dye (PDLC), their advantages, wide spectrum of usage of the fluorescent CLC systems. The device architecture of the CLC preparation using chiral dopants, improving the fluorescence by luminescence compounds etc. The chapter also discusses briefly the limitations, future aspects of fabricating NLC with chiral dopants, improving fluorescence characteristics of CLC etc.

Acknowledgements All authors are thankful to Chitkara University Research and Innovation Network, Chitkara University, Patiala, Panjab, India. Authors are also thankful to Centre for Nano and Material Sciences (CNMS), Jain (Deemed-to-be University), Ramanagaram, Kanakapura, Bangalore, Karnataka, India. KLE Technological University, Dr. MS Sheshgiri Campus, Udyambag, Belagavi, Karnataka, India for constant encouragement and support for the research.

Authors Contributions Dr. Vinayak Adimule wrote the chapter, handled revision and Dr. Santosh Nandi, Dr. Rangappa Keri contributed for editing the chapter, grammatical and typo corrections, and obtaining the permission for the figures.

Funding Authors declare that they have not received any funding from any source or institutions or organizations.

Conflict of Interest All authors declare that they do not have any conflict of interest.

References

1. Imanishi, M., Tomishima, Y., Itou, S., Hamashima, H., Nakajima, Y., Washizuka, K., Sakurai, M., Matsui, S., Imamura, E., Ueshima, K., Yamamoto, T.: Discovery of a novel series of biphenyl benzoic acid derivatives as potent and selective human β3-adrenergic receptor agonists with good oral bioavailability. Part I. J. Med. Chem. **51**(6), 1925–1944 (2008)
2. Muhammad, K., Nazir, S., Hameed, S., Bechtold, I.H.: Mesomorphic and fluorescence properties of methyl 4-(4-alkoxystyryl) benzoates. Liq. Cryst. **43**(7), 863–873 (2016)
3. Fuji, K.: Asymmetric creation of quaternary carbon centers. Chem. Rev. **93**(6), 2037–2066 (1993)
4. Rodríguez-Salamanca, P., Fernández, R., Hornillos, V., Lassaletta, J.M.: Asymmetric synthesis of axially chiral C−N atropisomers. Chem. A Eur. J. **28**(28), e202104442 (2022)
5. Guo, J., Chen, F., Qu, Z., Yang, H., Wei, J.: Electrothermal switching characteristics from a hydrogen-bonded polymer network structure in cholesteric liquid crystals with a double-handed circularly polarized light reflection band. J. Phys. Chem. B **115**(5), 861–868 (2011)
6. Chepeleva, D.S., Yakovleva, A.S., Murauski, A.A., Kukhta, I.N., Muravsky, A.A.: Phototunable selective reflection of cholesteric liquid crystals. Доклады Белорусского государственного университета информатики и радиоэлектроники **S7**(125), 28–31 (2019)
7. Yoshida, H., Lee, C.H., Matsuhisa, Y., Fujii, A., Ozaki, M.: Bottom-up fabrication of photonic defect structures in cholesteric liquid crystals based on laser-assisted modification of the helix. Adv. Mater. **19**(9), 1187–1190 (2007)
8. Furumi, S.: Recent progress in chiral photonic band-gap liquid crystals for laser applications. Chem. Rec. **10**(6), 394–408 (2010)
9. Dogra, A.R., Kumar, P., Sharma, V., Malik, P., Khanra, P., Kapoor, M.: Deposition of nanoparticles layer on ITO substrate and its application for alignment control of liquid crystal. In: AIP Conference Proceedings, vol. 2357, no. 1. AIP Publishing (2022)
10. Sharma, V., Kumar, P., Chinky, C., Malik, P., Raina, K.K.: Morphological and electro optic studies of polymer dispersed liquid crystal in reverse mode. In: AIP Conference Proceedings, vol. 1953, no. 1. AIP Publishing (2018)
11. Dogra, A.R., Sharma, V., Gahrotra, R., Khanra, P., Dogra, A.: Thickness dependent electro-optical behaviour of homeotropically aligned nematic liquid crystal. ECS Trans. **107**(1), 5021 (2022)
12. Matharu, A.S., Jeeva, S., Ramanujam, P.S.: Liquid crystals for holographic optical data storage. Chem. Soc. Rev. **36**(12), 1868–1880 (2007)
13. Lin, Y.H., Ren, H., Wu, S.T.: Polarisation-independent liquid crystal devices. Liq. Cryst. Today **17**(1–2), 2–8 (2009)
14. Xianyu, H., Faris, S., Crawford, G.P.: In-plane switching of cholesteric liquid crystals for visible and near-infrared applications. Appl. Opt. **43**(26), 5006–5015 (2004)
15. Woliński, T.R., Bock, W.J.: Cholesteric liquid crystal sensing of high hydrostatic pressure utilizing optical fibers. Mol. Cryst. Liq. Cryst. **199**(1), 7–17 (1991)
16. Leal-Junior, A., Soares, M.S., de Almeida, P.M., Marques, C.: Cholesteric liquid crystals sensors based on nanocellulose derivatives for improvement of quality of human life: a review. Adv. Sens. Res. **2**(10), 2300022 (2023)
17. Kozlovsky, M.V.: Chiral polymers with photoaffected phase behaviour for photo-optical and optoelectronic applications. Synth. Met. **127**(1–3), 67–70 (2002)
18. Iwan, A.: An overview of LC polyazomethines with aliphatic–aromatic moieties: thermal, optical, electrical and photovoltaic properties. Renew. Sustain. Energy Rev. **52**, 65–79 (2015)
19. Sage, I.: Thermochromic liquid crystals in devices. Liquid Cryst. Appl. Uses 301–343 (1992)
20. Yu, R., Zhao, J., Wu, L., Li, Y., Liu, W., Yang, Y.: Colourful patterns prepared using a cholesteric liquid crystal mixture with both thermochromic and photochromic properties. Liq. Cryst. **50**(11–12), 1951–1959 (2023)
21. Zhang, L.Y., Gao, Y.Z., Song, P., Wu, X.J., Yuan, X., He, B.F., Chen, X.W., Hu, W., Guo, R.W., Ding, H.J., Xiao, J.M.: Research progress of cholesteric liquid crystals with broadband

reflection characteristics in application of intelligent optical modulation materials. Chin. Phys. B **25**(9), 096101 (2016)
22. Weyer, L.G.: Near-infrared spectroscopy of organic substances. Appl. Spectrosc. Rev. **21**(1–2), 1–43 (1985)
23. Zhao, Y.S., Fu, H., Peng, A., Ma, Y., Xiao, D., Yao, J.: Low-dimensional nanomaterials based on small organic molecules: preparation and optoelectronic properties. Adv. Mater. **20**(15), 2859–2876 (2008)
24. Liu, P., Zhang, Y., Martin, S.T.: Complex refractive indices of thin films of secondary organic materials by spectroscopic ellipsometry from 220 to 1200 nm. Environ. Sci. Technol. **47**(23), 13594–13601 (2013)
25. McConney, M.E., Rumi, M., Godman, N.P., Tohgha, U.N., Bunning, T.J.: Photoresponsive structural color in liquid crystalline materials. Adv. Opt. Mater. **7**(16), 1900429 (2019)
26. Kim, J.B., Lee, S.Y., Lee, J.M., Kim, S.H.: Designing structural-color patterns composed of colloidal arrays. ACS Appl. Mater. Interfaces **11**(16), 14485–14509 (2019)
27. Gevorgyan, A.H.: Optical properties of a stack of right-and left-handed layers of a cholesteric liquid crystal. Opt. Spectrosc. **113**, 141–152 (2012)
28. Kipfer, P., Klappert, R., Herzig, H.P., Grupp, J., Dändliker, R.: Improved red color with cholesteric liquid crystals in Bragg reflection mode. Opt. Eng. **41**(3), 638–646 (2002)
29. Agez, G., Bitar, R., Mitov, M.: Color selectivity lent to a cholesteric liquid crystal by monitoring interface-induced deformations. Soft Matter **7**(6), 2841–2847 (2011)
30. Boyon, C., Soldan, V., Mitov, M.: Bioinspired, cholesteric liquid-crystal reflectors with time-controlled coexisting chiral and achiral structures. ACS Appl. Mater. Interfaces. **13**(25), 30118–30126 (2021)
31. Huang, Y., Zhou, Y., Wu, S.T.: Broadband circular polarizer using stacked chiral polymer films. Opt. Express **15**(10), 6414–6419 (2007)
32. Mitov, M., Boyon, C., Soldan, V.: Self-assembly of twisted monolayer cholesteric films via surface tension local anisotropy: implications for multicolor optical tags. ACS Appl. Nano Mater. **5**(8), 10560–10571 (2022)
33. Ma, L., Zhao, W., Liang, Z., Wang, X., Xie, L., Jiao, L., Zhou, T.: An investigation on the mechanical property changing mechanism of high speed steel by pulsed magnetic treatment. Mater. Sci. Eng., A **609**, 16–25 (2014)
34. Liu, J., Song, Z.P., Sun, L.Y., Li, B.X., Lu, Y.Q., Li, Q.: Circularly polarized luminescence in chiral orientationally ordered soft matter systems. Responsive Mater. **1**(1), e20230005 (2023)
35. Kacher, J., Landon, C., Adams, B.L., Fullwood, D.: Bragg's law diffraction simulations for electron backscatter diffraction analysis. Ultramicroscopy **109**(9), 1148–1156 (2009)
36. Andruski, J.E., Costello, J.: Using polynomial equations to model pitch contour shape in lexical tones: an example from Green Mong. J. Int. Phon. Assoc. **34**(2), 125–140 (2004)
37. Li, H.H.: Refractive index of alkaline earth halides and its wavelength and temperature derivatives. J. Phys. Chem. Ref. Data **9**(1), 161–290 (1980)
38. Zhang, W., Froyen, A.A., Schenning, A.P., Zhou, G., Debije, M.G., de Haan, L.T.: Temperature-responsive photonic devices based on cholesteric liquid crystals. Adv. Photonics Res. **2**(7), 2100016 (2021)
39. Chu, G.: Controlled self-assembly of cellulose nanocrystal as custom-tailored photonics and complex soft matter. Acc. Mater. Res. (2024)
40. Concellón, A.: Liquid crystal emulsions: a versatile platform for photonics, sensing, and active matter. Angew. Chem. Int. Ed. **62**(51), e202308857 (2023)
41. Zhang, J., Zhang, Y., Yang, J., Wang, X.: Beyond color boundaries: pioneering developments in cholesteric liquid crystal photonic actuators. Micromachines **15**(6), 808 (2024)
42. Yao, W., Lan, R., Li, K., Zhang, L.: Multiple anti-counterfeiting composite film based on cholesteric liquid crystal and QD materials. ACS Appl. Mater. Interfaces **13**(1), 1424–1430 (2021)
43. Wang, Q., Zhang, X., Zhou, D., Chen, Y., Qu, C., Lu, C., Liu, Y.: Cholesteric liquid crystal doped with spiropyran and zinc ions for multi-state anti-counterfeiting. J. Mater. Chem. C **12**(20), 7225–7232 (2024)

44. Li, H., Gui, X., Wan, Z., Qi, Y., Wang, S., Zhang, H., Niu, M., Guo, Y.: Efficient preparation of cellulose nanocrystals and regulation of cholesteric liquid crystals for advanced anti-counterfeiting. J. Appl. Polym. Sci. e56043 (2024)
45. Shi, Y., Han, J., Jin, X., Miao, W., Zhang, Y., Duan, P.: Chiral luminescent liquid crystal with multi-state-reversibility: breakthrough in advanced anti-counterfeiting materials. Adv. Sci. **9**(20), 2201565 (2022)
46. Zhou, Y., Bukusoglu, E., Martínez-González, J.A., Rahimi, M., Roberts, T.F., Zhang, R., Wang, X., Abbott, N.L., De Pablo, J.J.: Structural transitions in cholesteric liquid crystal droplets. ACS Nano **10**(7), 6484–6490 (2016)
47. Coates, D.: Development and applications of cholesteric liquid crystals. Liq. Cryst. **42**(5–6), 653–665 (2015)
48. Gahrotra, R., Sharma, V., Dogra, A.R., Malik, P., Kumar, P.: Temperature dependent morphological and electro-optical characteristics of dye doped cholesteric liquid crystal. ECS Trans. **107**(1), 5481 (2022)
49. Dogra, A.R., Sharma, V., Malik, P., Kumar, P.: In-situ homeotropic alignment of dye doped liquid crystal molecules on multilayered self assembled nanoparticles in confined cells for next generation display devices. Opt. Mater. **154**, 115731 (2024)
50. Sharma, V., Kumar, P.: Electro-optically oriented Kerr and orientational phase study of normal mode polymer dispersed liquid crystals–effect of dispersion of nanoparticles. J. Mol. Liq. **348**, 118030 (2022)
51. Sharma, V., Kumar, P., Sharma, A., Raina, K.K.: Droplet configuration control with orange azo dichroic dye in polymer dispersed liquid crystal for advanced electro-optic characteristics. J. Mol. Liq. **233**, 122–130 (2017)
52. Manhas, N., Adimule, V., Lakshminarayana, P., Keri, R., Kumar, L.S. and Bathula, C.: Utilization of CO_2^+ doped Eu_2O_3 nanostructures and their effect on the dielectric and photoisomerization of liquid crystals. Phys. Fluids **35**(10) (2023)
53. Qi, Y., Zhang, S.: Structural Color Due to Self-Assembly. Functional Materials from Colloidal Self-Assembly, pp. 183–235 (2022)
54. Gollapelli, B., Potu, S., Rajaboina, R., Vallamkondu, J.: Optical anti-counterfeiting with cholesteric liquid crystal emulsions: preparation, properties, and applications. Mater. Adv. (2024)
55. Blumstein, A., Vilasagar, S., Ponrathnam, S., Clough, S.B., Blumstein, R.B., Maret, G.: Nematic and cholesteric thermotropic polyesters with azoxybenzene mesogenic units and flexible spacers in the main chain. J. Polym. Sci. Polym. Phys. Ed. **20**(5), 877–892 (1982)
56. Katsonis, N., Lacaze, E., Ferrarini, A.: Controlling chirality with helix inversion in cholesteric liquid crystals. J. Mater. Chem. **22**(15), 7088–7097 (2012)
57. Singh, S.: Phase transitions in liquid crystals. Phys. Rep. **324**(2–4), 107–269 (2000)
58. Adimule, V., Bathula, C., Sharma, K., Keri, R.: Study of structural, optical, dielectric properties poly-3-butyl thiophene reinforced with Co_3O_4 nanoparticles. Nano-Struct. Nano-Obj. **38**, 101150 (2024)
59. Nandi, S.S., Adimule, V., Kadapure, S.A., Kerur, S.S.: Rare earth based nanocomposite materials for prominent performance supercapacitor: a review. Appl. Mech. Mater. **908**, 3–18 (2022)
60. Renn, S.R., Lubensky, T.C.: Abrikosov dislocation lattice in a model of the cholesteric–to–smectic-a transition. Phys. Rev. A **38**(4), 2132 (1988)
61. Adimule, V., Sharma, V., Kumar, P., Keri, R., Sharma, K.: Perylene-3, 4, 9, 10-tetracarboxylic dianhydride dye intercalated liquid crystal molecules: synthesis, spectroscopic characterizations and photoluminescence dynamics. J. Mol. Liq. 125990 (2024)
62. Nagalakshmi, G., Nandeesh, I.M., Yallur, B.C., Adimule, V., Batakurki, S.: Synthesis and optical properties of copper terephthalate metal organic frame works. Eng. Chem. **2**, 3–11 (2023)
63. Adimule, V., Keri, R., Rajendrachari, S., Kendrekar, P., Kulkarni, C.V.: Highly photoluminescence and wide band gap insulating metal hybrid nanoparticles array of samarium-doped SrO: CoO: synthesis, characterizations and sensor characteristics. J. Mater. Sci.: Mater. Electron. **34**(5), 442 (2023)

64. Jang, J.H., Park, S.Y.: PH-responsive cholesteric liquid crystal double emulsion droplets prepared by microfluidics. Sens. Actuators, B Chem. **241**, 636–643 (2017)
65. Gollapelli, B., Suguru Pathinti, R., Vallamkondu, J.: Carbon quantum dots doped cholesteric liquid crystal films and microdroplets for anti-counterfeiting. ACS Appl. Nano Mater. **5**(8), 11912–11922 (2022)
66. Bobrovsky, A., Mochalov, K., Oleinikov, V., Sukhanova, A., Prudnikau, A., Artemyev, M., Shibaev, V., Nabiev, I.: Optically and electrically controlled circularly polarized emission from cholesteric liquid crystal materials doped with semiconductor quantum dots. Adv. Mater. **24**(46), 6216 (2012)
67. Adimule, V., Yallur, B.C., Keri, R., Bathula, C., Batakurki, S.: Microstructure, photoluminescence and electrical properties of SmxGd (1–x): SrO hybrid nanomaterials synthesized via facile coprecipitation method. Electron. Mater. Lett. **19**(3), 278–297 (2023)
68. Challa, M., Yallur, B.C., Ambika, M.R., Adimule, V.: Influence of nano particles on optical properties of Cu-MOFs. Adv. Mater. Res. **1173**, 23–33 (2022)
69. Adimule, V., Yallur, B.C., Gowda, A.H.: Advanced sensors based on carbon nanomaterials. In: Carbon Nanomaterials-Based Sensors, pp. 259–268. Elsevier (2022)
70. Adimule, V., Nandi, S., Sharma, K., Keri, R.: Microstructure, zeta potential and enhanced photocatalytic dye degradation performances of mesoporous flower like $Nb_2O_5@CeO_2$ bimetallic oxide nanomaterials decorated with reduced graphene oxide. Inorg. Chem. Commun. 112592 (2024)
71. Liu, X., Qin, L., Zhan, Y., Chen, M., Yu, Y.: Phototuning of structural colors in cholesteric liquid crystals. Acta Chim. Sinica **78**(6), 478 (2020)
72. Sharma, V., Kumar, P., Chinky, C., Gahrotra, R., Raina, K.K., Malik, P.: Effect of nano particles on electro optic properties of polymer dispersed liquid crystal in normal mode. In: AIP Conference Proceedings, vol. 2142, no. 1. AIP Publishing (2019)
73. Qin, J., Wang, X.Q., Yuan, C., Zheng, Z., Shen, D.: Light-rewritable geometric phase and reflectance modulations enabled by pattern-aligned photoresponsive liquid crystal superstructures. Liq. Cryst. **47**(2), 255–262 (2020)

Recent Advances in Optical Storage and Display Applications of Cholesteric Liquid Crystals

Gangadhar Bagihalli and Bojappa Ginni

Abstract Cholesteric liquid crystals (CLCs) have garnered significant attention in recent years due to their unique optical properties, which make them well suited for a wide range of advanced applications in both optical storage and display technologies. This chapter reviews the recent advances in the field, focusing on the use of CLCs for high-density optical data storage, color displays, and dynamic, tunable reflectors. The helical structure of CLCs enables selective reflection of circularly polarized light, which can be harnessed for various photonic applications, including 3D displays and anti-counterfeiting measures. Advances in material synthesis, chirality control, and polymer-stabilized systems have allowed for improved performance in terms of contrast, color purity, and response times. The chapter also explores novel methods for achieving multi-stable and rewritable optical storage through CLCs, offering potential for next-generation data storage devices. Additionally, innovations in light modulation techniques and flexible display technologies further expand the versatility of CLC-based systems. By examining these cutting-edge developments, this chapter gives a comprehensive overview of the role of CLCs in revolutionizing optical technologies, while highlights the ongoing challenges and future directions in the field.

Keywords CLCs · Optical data storage · Selective light reflection · Photonic applications

G. Bagihalli (✉)
Department of Engineering Chemistry, KLE Institute of Technology, Airport Road, Gokul, Hubballi 580030, Karnataka, India
e-mail: g.bagihalli@gmail.com

B. Ginni
VTU Recognized Research Centre in Chemistry, KLE Institute of Technology, Hubballi, Karnataka, India

P. Kumar et al. (eds.), *Cholesteric Liquid Crystals*, Engineering Materials,
https://doi.org/10.1007/978-981-96-3821-5_2

1 Introduction

Cholesteric liquid crystals (CLCs), also called chiral nematic liquid crystals (LCs), it can be representing a unique class of liquid crystalline materials characterized by their helical molecular arrangement. Unlike nematic LCs, molecules align in parallel but lack positional order, CLCs exhibit a periodic helical structure, with the molecules twisting along an axis. helical arrangement is defined by its pitch the distance over which the molecular orientation completes a full 360° twist, imparts distinctive optical properties to CLCs, such as selective reflection and circular polarization of light [1].

CLCs are created at the molecular level by integrating chiral doping impurities into nematic LC or employing mesogens that are intrinsically chiral. The molecules create helical structure as a product of this chirality. The wavelength range of light is selectively reflected by the CLCs. The CLCs were found by using pitch of the helix, which is affected by the chiral dopant's composition and concentration. CLCs helical axis, which is perpendicular to the reactant, enables varying light interaction is depends on the composition of the material and outside stimuli [2, 3].

One of the different characteristics of CLCs is their capacity to selectively reflect light within a specific wavelength range that corresponds to the pitch of the helix. This characteristic results from Bragg reflection, in which the periodic structure of the helical alignment causes constructive interference for specific wavelengths of light. In relation to the handedness of the helical structure, the reflected light is circularly polarized. Birefringence is one more characteristic of CLCs, where the refractive index varies in responds to the helical axis-related direction of light propagation. CLCs are highly versatile for technological applications due to these type properties, as well as their thermotropic and electro-optic responsiveness, which enable dynamic modulation of their optical behaviour. Because of its special optical properties, CLCs will become essential in the fields of photonics and optoelectronics. For uses like adjustable optical filters and security features, their capacity for selective reflection is needed. with they have well developed applications in adaptive optics, sensors, and smart coatings due to their sensitivity to external stimuli with light, temperature. CLCs provide high-density data encoding capabilities in optical storage technologies. By using their pitch dependent optical behavior in display applications, they will give different color rendering that is energy efficient. CLCs' potential in cutting-edge applications like augmented reality and virtual reality has been further increased by its integration with present materials and technologies like meta surfaces and nanostructure [4, 5].

This chapter will provide an overview of good achievements in optical storage and display applications using CLCs. This chapter begins with an overview of the structure and properties of CLCs. It will discuss the CLCs mechanisms and usage in data storage and display technologies. It will also lead to material advances and production approaches that improve the performance of CLCs (Fig. 1). Finally, the chapter will discuss the problems, future directions, and revolutionary possibilities of CLCs in the continuously changing world optical technology.

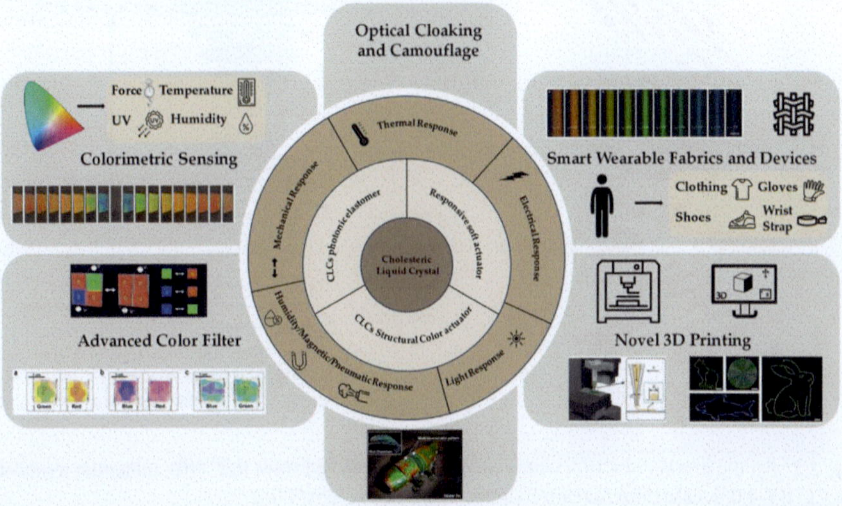

Fig. 1 The overview of classification, stimulus response, and potential applications of the CLCs (Reproduced from Ref. [6], under the terms of the CC BY 4.0 license, http://creativecommons.org/licenses/by/4.0/.)

2 Fundamental Concepts of Cholesteric Liquid Crystals (CLCs)

CLCs also known as chiral nematic LC are a fascinating phase of LCs distinguished by their helical molecular arrangement. This unique structure is because of the presence of chirality in the system or from inherently chiral molecules or from the introduction of chiral doping impurities into nematic LCs. The helical configuration imparts CLCs with good optical and electro-optical properties making them exceedingly important for improvable technological applications.

2.1 Helical Arrangement and Pitch of CLCs

In CLCs the molecules exhibit a periodic twisting orientation along an axis known as the helical axis. This periodicity, referred to as the pitch denoted by P represents the distance over which the molecular alignment completes a full 360° twist (Fig. 2). The pitch of the helix is a critical parameter that governs the optical behavior of CLCs. By introduction of chirality convinces a torque that will cause the nematic molecular alignment to twist, shaping a helical structure. The handedness of the helix left-handed or right-handed is determined by the chirality of the dopant or the molecular structure of the LC [7]. The pitch is influenced by factors such as:

A. The concentration and the strength of the chiral dopant.

Fig. 2.2 An illustration of CLCs with a pitch (*p*) (Reproduced from Ref. [10], under the terms of the CC BY 4.0 license, http://creativecommons.org/licenses/by/4.0/.)

B. Temperature, as thermal expansion or contraction affects molecular spacing.
C. External fields, which can modify the helical twist or disrupt the alignments [8, 9].

2.2 *Optical Behavior*

The relationship between pitch and reflection wavelength enables dynamic tunability of the optical properties of CLCs. Changes in pitch, convinced by the external stimuli those are:

a. **Temperature**: Thermal expansion or contraction adjusts the pitch, shifting the reflection wavelength (Fig. 3). The CLCs transited from the anisotropic state to the isotropic state upon heating. Specifically, its spiral structure left after being heated to 70 °C, and it achieved isotropy and displayed as a transparent liquid [11].

b. **Electric Fields**

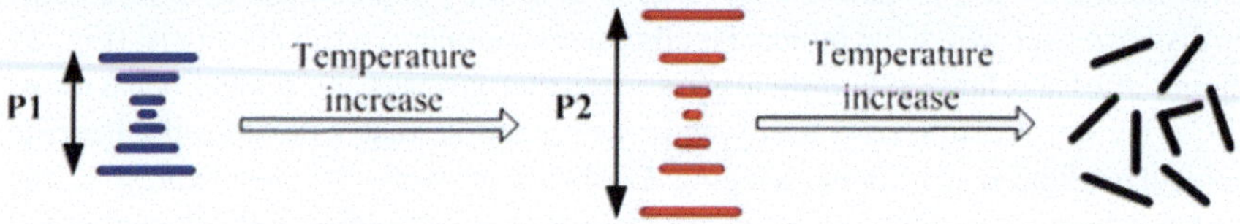

Fig. 3 The effect of temperature on the helical structure of CLCs (Reproduced from Ref. [11], under the terms of the CC BY 4.0 license, http://creativecommons.org/licenses/by/4.0/.)

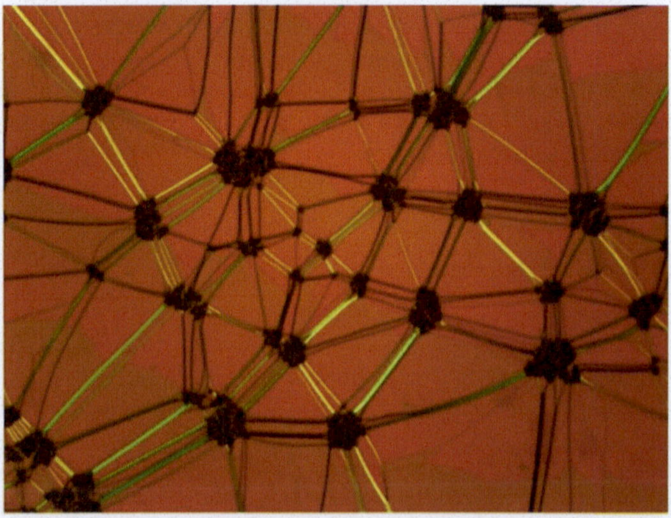

Fig. 4 CLCs exhibit the strongest selective reflection of incident light wavelength, especially when the incident light is projecting along its spiral axis (Reproduced from Ref. [12], under the terms of the CC BY 4.0 license, http://creativecommons.org/ licenses/by/4.0/.)

Realignment of molecules within the helical structure changes the pitch or disrupts the helix entirely. Photo-responsive impurities can convince the pitch variations upon illumination The typical texture of the CLCs can be as shown in Fig. 4.

This tunable nature is critical for the use of CLCs in devices such as tunable filters, reflecting displays, and adaptive optics, which require fine control over light-matter interactions. CLCs can be tailored for specific optical and photonic applications by understanding the interaction between the helical structure, pitch, and reflected wavelength. These ideas form the base for their application in present optical storage and display technologies.

CLCs are advantageous in optical data storage because of their changeable pitch, specific light reflection, and dynamic adaptation to external stimuli such as light. Their helical shape, combined with the use of photo responsive materials, gives the unique techniques of data encoding, storage, and retrieval [12–16]. The mechanism involved in the optical data storage is certain light wavelengths can be utilized for the change the helical pitch of CLCs. Changes in the molecular orientation inside the CLC matrix caused by light exposure cause pitch to fluctuate. These pitch shifts enable information to be encoded as colour or wavelength changes by changing the reflected wavelength. UV and visible light can induce reversible or irreversible pitch modifications and thermal relaxation returns the pitch to its original state, providing a mechanism for erasable or rewritable storage [17–20].

Using light-sensitive dopants such as photo responsive materials, including azobenzene derivatives, can be used to precisely regulate the characteristics of CLCs.

When azobenzene molecules are revealed to light, they undergo *cis–trans* isomerization, which causes them to alter shape and alignment. This realignment allows the use of particular light wavelengths to write and erase data by changing the helical pitch. Multi-state or multi-wavelength data increased by additional dopants such as spiropyrans, increases storage density.

3 Technological Developments of CLCs

CLCs can store data in several layers by storing different pitch values, each representing a distinct wavelength. This multi-layer technique increases data capacity by storing and reading information at different depths within the material. Laser-based writing and reading devices use CLCs' selective reflection properties to handle multi-wavelength data [21].

The spatial resolution of pitch alterations allows for the encoding of data at sub-micrometre levels. Advances in laser technology and nanoengineering have permitted the precise manipulation of CLC formations, allowing for ultra-high-density optical storage. The capacity to encode several states at a single place increases data density over binary systems [22].

4 Challenges and Solutions of CLCs in Technological Developments of CLCs

4.1 Stability and Repeatability of Recorded Data

a. **Challenge**: Environmental conditions include temperature, humidity, and UV exposure can deteriorate the structure of CLCs over time, causing data loss or mistakes.
b. **Solutions**: **Polymer stabilization**: Embedding CLCs in polymer networks improves structural rigidity while maintaining pitch stability.
c. **Protective coatings**: It can significantly enhance the stability and repeatability of the recorded data in CLCs by shielding materials reduce environmental impact [23].

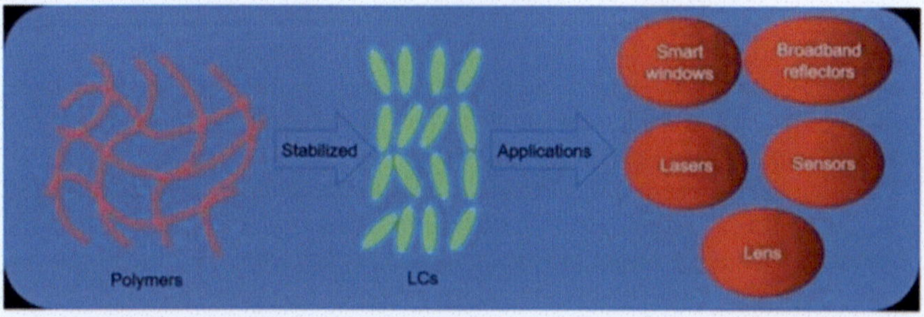

Fig. 5 Polymer stabilized liquid crystals (Reproduced from Ref. [28], under the terms of the CC BY 4.0 license, http://creativecommons.org/licenses/by/4.0/.)

4.2 Material Engineering to Improve Performance

a. **Challenge**: Hybrid materials are being developed that combine CLCs with nanoparticles or quantum dots (QDs) to improve functionality. Use of sophisticated photo-responsive dopants with faster response times and higher sensitivity. Engineering materials with increased tunability, allowing for multi-state encoding and speedier data retrieval [3, 24].

4.3 Environmental Stability

a. **Challenge**: CLCs are susceptible to environmental conditions like temperature, UV light, and humidity, which can break their helical structure and reduce their performance over time. For example, extended UV exposure can modify the molecular structure, resulting in a loss of selective reflection or colour shifts.
b. **Polymer Stabilization**: Embedding CLCs in a polymer matrix developed their thermal and mechanical stability (Fig. 5). Under various environmental conditions, polymer stabilized CLCs retain their helical structure intact.
c. **UV Protective Coatings**: Adding UV-absorbing coatings or inserting photostable molecules can help to decrease the impacts of UV light.
d. **Humidity Resistance**: Thin protective coatings and other encapsulation techniques reduce moisture penetration while maintaining the characteristics of CLCs [25, 26].

4.4 Scalability and Fabrication Challenges

a. **Challenge**: Fabricating large-area, defect-free CLCs films with uniform optical properties is difficult, particularly for industrial-scale production. Misalignment, thickness variations, and defects can reduce the performance and consistency of CLCs -based devices.

b. **Solutions**: Large-scale production with greater control over alignment and consistency can be made possible by techniques like self-assembly, roll-to-roll processing, and inkjet printing.
c. **Layer-by-Layer Assembly**: Precise deposition techniques ensure consistent thickness and minimize defects in multilayer CLCs systems.
d. **Automation and Quality Control**: Automated systems for real-time monitoring and defect correction during fabrication can improve yield and quality [28–30].

4.5 Limited Material Diversity

a. **Challenge**: The availability of materials with tuneable properties, environmental stability, and compatibility with modern fabrication techniques is limited.
b. **Development of New Mesogens**: The variety of possible materials is increased by creating new LC molecules with enhanced chiral and thermal characteristic.
c. **Hybrid and Composite Materials**: Combining CLCs with organic and inorganic nanomaterials enhances functionality and widens application potential.
d. **Green Chemistry Approaches**: Developing eco-friendly, sustainable materials for CLCs addresses environmental concerns and expands material diversity [31, 32].

5 Display Applications

CLCs are critical in the development of display systems because of their remarkable selective light-reflection capabilities and tuneable optical properties. CLCs offer a number of low-power, high-resolution display options, making them a viable alternative to traditional display technologies in some applications.

5.1 CLCs-Based Display Technologies

a. **Reflective Displays**: CLCs function similarly to electronic paper in that they reflect light from the environment rather than emitting it. These displays use the selective reflection property of CLCs to produce vivid, full-color visuals without a backlight.
b. **Tuneable Color Displays**: The pitch of CLCs can be dynamically adjusted to reflect specific wavelengths of light, enabling tuneable color displays.

In CLCs displays can produce a wide range of colors by changing the helical pitch in response to temperature, electric fields, or light exposure [33]. Recent Advances

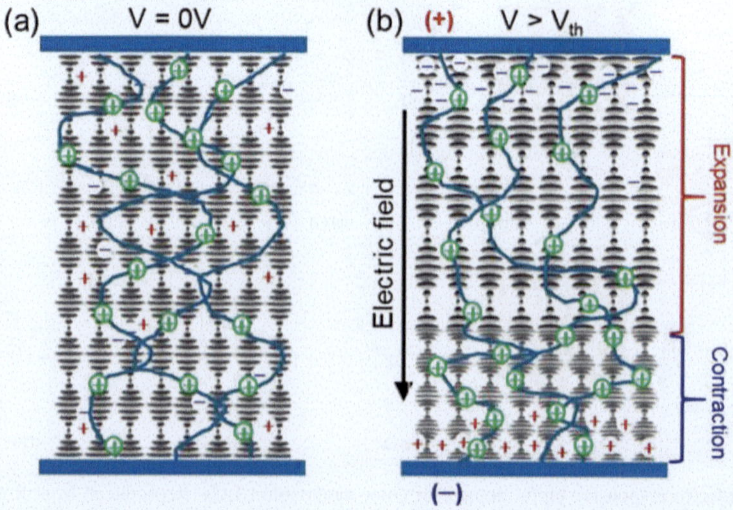

Fig. 6 Schematic of the polymer network in a deformable polymer-stabilized CLCs (PSCLC) with $\Delta\varepsilon < 0$: **a** with no applied field, **b** with a DC field applied between the top and bottom substrates. The blue lines represent the polymer network, the grey horizontal bars are the low-molecular-weight CLC (CLC) molecules, + and – are the free cationic and anionic impurities and are trapped cationic impurities (Reproduced from Ref. [37], under the terms of the CC BY 4.0 license, http://creativecommons.org/licenses/by/4.0/.)

in CLCs based on usage of polymer stabilized cholesteric textures which improve their mechanical and thermal properties and stabilization allows for sharper colour reflection, increased durability, and improved response times in tunable displays (Fig. 6) [34, 35].

Enhanced materials and structural designs have led to broader viewing angles without significant colour distortion. Adding light-scattering layers or optimising refractive indices to reflect displays makes them more competitive with emissive technology, such as OLEDs.

5.2 *Advantages Over Competing Technologies*

b. **Low Power Consumption**: Unlike emissive displays CLCs-based displays do not require a backlight and hence power is consumed only during content updates, making them suitable for applications requiring prolonged operation on limited energy.
c. **Wide Color Gamut**: CLCs displays offer vibrant and highly saturated colors because of their ability to selectively reflect specific wavelengths (Fig. 7).They outperform certain conventional displays in achieving vividness without excessive power demands [37, 38].

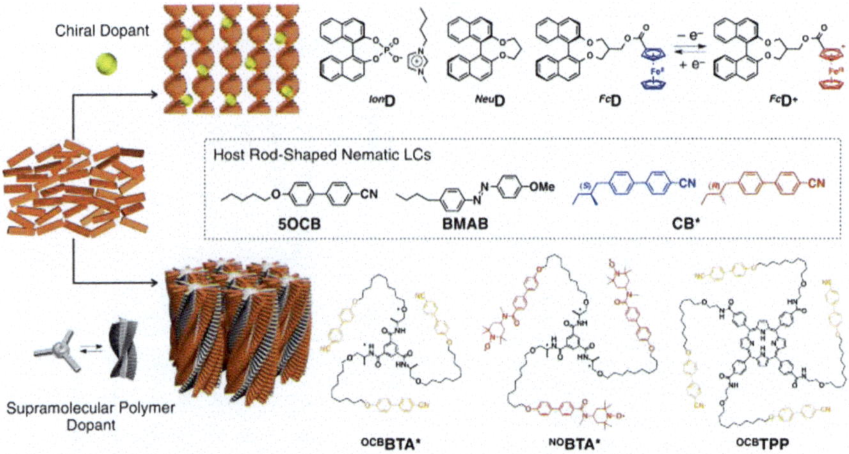

Fig. 7 Electro responsive chiral dopant for color-modulable CLCs (Reproduced from Ref. [40], under the terms of the CC BY 4.0 license, http://creativecommons.org/licenses/by/4.0/.)

d. **Eco-Friendliness**: With fewer energy requirements and the potential for sustainable materials, CLCs contribute to greener display technologies.

5.3 *Challenges in Display Applications of CLCs*

a. **Challenge**: CLCs often exhibit slower response times compared to traditional LCDs or OLEDs, particularly when tuned thermally or mechanically.
b. **Solutions**: Incorporating highly responsive dopants. The most characteristic feature of CLCs is selective reflection due to internal helical ordering. When the wavelength and the corresponding helical pitch fall into the range of the visible region, a bright reflection color can be observed. Meanwhile, the cholesteric phase is formed not only from the cholesterol derivative but also from a nematic LC doped with a chiral molecule (dopant) [39].

Employing advanced electrode designs and fast-switching materials [40, 41]

5.4 *Environmental Stability and Durability*

c. **Challenge**: The optical and mechanical properties of CLCs can degrade under prolonged exposure to UV light, heat, or moisture.
d. **Solutions**: Polymer stabilization to protect the helical structure. UV-blocking coatings or encapsulation to enhance environmental resistance. Engineering hybrid CLCs materials with improved thermal and chemical stability [42].

5.5 Advances in Materials and Fabrication

The performance and applicability of CLCs have been significantly increased through innovations in material development and fabrication methods. These advances address key challenges such as stability, scalability, and environmental impact, paving the way for next-generation optical and photonic technologies [43–45].

5.6 Development of New CLCs Mixtures and Dopants

Tailored CLC mixtures with optimized birefringence, pitch tunability, and thermal stability improve device performance. Introduction of chiral impurities with higher pitch-tuning sensitivity give a chance over more precise control over selective reflection properties. Novel photoactive dopants, such as azobenzene derivatives, allow dynamic modulation of CLCs pitch in response to light stimuli, enhancing functionality in tunable filters and smart displays [46, 47].

5.7 Incorporation of Nanoparticles and Nanomaterials

Nanoparticles placed in CLC matrices improve optical properties such as reflectivity, brightness, and colour saturation. CLC alignment is directed by nanostructured surfaces and templates, resulting in defect-free films and more consistent performance. Graphene and other 2D materials are added to devices made of CLCs to improve their thermal stability and electrical conductivity [48–50].

5.8 Layer-By-Layer Assembly for Precise Control

This method offers nanoscale precision in fabricating multilayer CLCs systems, which is required for creating complex devices with distinct reflection wavelengths or multi-functional properties. Layer-by-layer approaches are extremely important in optical storage and multi-wavelength reflective display technologies, which require perfect pitch and alignment [51].

Sustainability and Green Chemistry

5.9 Sustainable Processing Techniques

The use of low-energy production processes, such as photopolymerization at room temperature, lowers energy usage.

Recycling and reuse of chiral dopants and substrates contribute to a circular material economy in CLCs production [52, 53].

6 Challenges and Future Directions of CLCs

While CLCs have shown great promise in a variety of optical and photonic applications, a number of obstacles remain. To address these restrictions, continued innovation in materials research, fabrication processes, and interdisciplinary collaboration will be necessary. At the same time, growing research trends and prospects point to a promising future for CLC technology [54, 55].

6.1 Long-Term Stability Under Varying Environmental Conditions

Environmental factors including temperature, humidity, and UV radiation can cause CLC materials to deteriorate with the time. These instabilities have a negative impact on the long-term survivability of CLC-based devices, especially when used outdoors or under high stress. CLCs' slow response to outside stimuli restricts their application in adaptive optics and fast-switching displays [7].

6.2 Opportunities for Growth of CLCs

The ability of CLCs to selectively reflect light at specific wavelengths, coupled with their tunable nature, makes them ideal candidates for high-density holographic data storage. Photo responsive materials and multi-layered CLC devices have the potential to boost storage capacity and retrieval speeds even more were reported the responsive optical labels based on motor-doped CLC microdroplets that were prepared by a microfluidic technique (Fig. 8).

The inherent flexibility of CLCs films makes them suitable for integration into bendable, lightweight, and energy-efficient displays for wearable technology. Potential applications include smart textiles, flexible e-paper, and adaptive color displays for consumer electronics.

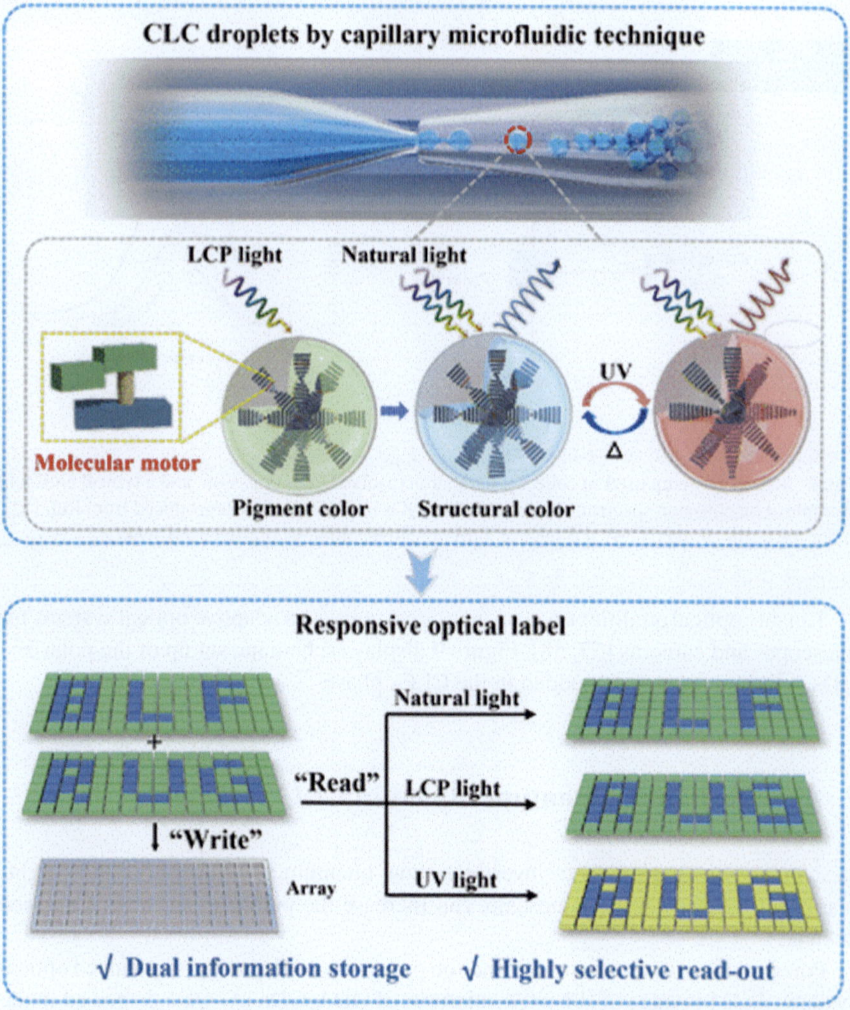

Fig. 8 Representative scheme of a dual–level responsive optical label by molecular motor doped CLC microdroplets. The label is able to carry two levels of distinct information that can be selectively read-out by different stimuli. Letters "BLF" is shown under natural white light and letters "RUG" are observed instead when the label is placed under LCP or UV light. Role in Flexible and Wearable Displays (Reproduced from Ref. [56], under the terms of the CC BY 4.0 license, http://creativecommons.org/licenses/by/4.0/.)

6.3 Energy-Saving Applications in Smart Windows and Adaptive Optics

CLCs are suitable for energy-efficient smart windows in automobiles and buildings because of their ability to dynamically adjust light transmission and reflection.

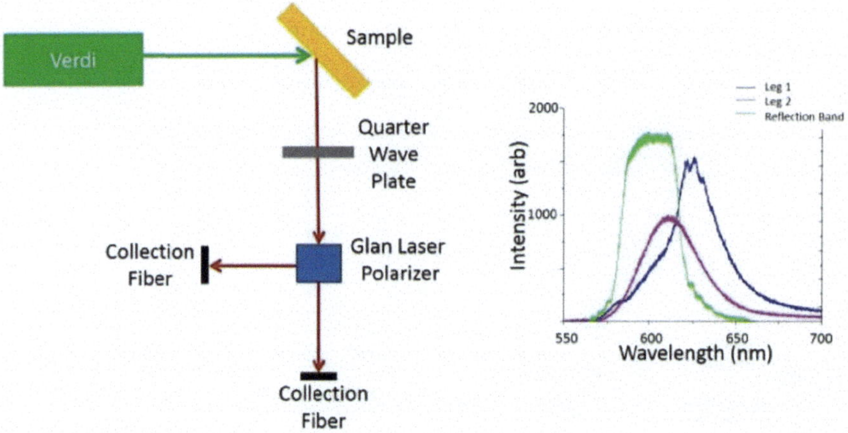

Fig. 9 Schematic setup used to collect polarization resolved QDs emission, and a typical emission example of an emission spectrum from embedded QDs in CLCs phase (Reproduced from Ref. [12], under the terms of the CC By 4.0 license, http://creativecommons.org/licenses/by/4.0/.)

Tunable optical qualities open up new opportunities in adaptive optical systems for telescopes and cameras [57, 58]. Figure 9 displays schematic set up of the polarized QDs emission spectra embedded in the CLCs phase.

7 Exploration of Quantum Effects in CLC Systems

Researchers are beginning to investigate how quantum phenomena, such as light-matter interactions at the nanoscale, can Increase the performance of CLCs-based devices.

Potential uses for quantum information processing and quantum-enhanced optical systems. POM images of 0.1 wt.% mixtures of ODA-QDs LC QDs dispersed in the CLCs phase and their fluorescence images of QDS embedded in the fluid cavities shown in Fig. 10. Figure 11 unveils fluorescence microscope images of QDs in micro-shells, flat membrane generated by phase transitions, single QDs, and their crumpled membrane structure.

7.1 *Interdisciplinary Approaches in Material Science and Optics*

Collaboration between chemists, physicists, and engineers is driving the development of hybrid materials that combine CLCs with nanoparticles, quantum dots, and advanced polymers. The photoluminescence (PL) spectra of 0.02 wt.% of ODA-QDs

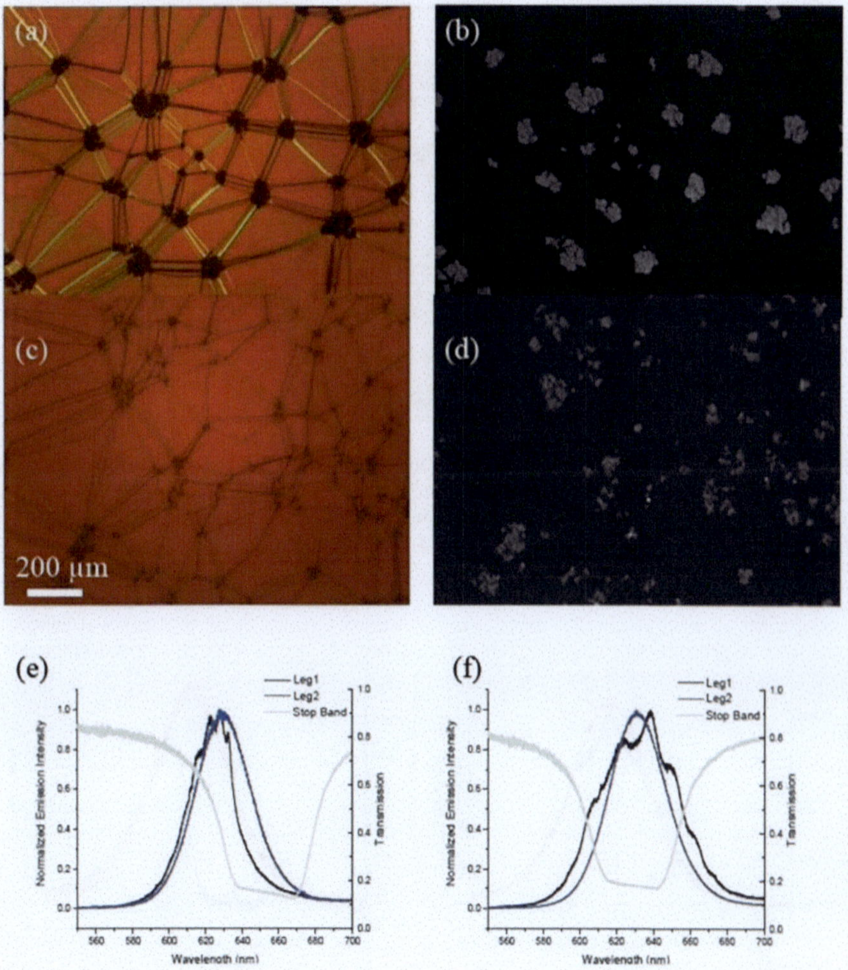

Fig. 10 POM images of 0.1 wt.% mixtures of **a** ODA–QDs and **c** functionalized LC–QDs dispersed in CLCs and fluorescence microscopy images **b** and **d**–**f** show emission spectra for QDs embedded in fluid cavities (Reproduced from Ref. [12], under the terms of the CC BY 4.0 license, http://creativecommons.org/licenses/by/4.0/.)

and QDs dispersion in CLCs film are shown in Fig. 12. Polymer dispersed CLS were found to be electrically switchable materials four decades ago (Fig. 13).

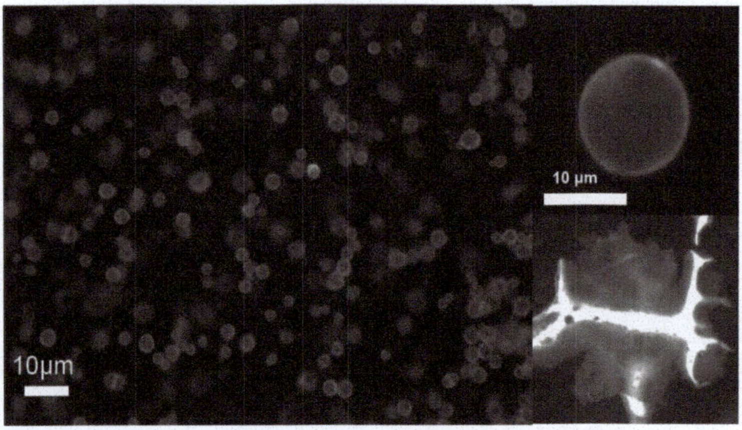

Fig. 11 Fluorescence microscope images of QDs in micro-shells, or "vesicles" and a flat membrane generated by phase transition templating. **a** Multiple shells suspended in the nematic phase, **b** a single QD shell and **c** a crumpled membrane structure (Reproduced from Ref. [12], under the terms of the CC BY 4.0 license, http://creativecommons.org/licenses/by/4.0/.)

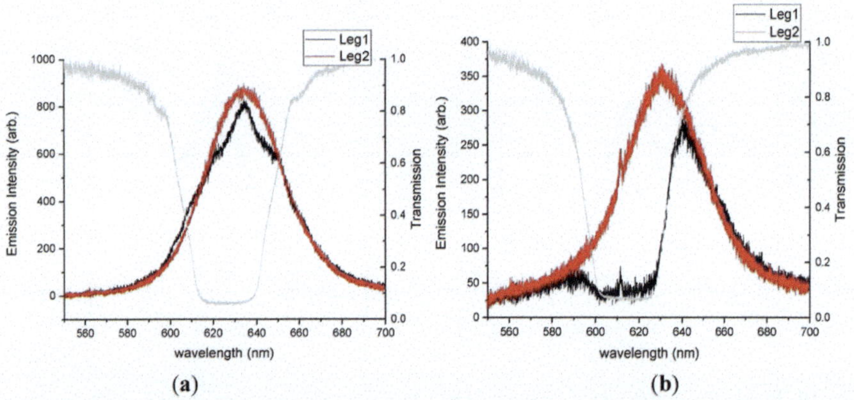

Fig. 12 PL spectra for 0.02 wt.% ODA-QDs. **a** QD dispersion in CLCs. **b** QDs located beneath CLC film (Reproduced from Ref. [12], under the terms of the CC BY 4.0 license, http://creativecommons.org/licenses/by/4.0/.)

7.2 High-Performance CLC Materials

Research on solvent free manufacturing processes and biodegradable mesogens is advancing. The development of materials with higher stability, shorter response times, and greater tunability will broaden the spectrum of viable applications. The integration of CLCs with cutting-edge fields such as metamaterials and photonic crystals opens up new functional possibilities.

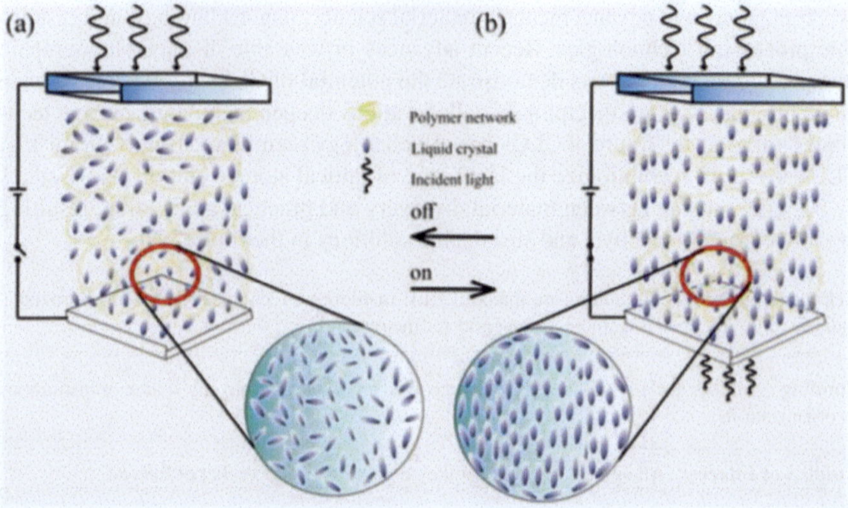

Fig. 13 Operating principle of a common polymer dispersed liquid crystal. **a** Off-state, **b** on-state (Reproduced from Ref. [60], under the terms of the CC BY 4.0 license, http://creativecommons.org/licenses/by/4.0/.) Interdisciplinary work is also leading to better theoretical models for predicting and optimizing CLC behavior under complex conditions [60, 61]

7.3 Expanding Application Domains

Beyond displays and optical storage, CLCs can play a transformative role in fields like bio photonics (e.g., biosensors), energy harvesting, and environmental monitoring.

Emerging applications in holography, augmented reality, and quantum optics hold tremendous promise [62, 63].

8 Conclusion

In this book chapter, CLCs are introduced as a foundation for modern optical technologies, providing variety types of features such as particular reflection, tunable optical response, and energy efficiency. Recent advances in materials and production techniques have good expanded their applications, particularly in optical storage and display technologies. Light-sensitive doping impurities and photo induced pitch alterations are two optical storage approaches that have appeared guarantee for multi-wavelength, high-density data storage. CLCs are difficult to utilize in a variety of classic display technologies because to developments in display applications, such as polymer-stabilized cholesteric texturing, which has enhanced brightness, viewing angles, and environmental endurance. Notwithstanding these difficulties, scalability, stability, and reaction time optimization are among the constraints being overcome by recent advancements in material science, nanotechnology, and sustainable processing technologies. Recent advances in wearable displays, holographic storage, and quantum optics demonstrate the potential of CLCs to transform a wide range of areas. As interdisciplinary collaborations deepen and manufacturing techniques improve, the future of CLCs-based technology seems increasingly promising. CLCs are set to revolutionize the landscape of optical storage and display systems by bridging the gap between material discovery and practical application, resulting in more efficient, adaptive, and sustainable solutions in the future.

Acknowledgements All authors are thankful KLE Institute of Technology Hubballi, Karnataka, India for constant encouragement and support for the research.

Funding Authors declare that they have not received any funding from any source or institutions or organizations.

Conflict of Interest All authors declare that they do not have any conflict of interest.

References

1. Raghavendra, G.M., Varaprasad, K., Jayaramudu, T.: Biomaterials: design, development and biomedical applications. Nanotechnol. Appl. Tissue Eng. 21–44 (2015). https://doi.org/10.1016/B978-0-323-32889-0.00002-9
2. Meyer, R.B.: Distortion of a cholesteric structure by a magnetic field. Appl. Phys. Lett. **14**(7), 208–209 (1969). https://doi.org/10.1063/1.1652780
3. Manhas, N., Adimule, V., Lakshminarayana, P., Keri, R., Kumar, L.S., Bathula, C.: Utilization of Co^{2+} doped Eu_2O_3 nanostructures and their effect on the dielectric and photoisomerization of liquid crystals. Phys. Fluids **35**(10) (2023). https://doi.org/10.1063/5.0171578/2916069
4. Sinha, N., Ray, G., Bhandari, S., Godara, S., Kumar, B.: Synthesis and enhanced properties of cerium doped ZnO nanorods. Ceram. Int. **40**(8), 12337–12342 (2014). https://doi.org/10.1016/J.CERAMINT.2014.04.079
5. Sharma, V., Kumar, P., Chinky, C., Gahrotra, R., Raina, K.K., Malik, P.: Effect of nano particles on electro optic properties of polymer dispersed liquid crystal in normal mode. In: AIP Conference Proceedings, vol. 2142 (2019). https://doi.org/10.1063/1.5122510
6. Zhang, J., Zhang, Y., Yang, J., Wang, X.: Beyond color boundaries: pioneering developments in cholesteric liquid crystal photonic actuators. Micromachines **15**(6), 808 (2024). https://doi.org/10.3390/MI15060808
7. Jiang, Q., et al.: Light regulation and long-lived stability of RGB colors in cholesteric liquid crystal physical gels via a mixing strategy. Soft Matter **17**(11), 3216–3221 (2021). https://doi.org/10.1039/D0SM02283G
8. Liu, B., Zheng, Z., Chen, X., Shen, D.: Low-voltage-modulated laser based on dye-doped polymer stabilized cholesteric liquid crystal. Opt. Mater. Express **3**(4), 519 (2013). https://doi.org/10.1364/OME.3.000519
9. Liu, B., et al.: Dynamically actuated soft heliconical architecture via frequency of electric fields. Nat. Commun. **13**(1) (2022). https://doi.org/10.1038/S41467-022-30486-2
10. Kemiklioglu, E., Kemiklioglu, E.: From a chiral molecule to blue phases. Liq. Cryst. Recent Adv. Fundam. Device Technol. (2017). https://doi.org/10.5772/INTECHOPEN.70555
11. Advances in cholesteric liquid crystals. MDPI (2020). https://doi.org/10.3390/books978-3-03928-229-6

12. Rodarte, A.L., Cisneros, F., Hein, J.E., Ghosh, S., Hirst, L.S.: Quantum dot/liquid crystal nanocomposites in photonic devices. Photonics **2**(3), 855–864 (2015). https://doi.org/10.3390/photonics2030855
13. Jeong, M.Y., Choi, H.J., Kwak, K., Yu, Y.: Multifunctional optical device with a continuous tunability over 500 nm spectral range using polymerized cholesteric liquid crystals. Polymers **13**(21), 3720 (2021). https://doi.org/10.3390/POLYM13213720
14. Dreher, R., Meier, G.: Optical properties of cholesteric liquid crystals. Phys. Rev. A **8**(3), 1616 (1973). https://doi.org/10.1103/PhysRevA.8.1616
15. Dolganov, P.V., Baklanova, K.D., Dolganov, V.K.: Optical properties and photonic density of states in one-dimensional and three-dimensional liquid-crystalline photonic crystals. Liq. Cryst. **47**(2), 231–237 (2020). https://doi.org/10.1080/02678292.2019.1641636
16. Adimule, V., Bathula, C., Sharma, K., Keri, R.: Study of structural, optical, dielectric properties poly-3-butyl thiophene reinforced with CO_3O_4 nanoparticles. Nano-Struct Nano-Obj. **38**, 101150 (2024). https://doi.org/10.1016/J.NANOSO.2024.101150
17. Adimule, V., Sharma, V., Kumar, P., Keri, R., Sharma, K.: Perylene-3,4,9,10-tetracarboxylic dianhydride dye intercalated liquid crystal molecules: synthesis, spectroscopic characterizations and photoluminescence dynamics. J. Mol. Liq. **413** (2024). https://doi.org/10.1016/J.MOLLIQ.2024.125990
18. Chychłowski, M.S., Kajkowska, M., Jankiewicz, B., Bartosewicz, B., Woliński, T.R., Lesiak, P.: Photopolymerization of 1D photonic structures induced by nematic-isotropic phase transition in liquid crystal. Soft Matter **19**(18), 3398–3404 (2023). https://doi.org/10.1039/D3SM00173C
19. Dolganov, P.V., Dolganov, V.K.: Photon density of states in a cholesteric photonic crystal. JETP Lett. **108**(3), 170–174 (2018). https://doi.org/10.1134/S0021364018150043
20. Sharma, V., Kumar, P.: Electro-optically oriented Kerr and orientational phase study of normal mode polymer dispersed liquid crystals—effect of dispersion of nanoparticles. J. Mol. Liq. **348**, 118030 (2022). https://doi.org/10.1016/J.MOLLIQ.2021.118030
21. Dogra, A.R., Sharma, V., Gahrotra, R., Khanra, P., Dogra, A.: Thickness dependent electro-optical behaviour of homeotropically aligned nematic liquid crystal. ECS Trans. **107**(1), 5021–5025 (2022). https://doi.org/10.1149/10701.5021ECST
22. Inoue, Y., Sasaki, S., Moritake, H.: High-quality tuning of cholesteric liquid crystal lasers based on polymer composite system. J. Appl. Phys. **127**(8) (2020). https://doi.org/10.1063/1.5136048
23. Wang, K., et al.: Research progress of electrically driven multi-stable cholesteric liquid crystals. Materials (Basel). **17**(1) (2024). https://doi.org/10.3390/MA17010136
24. Ortega, J., Folcia, C.L., Etxebarria, J.: Upgrading the performance of cholesteric liquid crystal lasers: improvement margins and limitations. Materials **11**(1), 5 (2017). https://doi.org/10.3390/MA11010005
25. Moss, M.R.: Environmental stability. Environ. Geol. 227–229 (1999). https://doi.org/10.1007/1-4020-4494-1_126
26. Fan, F.R., Tian, Z.Q., Lin Wang, Z.: Flexible triboelectric generator. Nano Energy **1**(2), pp. 328–334 (2012). https://doi.org/10.1016/j.nanoen.2012.01.004
27. Ye, Y., Guo, L., Zhong, T.: A review of developments in polymer stabilized liquid crystals. Polymers **15**(13), 2962 (2023). https://doi.org/10.3390/POLYM15132962
28. Hayat, A., et al.: Recent advances, properties, fabrication and opportunities in two-dimensional materials for their potential sustainable applications. Energy Storage Mater. **59**, 102780 (2023). https://doi.org/10.1016/J.ENSM.2023.102780
29. Jeong, M.Y., Wu, J.W.: Temporally stable and continuously tunable laser device fabricated using polymerized cholesteric liquid crystals. Jpn. J. Appl. Phys. **51**(8) (2012). https://doi.org/10.1143/JJAP.51.082702
30. Balamurugan, R., Liu, J.H.: A review of the fabrication of photonic band gap materials based on cholesteric liquid crystals. React. Funct. Polym. **105**, 9–34 (2016). https://doi.org/10.1016/j.reactfunctpolym.2016.04.012
31. Dreher, R., Meier, G., Saupe, A.: Selective reflection by cholesteric liquid crystals. Mol. Cryst. **13**(1), 17–26 (1971). https://doi.org/10.15354/SI.22.RE051

32. Sharma, V., Kumar, P., Sharma, A., Chinky, C., Raina, K.K.: Droplet configuration control with orange azo dichroic dye in polymer dispersed liquid crystal for advanced electro-optic characteristics. J. Mol. Liq. **233**, 122–130 (2017). https://doi.org/10.1016/J.MOLLIQ.2017.03.012
33. Cholesteric Liquid Crystal The Future of Display Technology. Accessed: 09 Dec 2024. [Online]. Available: https://www.dakenchem.com/cholesteric-liquid-crystal/
34. Wang, T., Zhao, J., Wu, L., Liu, W., Li, Y., Yang, Y.: Polymer-stabilized cholesteric liquid crystal films with double reflection bands prepared based on the competition between photopolymerization and photoisomerization. ACS Appl. Mater. Interfaces **15**(37), 44314–44321 (2023). https://doi.org/10.1021/ACSAMI.3C09576
35. Chen, X., et al.: Broadband reflection of polymer-stabilized chiral nematic liquid crystals induced by a chiral azobenzene compound. Chem. Commun. **50**(6), 691–694 (2014). https://doi.org/10.1039/C3CC47438K
36. Lee, K.M., Crenshaw, E.P., Rumi, M., White, T.J., Bunning, T.J., McConney, M.E.: Effect of cell thickness on the electro-optic response of polymer stabilized cholesteric liquid crystals with negative dielectric anisotropy. Materials **13**(3), 746 (2020). https://doi.org/10.3390/MA13030746
37. Ozols, A., Linina, E., Zabels, R., Komitov, L.: Evaluation of the characteristics of cholesteric liquid crystal diffuser element applied in multi-focal display architectures. Crystals **12**(5), 733 (2022). https://doi.org/10.3390/CRYST12050733
38. Hu, X., Hua, H.: High-resolution optical see-through multi-focal-plane head-mounted display using freeform optics. Opt. Express **22**(11), 13896 (2014). https://doi.org/10.1364/OE.22.013896
39. Itoh, Y., Morishita, D.: Development of electroresponsive functional soft materials by electroresponsive dopants. Polym. J. **55**(10), 1035–1048 (2023). https://doi.org/10.1038/s41428-023-00805-5
40. Kim, J.K., Joo, S.H., Song, J.K.: Reflective-emissive photoluminescent cholesteric liquid crystal display. Appl. Opt. **52**(34), 8280–8286 (2013). https://doi.org/10.1364/AO.52.008280
41. Zhang, X., et al.: Liquid crystal-templated chiral nanomaterials: from chiral plasmonics to circularly polarized luminescence. Light Sci. Appl. **11**(1) (2022). https://doi.org/10.1038/S41377-022-00913-6
42. Sun, C., Lu, J.: Temperature self-adaptive and color-adjustable smart window based on templated cholesteric liquid crystals. Polymers (Basel) **16**(1) (2024). https://doi.org/10.3390/POLYM16010082
43. Ryabchun, A., Bobrovsky, A.: Cholesteric liquid crystal materials for tunable diffractive optics. Adv. Opt. Mater. **6**(15) (2018). https://doi.org/10.1002/ADOM.201800335
44. Mitov, M.: Cholesteric liquid crystals with a broad light reflection band. Adv. Mater. **24**(47), 6260–6276 (2012). https://doi.org/10.1002/ADMA.201202913
45. Gahrotra, R., Sharma, V., Dogra, A.R., Malik, P., Kumar, P.: Temperature dependent morphological and electro-optical characteristics of dye doped cholesteric liquid crystal. ECS Trans. **107**(1), 5481–5486 (2022). https://doi.org/10.1149/10701.5481ECST
46. Shaban, H., Wu, P.C., Jia, Y.F., Lee, W.: Overlooked ionic contribution of a chiral dopant in cholesteric liquid crystals. Materials **17**(20), 5080 (2024). https://doi.org/10.3390/MA17205080
47. Froyen, A.A.F., Debije, M.G., Schenning, A.P.H.J.: Polymer dispersed cholesteric liquid crystal mixtures for optical time-temperature integrators. Adv. Opt. Mater. **10**(22), 2201648 (2022). https://doi.org/10.1002/ADOM.202201648
48. Mitov, M., Bourgerette, C., de Guerville, F.: Self-assembly of metallic nanoparticles in a cholesteric liquid crystal. Nanofabrication Technol. **5220**, 20 (2003). https://doi.org/10.1117/12.503558
49. Pathinti, R.S., Tatipamula, A.K., Vallamkondu, J.: ZnO nanoparticles dispersed cholesteric liquid crystal based smart window for energy saving application. J. Alloys Compd. **963**, 171198 (2023). https://doi.org/10.1016/J.JALLCOM.2023.171198

50. Adimule, V., Yallur, B.C., Keri, R., Bathula, C., Batakurki, S.: Microstructure, photoluminescence and electrical properties of SmxGd(1–x): SrO hybrid nanomaterials synthesized via facile coprecipitation method. Electron. Mater. Lett. **19**(3), 278–297 (2023). https://doi.org/10.1007/S13391-022-00394-0/METRICS
51. Yu, C., et al.: Photopolymerizable biomaterials and light-based 3d printing strategies for biomedical applications. Chem. Rev. **120**(19), 10695–10743 (2020). https://doi.org/10.1021/ACS.CHEMREV.9B00810
52. Peng, Y., Wei, X., Liang, Y., Wang, X., Chen, S., Niu, X.: Advances in structural color composite films based on cellulose nanocrystals. Ind. Crops Prod. **221** (2024). https://doi.org/10.1016/J.INDCROP.2024.119294
53. Guo, M., et al.: Cellulose-derived cholesteric liquid crystal: towards smart and sustainable colorimetric sensors. Ind. Crops Prod. **192**, 116007 (2023). https://doi.org/10.1016/J.INDCROP.2022.116007
54. Liu, J., Tai, W., Wang, D., Su, J., Yu, L.: Cholesteric liquid crystal photonic hydrogel films immobilized with urease used for the detection of Hg^{2+}. Chemosensors **10**(4), 140 (2022). https://doi.org/10.3390/CHEMOSENSORS10040140
55. Murphy, C., Abbott, N.: Liquid crystals with reduced toxicity and applications thereof. Art. no. US7303694 B2 (2007)
56. Li, J., et al.: Dynamic control of multiple optical patterns of cholesteric liquid crystal microdroplets by light-driven molecular motors. CCS Chem. **6**(2), 427–438 (2024). https://doi.org/10.31635/CCSCHEM.023.202303171
57. Liu, H., et al.: Triboelectric-optical responsive cholesteric liquid crystals for self-powered smart window, E-paper display and optical switch. Sci. Bull. **66**(19), 1986–1993 (2021). https://doi.org/10.1016/J.SCIB.2021.05.016
58. Zhang, H., Miao, Z., Shen, W.: Development of polymer-dispersed liquid crystals: from mode innovation to applications. Compos. Part A Appl. Sci. Manuf. **163** (2022). https://doi.org/10.1016/j.compositesa.2022.107234
59. Saeed, M.H., et al.: Recent advances in the polymer dispersed liquid crystal composite and its applications. Molecules **25**(23) (2020). https://doi.org/10.3390/MOLECULES25235510
60. Omar, Y., Plenio, M.B., Huelga, S., Rasetti, M.: Quantum effects in biology and their applications to light harvesting and sensing. Procedia Comput. Sci. **7**, 92–95 (2011). https://doi.org/10.1016/J.PROCS.2011.12.028
61. Gollapelli, B., Suguru Pathinti, R., Vallamkondu, J.: Carbon quantum dots doped cholesteric liquid crystal films and microdroplets for anti-counterfeiting. ACS Appl. Nano Mater. **5**(8), 11912–11922 (2022). https://doi.org/10.1021/ACSANM.2C02933
62. Leal-Junior, A., Soares, M.S., de Almeida, P.M., Marques, C.: Cholesteric liquid crystals sensors based on nanocellulose derivatives for improvement of quality of human life: a review. Adv. Sens. Res. **2**(10), 2300022 (2023). https://doi.org/10.1002/ADSR.202300022
63. Zhang, Y., Yang, H., Chen, Y., Yu, H.: Progress in fabrication and applications of cholesteric liquid crystal microcapsules. Chem. A Eur. J. **30**(8), e202303198 (2024). https://doi.org/10.1002/CHEM.202303198

Recent Advances in the Optical Properties of Cholesteric Liquid Crystals

Sushant Kakati, Raghavendra P. Bakale, S. N. Mathad, Sheela A. Sangam, and Gangadhar Bagihalli

Abstract Recent advances in the optical properties of cholesteric liquid crystals (CLCs) have significantly expanded their potential applications. The incorporation of chiral dopants has led to enhanced luminescence, making CLCs suitable for innovative applications in lighting and displays. Additionally, the pitch of the helical structure can be modulated to achieve tunable emission wavelengths, rendering CLCs suitable for optical filters and sensors. CLCs also exhibit polarized emission, making them an attractive choice for polarization-sensitive applications such as optical storage and imaging. Furthermore, the optical activity of CLCs has been harnessed for applications in optical rotation and circular dichroism spectroscopy. Researchers have also successfully engineered photonic bandgaps in CLCs, enabling applications in optical filtering and sensing. The optical properties of CLCs can be dynamically tuned by applying external stimuli such as temperature, electric fields, or light. The development of nanostructured CLCs has led to unique optical properties, including enhanced luminescence and tunable emission wavelengths. CLCs have also been explored for biological imaging applications, leveraging their optical activity and luminescence properties. They have been developed as optical sensors for detecting changes in temperature, concentration, and other environmental factors. Moreover, CLCs have been integrated into photonic devices such as lasers, modulators, and optical switches, enabling new functionalities and applications. This chapter provides a comprehensive review of the current state of research in CLCs, exploring

S. Kakati · S. N. Mathad (✉)
Department of Engineering Physics, KLE Institute of Technology, Airport Road, Gokul, Hubballi 580030, Karnataka, India
e-mail: physicssiddu@kleit.ac.in

R. P. Bakale (✉)
Department of Chemistry, Jain College of Engineering, Belagavi, Karnataka, India
e-mail: raghubakale@gmail.com

S. A. Sangam
Department of Chemistry, K.L.S Gogte Institute of Technology, Belagavi, Karnataka, India
e-mail: sasangam@git.edu

G. Bagihalli
Department of Engineering Chemistry, KLE Institute of Technology, Airport Road, Gokul, Hubballi 580030, Karnataka, India

P. Kumar et al. (eds.), *Cholesteric Liquid Crystals*, Engineering Materials,
https://doi.org/10.1007/978-981-96-3821-5_3

their optical properties and potential applications, and discussing future directions and challenges in the field.

Keywords Cholesteric Liquid Crystals (CLCs) · Optical Properties · Luminescence · Photonic Devices · Nanostructured Materials

1 Introduction

Cholesteric liquid crystals (CLCs) have emerged as a versatile and dynamic material platform [1, 2], revolutionizing the field of optics [3, 4] and photonics [5, 6]. Recent breakthroughs in understanding and manipulating their optical properties have unlocked a vast array of innovative applications [7, 8], transforming CLCs into a cornerstone of modern materials science. With their unique helical structure [9, 10] and tunable optical properties [11, 12], CLCs have become an attractive choice for researchers and engineers seeking to develop cutting-edge technologies [13, 14]. From enhanced luminescence [15, 16] and polarized emission [17, 18] to photonic bandgaps [19, 20] and dynamic tuning [21, 22], CLCs have demonstrated unparalleled potential in various fields, including lighting [23, 24], displays [25, 26], optical filtering [27, 28], sensing [29, 30], and biological imaging [31, 30]. The incorporation of chiral dopants [32, 33], nano structuring [34, 35], and external stimuli has further expanded the possibilities [36, 37], enabling CLCs to address complex challenges in optics [38], photonics [39], and biotechnology [40].

This chapter provides a comprehensive review of the current state of research in CLCs, delving into their optical properties [41], applications [42], and future directions are CLCs are tabulated in Table 1. By exploring the latest advances [43] and challenges in the field [44], this chapter aims to inspire new research avenues and foster collaboration among scientists, engineers, and industry experts, ultimately driving innovation and applications of CLCs in various disciplines as shown in Fig. 1.

2 Optical Properties of Cholesteric Liquid Crystals (CLCs)

The optical properties of CLCs are characterized by their unique helical structure [45], which makes them attractive for various applications. CLCs exhibit selective reflection [46], where they selectively reflect circularly polarized light depending on the helix pitch [47] and wavelength [48]. Additionally, they display circular dichroism [49], absorbing one circular polarization while transmitting the other. The chiral structure of CLCs also enables them to rotate the plane of polarized light, a phenomenon known as optical rotation [50]. Furthermore, CLCs exhibit birefringence [51], with different refractive indices for ordinary and extraordinary rays. CLCs can also exhibit luminescence [52], with emission wavelengths dependent on the molecular structure [53]. Notably, they emit polarized light, making

Table 1 Various aspects, properties and applications of CLCs

Aspect	Advances	Applications	Insights
Photonic band gaps	Enhanced tunability of photonic band gaps through external stimuli (electric, magnetic fields, temperature)	Optical filters, lasers, and reflective displays	Advances in stimuli-responsive materials
Circularly polarized light	Improved control of circularly polarized light via doping with nanoparticles or chiral additives	Circular polarizers, sensors, and optical data storage	Techniques for optimizing polarization properties
Structural colors	Refinement in color generation via layer thickness and pitch modulation	Decorative coatings, anti-counterfeiting	Bio-inspired designs mimicking beetle shells
Dynamic tuning	Development of real-time tunable devices using liquid crystal elastomers and photo-responsive materials	Smart windows, adaptive optics	Integration of liquid crystal with responsive polymers
Defect structures	Investigation of topological defects for unique optical effects	Metamaterials, advanced optics	Studies on disclination lines and defect-based systems
Chiral doping	Use of advanced dopants to achieve specific reflection wavelengths and reduced scattering losses	Display technologies, spectroscopic applications	Exploration of novel organic and inorganic chiral dopants
Laser emission	Development of CLC-based lasers with low-threshold, tunable wavelength emission	Tunable laser sources, optical communication	Inclusion of dyes and quantum dots for lasing
Thermal stability	Enhanced stability under thermal fluctuations through polymer stabilization	High-performance displays, rugged optical devices	Use of polymer-stabilized CLCs (PSCLCs)
Electro-optic effects	Faster response times and lower voltage operation via optimized molecular alignments	Modulators, shutters	Advances in material synthesis and alignment techniques
Bistable displays	Development of bistable cholesteric displays (BCD) for low-energy, high-contrast applications	E-readers, signage	Improvements in display bistability and refresh rates

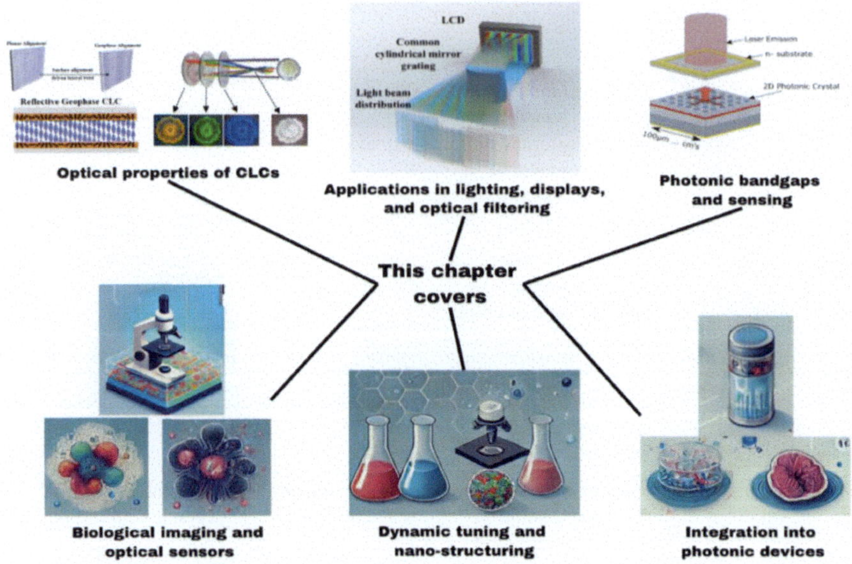

Fig. 1 Schematic of applications of CLCs

them useful for applications requiring polarized emission [54]. Moreover, CLCs can exhibit photonic bandgaps [55], prohibiting certain wavelengths from propagating [56]. Various factors influence the optical properties of CLCs, including helix pitch, molecular structure [57],, chiral dopant concentration [58], temperature [59], electric field [60], and nano structuring [61].

The authors Jung et al. demonstrated the thermo-responsive reconfigurability of chiral plasmonic films, which exhibit reversible changes in their chiroptical properties upon heating and cooling cycles. The complete disappearance of the CD response at elevated temperatures and its recovery upon cooling indicate that the chirality transfer from the chiral dopants to the plasmonic additives is not the primary factor responsible for the CD signal amplification. Instead, the enhanced light scattering and absorption preferentially toward the CPL with a specific orientational direction through the excitation of the LSPR of the assembled AuNRs play a crucial role in the enhancement of the chiroptical activity of the composite films. The high CD signal recovery of the chiral plasmonic films after consecutive heating and cooling cycles highlights their potential for applications in reversible optical storage, sensing, and other optoelectronic devices. Circular Dichroism (CD) spectra and POM images show thermal behavior and structural changes in R5011-doped chiral plasmonic films with varying fNR2 concentrations shown in Fig. 2. The unique optical properties of CLCs make them suitable for various applications, such as optical filters, polarized displays, optical sensors, biological imaging, lasers and optical devices. Understanding the optical properties of CLCs is crucial for optimizing their performance in these applications. By leveraging their optical properties, researchers and engineers can develop innovative technologies that harness the potential of CLCs.

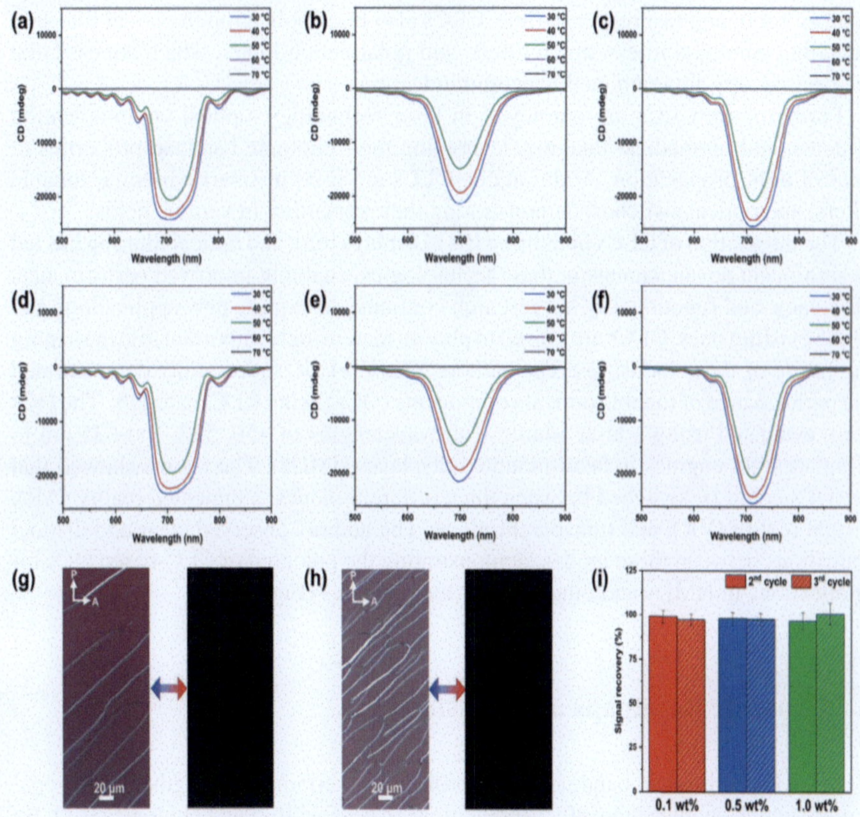

Fig. 2 Circular Dichroism (CD) Spectra and Polarized Optical Microscopy (POM) Images of R5011-Doped Chiral Plasmonic Films. CD spectra of R5011-doped chiral plasmonic films with varying concentrations of fNR2 (0.1, 0.5, and 1 wt.%) are shown during the 2nd and 3rd heating/cooling cycles (**a**–**f**). POM images of R5011-doped chiral plasmonic films with 0 and 0.1 wt.% fNR2 at $T_{cholesteric}$ and T_{iso} are also presented (**g**–**h**). The direction of the polarizer and analyzer is indicated by P and A, respectively. Additionally, the recovery of CD signals in R5011-doped chiral plasmonic films with different fNR2 concentrations after the 2nd and 3rd heating/cooling cycles (Reproduced from Ref. [60], under the terms of the CC BY 4.0 license, http://creativecommons.org/licenses/by/4.0/.)

3 Applications in Lighting, Displays, and Optical Filtering

CLCs have numerous applications in lighting [62], displays [63], and optical filtering due to their unique optical properties. In lighting, CLCs are used in reflective displays, backlighting, and color filters, enabling energy-efficient and high-color-accuracy solutions. Their ability to selectively reflect circularly polarized light makes them ideal for polarized displays, reducing glare and enhancing visibility. In displays, CLCs are utilized in liquid crystal displays (LCDs), electroluminescent displays, and flexible display, offering advantages such as low power consumption, high contrast ratio, and fast response time. CLCs also find application in optical filtering, including bandpass filters, notch filters, and polarization filters, which are essential in spectroscopy, imaging, and telecommunications.

Furthermore, CLCs are employed in laser technology, optical communication systems, and biomedical imaging, leveraging their photonic bandgap properties to control light propagation. Additionally, CLCs are used in smart windows, tunable filters, and optical switches, demonstrating their versatility in various fields.

The integration of CLCs in lighting [64], displays [65], and optical filtering has led to significant advancements in these technologies, enabling improved performance, efficiency, and functionality. As research continues to explore new applications and refine existing ones, CLCs are poised to play an increasingly important role in shaping the future of these industries. The authors Xia, Y. *et. al.* successfully demonstrated the performance of tunable optical components (OCs) using CLC materials. The OCs were evaluated using a laser source with wavelengths of 450, 532, and 635 nm to simulate light engines in head-mounted displays (HMDs). The results showed that the OCs could be switched between three different modes: augmented reality (AR), virtual reality (VR), and transparent mode. The authors observed clear and distinct transitions between these modes, demonstrating the potential of CLC-based OCs for applications in HMDs and other display technologies (Fig. 3).

4 Photonic Bandgaps and Sensing

CLCs exhibit photonic bandgaps [66], prohibiting certain wavelengths from propagating, which enables innovative applications in sensing [67] and photonics [68]. By precisely controlling the helix pitch and defect density [69], CLCs can be engineered to create tailored photonic bandgaps for specific wavelengths [70]. This unique property allows CLCs to selectively reflect or transmit light [71], making them ideal for optical sensing applications [72]. The authors Kocaman, C. *et. al.* investigated the reusability of CNC-capsulated CLC marbles through a three-step sequential experiment, exposing them to varying concentrations of toluene vapor. The results showed that the marbles exhibited reversible responses to toluene concentrations within the OSHA permission limits (<500 ppm), demonstrating their potential as VOC-responsive materials. However, exposure to high concentrations of toluene (>5000 ppm) led to the destruction of the cholesteric feature and the loss of responsiveness. The authors' findings suggest that CNC-capsulated CLC marbles can be reused as VOC-responsive materials within safe exposure limits, retaining their encapsulated responsive cholesteric droplet structure as shown in Fig. 4. CLCs-based sensors can detect changes in temperature [73], humidity [74], pressure [75], and chemical composition [76] with high sensitivity [77] and selectivity [78]. For instance, CLCs can be designed to shift their photonic bandgap in response to environmental changes [79], enabling real-time monitoring of physical or chemical parameters [80]. Additionally, CLCs' photonic bandgaps can be exploited for biosensing [81], allowing for label-free detection of biomolecules and cells [82].

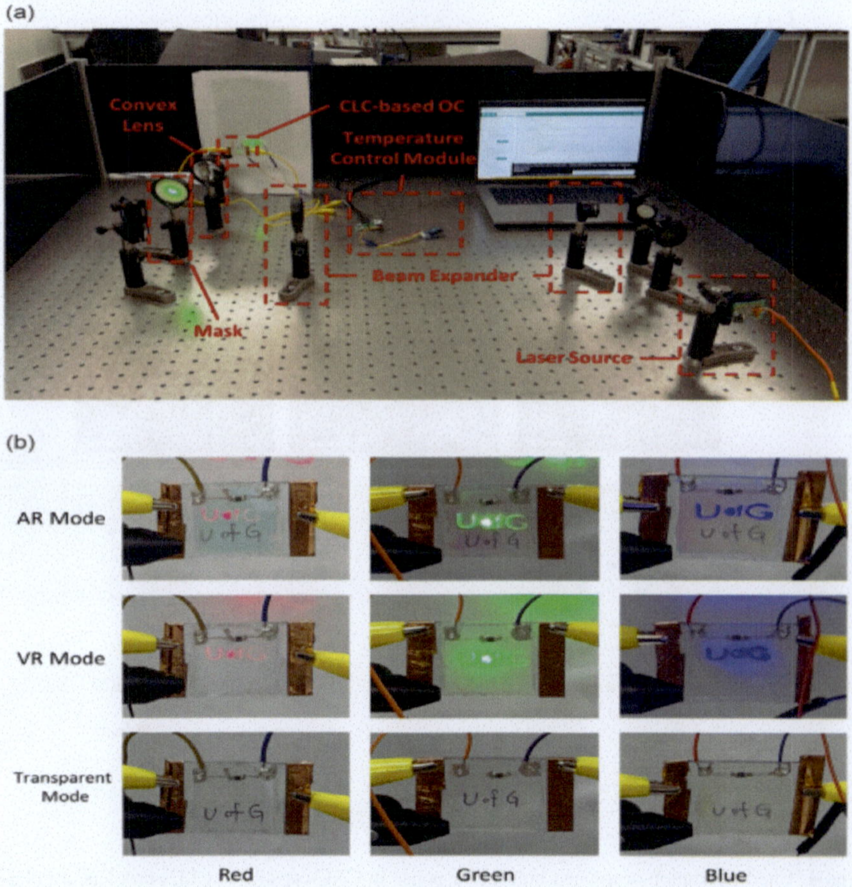

Fig. 3 **a** Optical system for projecting virtual images on cholesteric liquid crystal (CLC)-based optical components (OC). **b** Demonstration of the OC's switching capability between augmented reality (AR), virtual reality (VR), and transparent modes using RGB laser input (Reproduced from Ref. [65], under the terms of the CC BY 4.0 license, http://creativecommons.org/licenses/by/4.0/.)

As shown in Fig. 5, authors Li et al. demonstrate a novel dual-wavelength humidity monitoring system utilizing a single q-plate encoded CLC polymer (CLCP) film and two inexpensive laser diodes. The system enables real-time and visual monitoring of a broad range of humidity levels, from 10 to 95% RH. The reflected diffraction patterns of the film exhibit a distinct "8" pattern, which gradually appears and disappears as the humidity level changes. The intersection of the reflectance curves of the two wavelengths at 76% RH provides a clear visual indication of the humidity level. The results show that this system is capable of achieving extended-range humidity monitoring, making it a promising solution for various applications.

The photonic band gap (PBG) in photonic crystals arises from their periodic structures, creating a refractive index field that blocks certain wavelengths while allowing

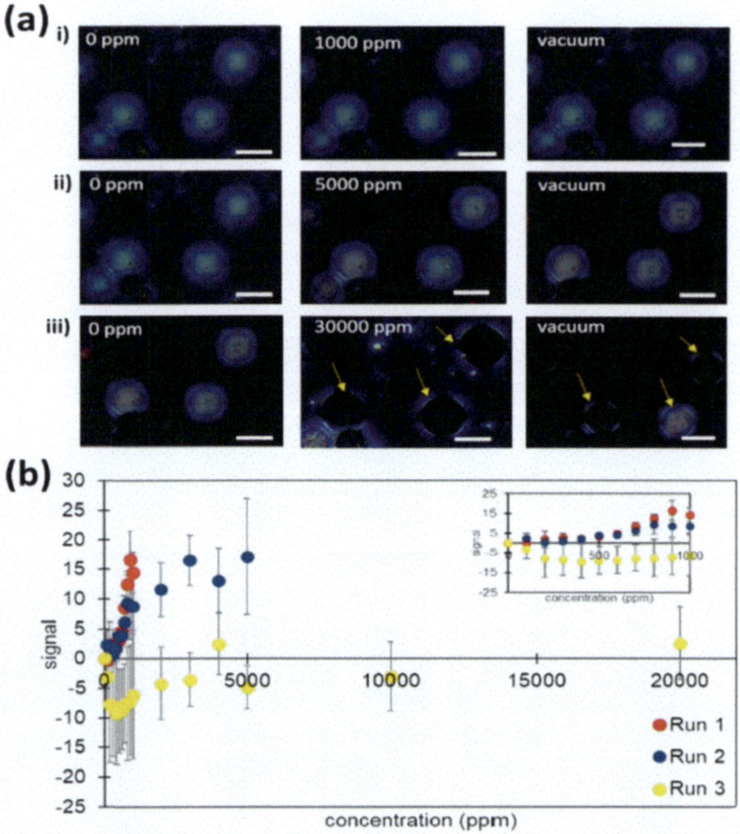

Fig. 4 Characterization of toluene-induced response in CNC-capsulated CLCs droplets. **a** Polarized light micrographs: reflection-mode images of CNC-capsulated CLC droplets synthesized from 23% chiral dopant and balanced E7, exposed to varying toluene concentrations. Images were captured at specified conditions during each run. Scale bar: 200 μm. **b** Quantitative analysis: plots showing the average response signals of three independent samples, calculated from images collected during each run (Reproduced from Ref. [73], under the terms of the CC BY 4.0 license, http://creativecommons.org/licenses/by/4.0/.)

others to propagate and various samples shown in Fig. 6. The integration of CLCs with other materials and technologies, such as nano-particles [83], polymers [84], and optical fibers [85], further expands their sensing capabilities. CLCs-based photonic bandgap structures also find applications in optical communication systems [86], laser technology [87], and imaging [88], where precise control over light propagation is crucial. By harnessing the photonic bandgap properties of CLCs, researchers and engineers can develop cutting-edge sensing technologies with unprecedented performance and functionality [89].

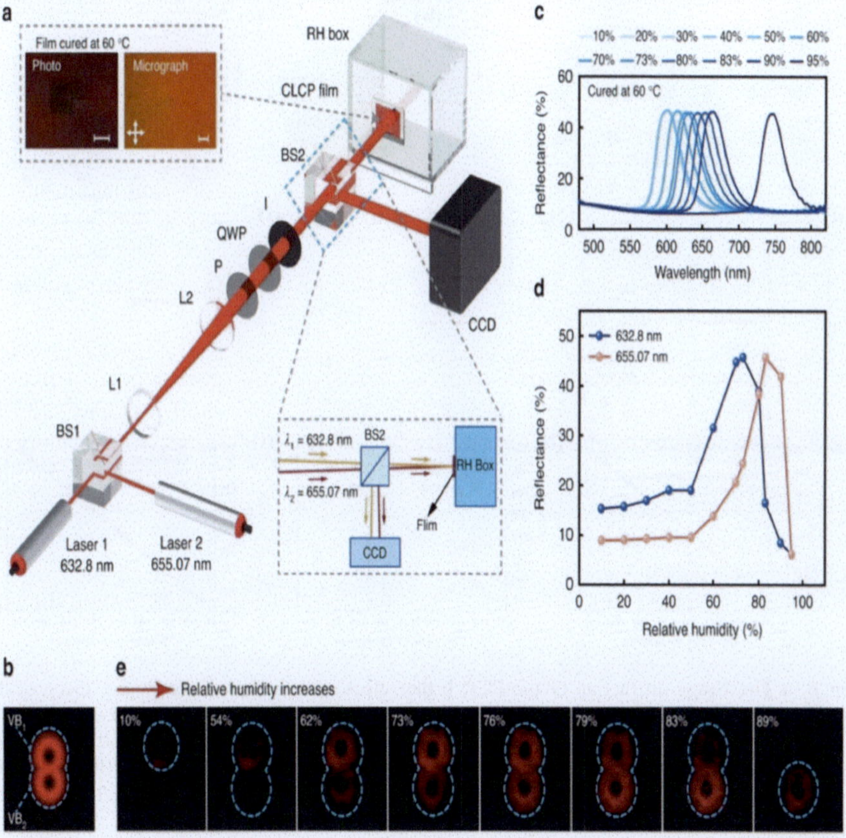

Fig. 5 Dual-wavelength humidity monitoring system. **a** Schematic diagram of the system, utilizing two lasers (Laser 1: 632.8 nm, 10 mW/cm^2; Laser 2: 655.07 nm, 10 mW/cm^2) to monitor humidity levels. Insets: Photograph (scale bar: 1 mm) and micrograph (scale bar: 100 μm) of the humidity-responsive CLCP film encoded with a single q-plate (q = + 0.5) pattern, cured at 60 °C. Bottom right inset: Schematic diagram of the optical path. **b** Schematic diagram of the reflected diffraction light field distribution. **c** Reflection spectra of humidity-responsive CLCP films cured at 60 °C. **d** Reflectance of the film at two wavelengths (λ1 = 632.8 nm and λ2 = 655.07 nm) across varying relative humidity (RH) levels. **e** Reflected diffraction patterns of a CLCP film encoded with a single q-plate, observed under the dual-wavelength humidity monitoring system across varying RH levels (Reproduced from Ref. [80], under the terms of the CC BY 4.0 license, http://creativecommons.org/licenses/by/4.0/.)

5 Biological Imaging and Optical Sensors

CLCs have emerged as versatile tools in biological imaging [90] and optical sensing [91], leveraging their unique optical properties to probe biological systems [92]. CLCs-based optical sensors can detect biomolecules [93], cells [94], and tissues with high sensitivity [95] and selectivity [96], enabling real-time monitoring of biological

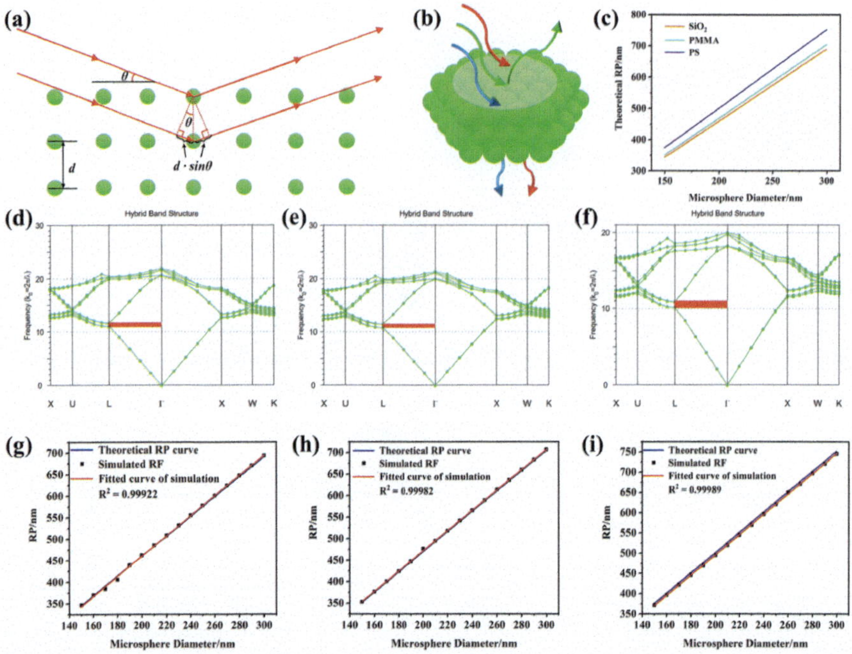

Fig. 6 **a** The Bragg diffraction of the PCs; **b** the selective reflection of the PCs; **c** theoretical diameter–reflection peak lines of the SiO_2 PCs, PMMA PCs and PS PCs; the Rsoft–Band solve simulation of 240 nm. **d** SiO_2 PCs, **e** PMMA PCs and **f** PS PCs; the comparison of theoretically calculated and simulated reflection peaks. **g** SiO_2 PCs, **h** PMMA PCs and **i** PS PCs (Reproduced from Ref. [85], under the terms of the CC BY 4.0 license, http://creativecommons.org/licenses/by/4.0/.)

processes [97]. By incorporating targeted ligands or receptors, CLCs can selectively bind to specific biomolecules, triggering changes in their optical properties [98]. Various biomedical applications of CLCs shown in Fig. 7.

In biological imaging, CLCs are used as contrast agents [99], enhancing the visualization of cellular structures and tissues [100]. Their ability to reflect circularly polarized light enables the differentiation of chiral molecules [101], facilitating the study of protein structures [102] and cellular metabolism [103]. CLCs-based probes have been developed for fluorescence microscopy [104], Raman spectroscopy [105], and optical coherence tomography [106], offering improved resolution [107] and contrast [108].

Moreover, CLCs are being explored for point-of-care diagnostics [109], detecting diseases such as cancer [110], infectious diseases [111], and neurological disorders [112]. Their integration with microfluidics and lab-on-a-chip devices enables the development of portable, low-cost biosensors [113]. The biocompatibility and non-toxicity of CLCs make them suitable for in vivo imaging [114] and sensing applications [115], opening up new avenues for understanding biological systems [116] and diagnosing diseases at the molecular level [117].

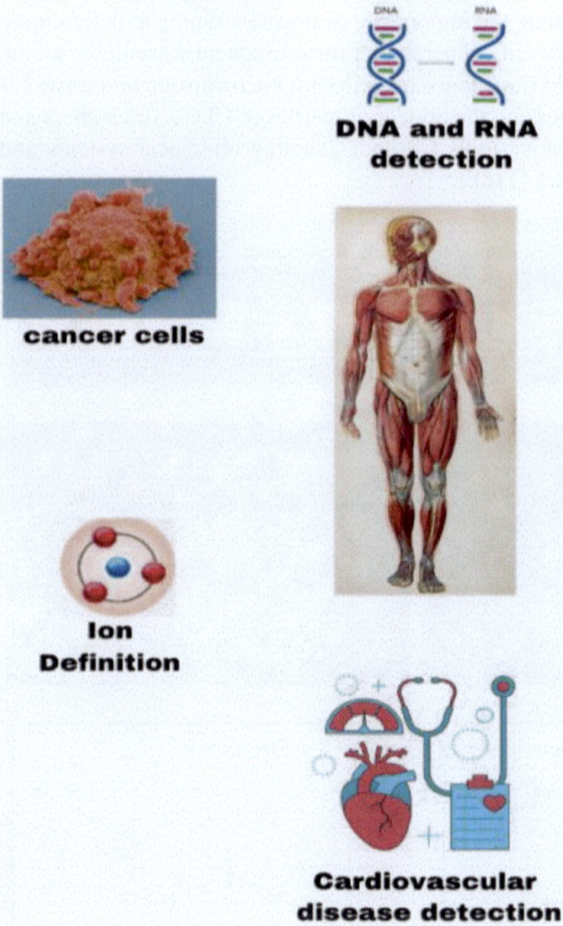

Fig. 7 Various applications of biomedical applications of CLCs

Key applications of CLCs in biological imaging and sensing include:

- Biomolecule detection
- Cellular imaging
- Tissue diagnostics
- Point-of-care testing
- In vivo imaging
- Biomedical research

The authors Wang et al. successfully fabricated hydroxypropyl cellulose (HPC) microcapsules using double-emulsion micro-fluidics, enabling precise control over core size and number. By adjusting the flow rates of internal, middle, and external phases, microcapsules with varying shell thickness were obtained which is shown in Fig. 8. The results demonstrate the importance of flow rate tuning in determining microcapsule shell thickness, a critical parameter for subsequent solvent extraction. This study provides a valuable framework for tailoring microcapsule properties for various applications. By harnessing the optical properties of CLCs, researchers and clinicians can develop innovative tools for understanding biological systems and improving healthcare outcomes [118].

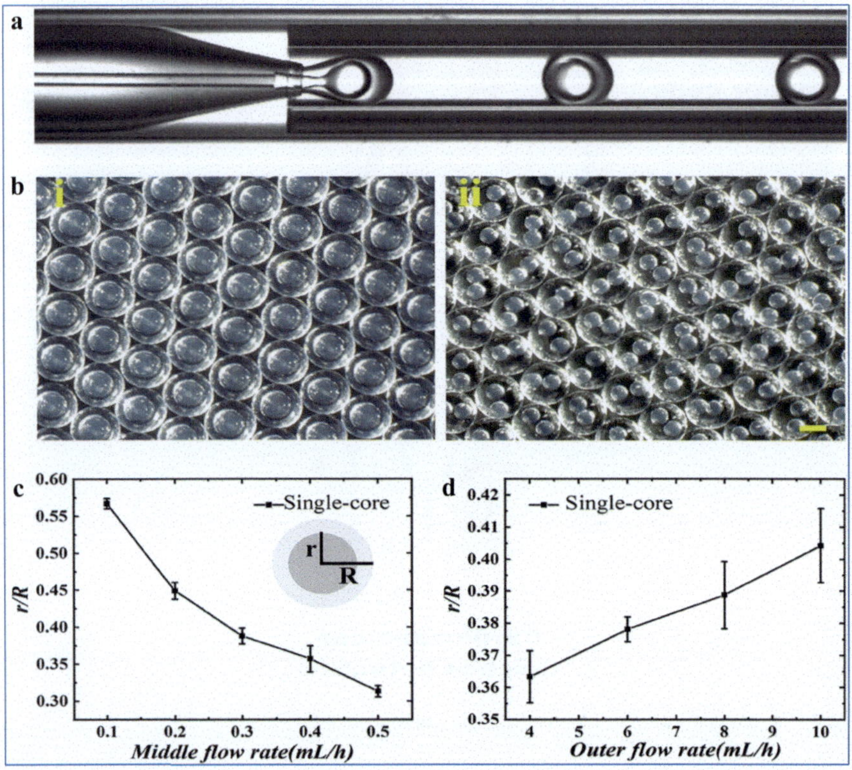

Fig. 8 **a** Real-time imaging of single-core double emulsion droplet formation. **b** Microscopic images of microcapsules under reflected light: (i) single-core microcapsules (ii) double-core microcapsules. **c** Relationship between r/R and middle phase flow rate (internal flow rate: 30 μL/h, external flow rate: 5 mL/h). **d** Relationship between r/R and external phase flow rate (internal flow rate: 30 μL/h, middle flow rate: 0.3 mL/h). Scale bar: 200 μm 119 (Reproduced from Ref. [], under the terms of the CC BY 4.0 license, http://creativecommons.org/licenses/by/4.0/.)

6 Dynamic Tuning and Nano Structuring

CLCs exhibit dynamic tuning [120] and nano structuring capabilities [121], enabling precise control over their optical properties [122]. External stimuli, such as temperature [123], electric fields [124], or light [125], can modulate the helix pitch [126] and defect density of CLCs [127], allowing for real-time adjustment of their photonic bandgaps [128] and optical properties [129]. Li et al. investigated the Evaporation-Induced Self-Assembly (EISA) process of CLC paints, revealing the dynamic evolution of structural color formation. The EISA process was driven by internal Marangoni flow, which aligned LC molecules and created a helical structure responsible for the coating's structural color. The authors demonstrated that the CLC paint's reflectivity increased significantly during the EISA process, reaching 42% in 60 s. After polymerization, the coating retained its high reflectivity and uniform structural color. This study provides valuable insights into the EISA process, enabling the development of advanced CLC materials with tailored optical properties which is shown in Fig. 9.

Nano structuring techniques, including nanolithography and self-assembly, enable the creation of complex CLC architectures with tailored optical properties [130]. This

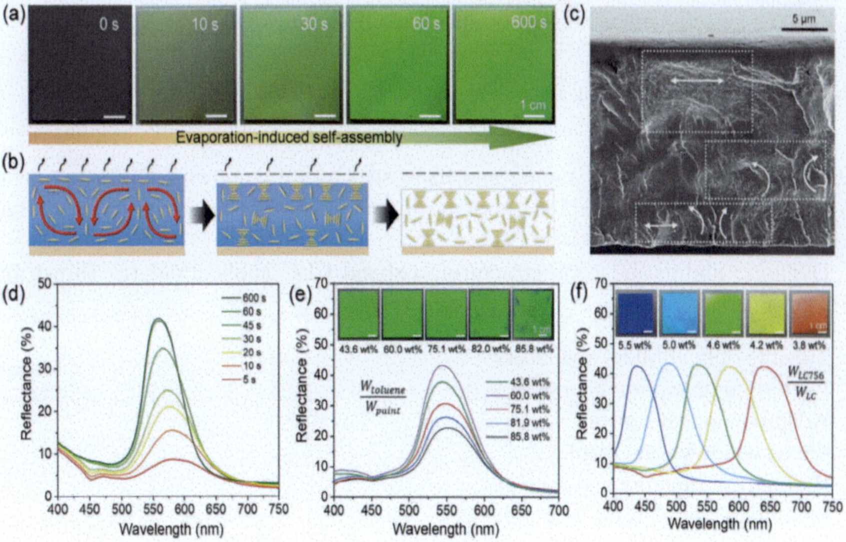

Fig. 9 EISA process and characterization of cholesteric liquid crystal (CLC) paints. **a** Evaporation-induced self-assembly (EISA) process of CLC paints. Scale bars: 1 cm. **b** Schematic illustration of internal Marangoni flow and molecular self-assembly during solvent evaporation. **c** Cross-sectional SEM images of CLCE coating. **d** Reflection spectra of CLCE coating during EISA process. **e** Reflection spectra of CLCE coatings with varying solvent amounts. Insets: corresponding photographs. Scale bars: 1 cm. **f** Reflection spectra of CLCE coatings with varying chiral dopant amounts. Insets: corresponding photographs. Scale bars: 1 cm (Reproduced from Ref. [121], under the terms of the CC BY 4.0 license, http://creativecommons.org/licenses/by/4.0/.)

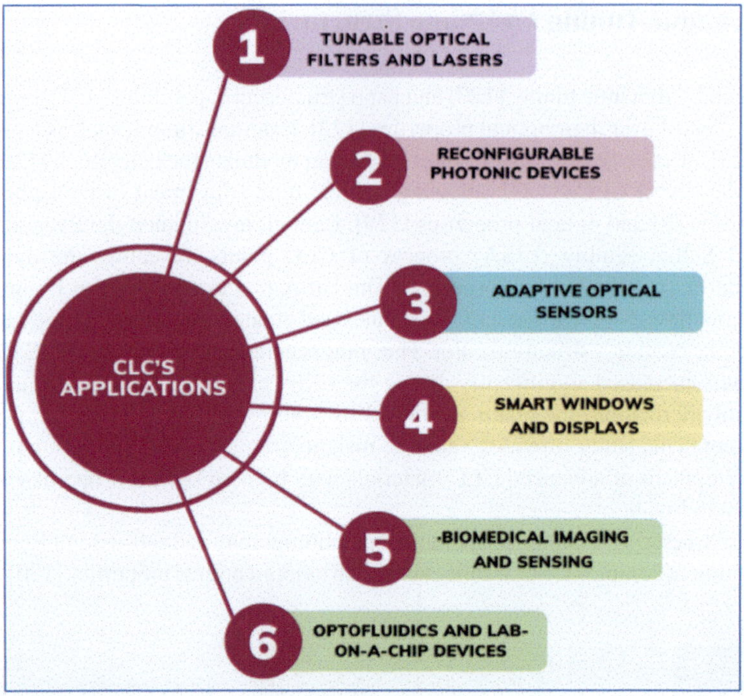

Fig. 10 Various dynamic tuning and nanostructured CLC applications

synergy between dynamic tuning and nano structuring empowers CLCs to adapt to changing environments [131], making them suitable for applications requiring flexibility and responsiveness [119].

Dynamic tuning and nano structuring expand CLC applications (as shown in Fig. 10) in:

- Tunable optical filters and lasers
- Reconfigurable photonic devices
- Adaptive optical sensors
- Smart windows and displays
- Biomedical imaging and sensing
- Optofluidic and lab-on-a-chip devices

Researchers and engineers can unlock the full potential of CLCs, driving innovation in photonics, biomedicine, and materials science.

7 Integration into Photonic Devices

CLCs are being integrated into photonic devices, leveraging their unique optical properties to enhance performance and functionality. CLCs are being incorporated into various devices, including lasers, optical switches, modulators, and photonic crystals, to achieve tunable emission, polarization control, and enhanced optical nonlinearity.

The integration of CLCs into photonic devices enables.

- Tunable lasers with precise wavelength control
- High-speed optical switches for telecommunications
- Polarization-sensitive modulators for optical communication
- Photonic crystals with adjustable bandgaps
- Reconfigurable optical interconnects

CLCs' compatibility with silicon photonics, fiber optics, and other photonic platforms facilitates seamless integration into existing systems. Researchers and engineers are exploring CLC-based photonic devices for applications in optical communication, sensing, imaging, and data processing.

Key benefits of CLC integration into photonic devices include:

- Enhanced optical properties
- Dynamic tuning and reconfigurability
- Compact and low-power designs
- Improved signal processing and transmission
- Increased functionality and versatility

By integrating CLCs into photonic devices, innovators can develop cutting-edge technologies that transform industries and revolutionize communication, healthcare, and information processing.

8 Future Directions and Challenges

As research in CLCs continues to advance, several future directions and challenges emerge. Future research will focus on exploring new materials and architectures, enhancing optical properties, and scaling up fabrication processes. Integrating CLCs with emerging technologies like metamaterials, graphene, and quantum dots will unlock novel applications. Furthermore, developing CLC-based devices for wearable technology, biomedical imaging, and environmental sensing will require innovative solutions.

Key challenges include:

- Improving stability and durability
- Scaling up fabrication while maintaining uniformity

- Developing precise control over optical properties
- Integrating CLCs with existing photonic platforms
- Addressing toxicity and biocompatibility concerns

Overcoming these challenges will require interdisciplinary collaborations between materials scientists, physicists, engineers, and biologists. The development of CLC-based technologies has the potential to transform various industries, from optics and photonics to healthcare and energy.

Future directions for CLC research include:

- Exploring new materials and architectures
- Enhancing optical properties
- Scaling up fabrication processes
- Integrating CLCs with emerging technologies
- Developing wearable and implantable devices
- Advancing biomedical imaging and sensing

By addressing the challenges and pursuing these future directions, researchers can unlock the full potential of CLCs and create transformative technologies.

9 Conclusion

In conclusion, recent breakthroughs in CLCs have revolutionized their optical properties, unlocking vast potential for innovative applications. The incorporation of chiral dopants, tunable emission wavelengths, polarized emission, and optical activity have made CLCs ideal for various uses, including lighting, displays, optical filters, sensors, optical storage, imaging, and biological imaging. Advances in nanostructured CLCs, dynamic tuning, and integration into photonic devices have further expanded their capabilities. This comprehensive review highlights the current state of CLC research, underscoring their transformative impact on materials science, optics, and photonics. As research continues to advance, CLCs are poised to enable groundbreaking applications in fields ranging from biotechnology to telecommunications.

Key Takeaways:

- Enhanced luminescence and tunable emission wavelengths
- Polarized emission and optical activity
- Applications in lighting, displays, optical filters, sensors, and biological imaging
- Dynamic tuning via external stimuli
- Integration into photonic devices
- Future directions: biotechnology, telecommunications, and materials science

This encapsulates the significance of CLC research, emphasizing its far-reaching implications and potential for innovative applications.

Acknowledgements Sushaant S.K., Shridhar Mathad and Gangagadhar Bagihalli are also thankful to KLEIT, Hubballi, Karnataka, India for constant encouragement and support for the research. Ragahvendra Bakale is thankful to Jain College of Engineering, Belagavi, Karnataka for the constant support & encouragement for the research.

Funding Authors declare that they have not received any funding from any source or institutions or organizations.

Conflict of Interest All authors declare that they do not have any conflict of interest.

References

1. Schwartz, M., Lenzini, G., Geng, Y., Rønne, P.B., Ryan, P.Y.A., Lagerwall, J.P.F.: Cholesteric liquid crystal shells as enabling material for information-rich design and architecture. Adv. Mater. **30**(30), 1707382 (2018). https://doi.org/10.1002/adma.201707382
2. Adimule, V., Yallur, B.C., Keri, R., et al.: Microstructure, photoluminescence and electrical properties of SmxGd(1–x): SrO hybrid nanomaterials synthesized via facile coprecipitation method. Electron. Mater. Lett. **19**, 278–297 (2023). https://doi.org/10.1007/s13391-022-00394-0
3. Zhang, J., Zhang, Y., Yang, J., Wang, X.: Beyond color boundaries: pioneering developments in cholesteric liquid crystal photonic actuators. Micro-machines **15**(6), 808 (2024). https://doi.org/10.3390/mi15060808
4. Adimule, V., Keri, R., Rajendrachari, S., Kendrekar, P., Kulkarni, C.V.: Highly photoluminescence and wide band gap insulating metal hybrid nanoparticles array of samarium-doped SrO: CoO: synthesis, characterizations and sensor characteristics. J. Mater. Sci. Mater. Electron. **34**(5), 442 (2023). https://doi.org/10.1016/j.mssp.2019.104722
5. Zia, A., Saeed, S., Man, T., Liu, H., Chen, C.X., Wan, Y.: Next-generation interfaces: integrating liquid crystal technologies in augmented and virtual reality–a review. Liq. Cryst. Rev. **12**(1), 30–56 (2024). https://doi.org/10.1080/21680396.2024.2401786
6. Challa, M., Yallur, B.C., Ambika, M.R., Adimule, V.: Influence of nano particles on optical properties of Cu-MOFs. Adv. Mater. Res. **1173**, 23–33 (2022). https://doi.org/10.4028/p-vn4hd4
7. Chen, X., Chen, J., Song, X., Du, T., Deng, X., Deng, Z., Hu, X.-G. et al.: Bioinspired mechanochromic liquid crystal materials: from fundamentals to functionalities and applications. Adv. Mater. 2403766 (2024). https://doi.org/10.1002/adma.202403766
8. Nandi, S.S, Adimule, V., Kadapure, S.A., Kerur, S.S.: Rare earth based nanocomposite materials for prominent performance supercapacitor: a review. Appl. Mech. Mater. **908**, 3–18 (2022). https://doi.org/10.4028/p-rff302
9. Yang, C., Wu, B., Ruan, J., Zhao, P., Shan, J., Zhang, R., Yoon, D.K., Chen, D., Liu, K.: Mechanochromic responses of cholesteric liquid crystal droplets with nanoscale periodic helical structures showing reversible and tunable structural color. ACS Appl. Polym. Mater. **4**(1), 463–468 (2021). https://doi.org/10.1021/acsapm.1c01362
10. Manhas, N., Adimule, V., Lakshminarayana, P., Keri, R., Kumar, L.S., Bathula, C.: Utilization of CO^{2+} doped Eu_2O_3 nanostructures and their effect on the dielectric and photoisomerization of liquid crystals. Phys. Fluids **35**(10) (2023). https://doi.org/10.1063/5.0171578
11. Gevorgyan, A.H., Vanyushkin, N.A., Efimov, I.M., Kamenev, A.O., Malinchenko, A.A., Golik, S.S., Oganesyan, K.B.: Theoretical study of a tunable optical device on the base of magnetoactive cholesteric liquid crystals. Opt. Mater. **156**, 115961 (2024). ISSN 0925-3467. https://doi.org/10.1016/j.optmat.2024.115961

12. Nagalakshmi, G., Nandeesh, I.M., Yallur, B.C., Adimule, V., Batakurki, S.: Synthesis and optical properties of copper terephthalate metal organic frame works. Eng. Chem. **2**, 3–11 (2023). https://doi.org/10.4028/p-fdqs03
13. Qu, R., Li, G.: Overview of liquid crystal biosensors: from basic theory to advanced applications. Biosensors **12**(4), 205 (2022). https://doi.org/10.3390/bios12040205
14. Adimule, V., Yallur, B.C., Gowda, A.H.J.: Advanced sensors based on carbon nanomaterials. In: Carbon Nanomaterials-Based Sensors, pp. 259–268. Elsevier (2022). https://doi.org/10.1016/B978-0-323-91174-0.00004-4
15. Lu, X., Zhou, Z., Ni, B., Li, H., Li, Y., Li, B., Liu, W., Wang, Y., Yang, Y.: Tuning the circularly polarized luminescence of polymer-stabilized cholesteric liquid crystal films using chiral dopants. J. Mater. Chem. C **10**(21), 8246–8253 (2022). https://doi.org/10.1039/D2TC00326K
16. Sharma, V., Kumar, P., Chinky, C., Gahrotra, R., Raina, K.K., Malik, P.: Effect of nano particles on electro optic properties of polymer dispersed liquid crystal in normal mode. In: AIP Conference Proceedings, vol. 2142, no. 1. AIP Publishing (2019). https://doi.org/10.1063/1.5122510
17. Risse, A.M., Schmidtke, J.: Angular-dependent spontaneous emission in cholesteric liquid-crystal films. J. Phys. Chem. C **123**(4), 2428–2440 (2019). https://doi.org/10.1021/acs.jpcc.8b11134
18. Adimule, V., Sharma, V., Kumar, P., Keri, R., Sharma, K.: Perylene-3, 4, 9, 10-tetracarboxylic dianhydride dye intercalated liquid crystal molecules: synthesis, spectroscopic characterizations and photoluminescence dynamics. J. Mol. Liq. **413**, 125990 (2024). https://doi.org/10.1016/j.molliq.2024.125990
19. Huang, Y., Zhou, Y., Doyle, C., Shin-Tson, W.: Tuning the photonic band gap in cholesteric liquid crystals by temperature-dependent dopant solubility. Opt. Express **14**(3), 1236–1242 (2006). https://doi.org/10.1364/OE.14.001236
20. Dogra, A.R., Kumar, P., Sharma, V., Malik, P., Khanra, P., Kapoor, M.: Deposition of nanoparticles layer on ITO substrate and its application for alignment control of liquid crystal. In: AIP Conference Proceedings, vol. 2357, no. 1. AIP Publishing (2022). https://doi.org/10.1063/5.0080664
21. White, T.J., McConney, M.E., Bunning, T.J.: Dynamic color in stimuli-responsive cholesteric liquid crystals. J. Mater. Chem. **20**(44), 9832–9847 (2010). https://doi.org/10.1039/C0JM00843E
22. Sharma, V., Kumar, P., Chinky, C., Malik, P., Raina, K.K.: Morphological and electro optic studies of polymer dispersed liquid crystal in reverse mode. In: AIP Conference Proceedings, vol. 1953, no. 1. AIP Publishing (2018). https://doi.org/10.1063/1.5032969
23. Bisoyi, H.K., Li, Q.: Light-directing chiral liquid crystal nanostructures: from 1D to 3D. Acc. Chem. Res. **47**(10), 3184–3195 (2014). https://doi.org/10.1021/ar500249k
24. Gahrotra, R., Sharma, V., Dogra, A.R., Malik, P., Kumar, P.: Temperature dependent morphological and electro-optical characteristics of dye doped cholesteric liquid crystal. ECS Trans. **107**(1), 5481 (2022). https://doi.org/10.1149/10701.5481ecst
25. Balamurugan, R., Liu, J.-H.: A review of the fabrication of photonic band gap materials based on cholesteric liquid crystals. Reactive Funct. Polym. **105**, 9–34 (2016). https://doi.org/10.1016/j.reactfunctpolym.2016.04.012
26. Adimule, V., Nandi, S., Sharma, K., Keri, R.: Microstructure, zeta potential and enhanced photocatalytic dye degradation performances of mesoporous flower like Nb_2O_5@CeO_2 bimetallic oxide nanomaterials decorated with reduced graphene oxide. Inorg. Chem. Commun. **166**, 112592 (2024). ISSN 1387-7003. https://doi.org/10.1016/j.inoche.2024.112592
27. Malik, P., Supreet, K.A., Castagna, R., Singh, G.: Recent advances and future perspectives on nanoparticles-controlled alignment of liquid crystals for displays and other photonic devices. Crit. Rev. Solid State Mater. Sci. **48**(1), 57–92 (2022). https://doi.org/10.1080/10408436.2022.2027226

28. Dogra, A.R., Sharma, V., Gahrotra, R., Khanra, P., Dogra, A.: Thickness dependent electro-optical behaviour of homeotropically aligned nematic liquid crystal. ECS Trans. **107**(1), 5021 (2022). https://doi.org/10.1149/10701.5021ecst
29. Liu, J., Song, Z.-P., Sun, L.-Y., Li, B.-X., Lu, Y.-Q., Li, Q.: Circularly polarized luminescence in chiral orientationally ordered soft matter systems. Responsive Mater. **1**(1), e20230005 (2023). https://doi.org/10.1002/rpm.20230005
30. Adimule, V., Bathula, C., Sharma, K., Keri, R.: Study of structural, optical, dielectric properties poly-3-butyl thiophene reinforced with CO_3O_4 nanoparticles. Nano-Struct. Nano-Objects **38**, 101150 (2024). https://doi.org/10.1016/j.nanoso.2024.101150
31. Wang, L., Urbas, A.M., Li, Q.: Nature-inspired emerging chiral liquid crystal nanostructures: from molecular self-assembly to DNA mesophase and nanocolloids. Adv. Mater. **32**(41), 1801335 (2020). https://doi.org/10.1002/adma.201801335
32. Kim, Y., Tamaoki, N.: Photoresponsive chiral dopants: light-driven helicity manipulation in cholesteric liquid crystals for optical and mechanical functions. Chem. Photo. Chem. **3**(6), 284–303 (2019). https://doi.org/10.1002/cptc.201900034
33. Dogra, A.R., Sharma, V., Malik, P., Kumar, P.: In-situ homeotropic alignment of dye doped liquid crystal molecules on multilayered self assembled nanoparticles in confined cells for next generation display devices. Opt. Mater. **154**, 115731 (2024). https://doi.org/10.1016/j.optmat.2024.115731
34. Bitar, R., Agez, G., Mitov, M.: Cholesteric liquid crystal self-organization of gold nanoparticles. Soft Matter **7**(18), 8198–8206 (2011). https://doi.org/10.1039/C1SM05628J
35. Sharma, V., Kumar, P.: Electro-optically oriented Kerr and orientational phase study of normal mode polymer dispersed liquid crystals–effect of dispersion of nanoparticles. J. Mol. Liq. **348**, 118030 (2022). https://doi.org/10.1016/j.molliq.2021.118030
36. Francés-Soriano, L., Zaballos-García, E., Pérez-Prieto, J.: Up-and down-shifting nanomaterials with chiroptical responses: origin of their chiroptical activity and applications. Adv. Opt. Mater. **11**(18), 2300337 (2023). https://doi.org/10.1002/adom.202300337
37. Sharma, V., Kumar, P., Sharma, A., Raina, K.K.: Droplet configuration control with orange azo dichroic dye in polymer dispersed liquid crystal for advanced electro-optic characteristics. J. Mol. Liq. **233**, 122–130 (2017). https://doi.org/10.1016/j.molliq.2017.03.012
38. Ryabchun, A., Bobrovsky, A.: Cholesteric liquid crystal materials for tunable diffractive optics. Adv. Opt. Mater. **6**(15), 1800335 (2018). https://doi.org/10.1002/adom.201800335
39. Zhang, W., Froyen, A.A.F., Schenning, A.P.H.J., Zhou, G., Debije, M.G., de Haan, L.T.: Temperature-responsive photonic devices based on cholesteric liquid crystals. Adv. Photonics Res. **2**(7), 2100016 (2021). https://doi.org/10.1002/adpr.202100016
40. Hussain, S., Zourob, M.: Solid-state cholesteric liquid crystals as an emerging platform for the development of optical photonic sensors. Small **20**(7), 2304590 (2024). https://doi.org/10.1002/smll.202304590
41. Wang, L., Liu, S., Luo, S., He, Z., Miao, Z.: Advances in the preparation and application of liquid crystal microcapsules. Liq. Cryst. 1–32 (2024). https://doi.org/10.1080/02678292.2024.2364789
42. Mitov, M.: Cholesteric liquid crystals with a broad light reflection band. Adv. Mater. **24**(47), 6260–6276 (2012). https://doi.org/10.1002/adma.201202913
43. Leal-Junior, A., Soares, M.S., de Almeida, P.M., Marques, C.: Cholesteric liquid crystals sensors based on nanocellulose derivatives for improvement of quality of human life: a review. Adv. Sens. Res. **2**(10), 2300022 (2023). .https://doi.org/10.1002/adsr.202300022
44. Belyakov, V.A., Dmitrienko, V.E., Orlov, V.P.: Optics of cholesteric liquid crystals. Soviet Phys. Uspekhi **22**(2), 64 (1979). https://doi.org/10.1070/PU1979v022n02ABEH005417
45. Mitov, M., Dessaud, N.: Going beyond the reflectance limit of cholesteric liquid crystals. Nat. Mater. **5**(5), 361–364 (2006). https://doi.org/10.1038/nmat1619
46. Zhou, H., Wang, H., He, W., Yang, Z., Cao, H., Wang, D., Li, Y.: Research progress of cholesteric liquid crystals with broadband reflection. Molecules **27**(14), 4427 (2022). https://doi.org/10.3390/molecules27144427

47. Palto, S.P., Barnik, M.I., Geivandov, A.R., Kasyanova, I.V., Palto, V.S.: Spectral and polarization structure of field-induced photonic bands in cholesteric liquid crystals. Phys. Rev. E **92**(3), 032502 (2015). https://doi.org/10.1103/PhysRevE.92.032502
48. Umanskii, B.A., Simdyankin, I.V.: Circular dichroism in cholesteric liquid crystals. Crystallogr. Rep. **64**, 437–442 (2019). https://doi.org/10.1134/S1063774519030283
49. Khan, M., Zhao, B., Zhang, M., Shah, S.N.A., Guo, Y., Qiongzheng, H.: Construction of cholesteric liquid crystal microlens array at air/CLC/air and water/CLC/air interfaces. Colloids Surf. A **683**, 133031 (2024). https://doi.org/10.1016/j.colsurfa.2023.133031
50. Chen, Y., Zhang, Y., Li, H., Li, Y., Zheng, W., Quan, Y., Cheng, Y.: Dynamic circularly polarized luminescence with tunable handedness and intensity enabled by achiral dichroic dyes in cholesteric liquid crystal medium. Adv. Mater. **34**(28), 2202309 (2022). https://doi.org/10.1002/adma.202202309
51. Wei, W., Farooq, M.A., Xiong, H.: Cholesteric liquid crystalline polyether with broad tunable circularly polarized luminescence. Langmuir **37**(40), 11922–11930 (2021). https://doi.org/10.1021/acs.langmuir.1c02144
52. Wu, Y., Li, M., Zheng, Z.-G., Zhen-Qiang, Y., Zhu, W.-H.: Liquid crystal assembly for ultra-dissymmetric circularly polarized luminescence and beyond. J. Am. Chem. Soc. **145**(24), 12951–12966 (2023). https://doi.org/10.1021/jacs.3c01122
53. Labeeb, A.M., Ward, Y.A., Fikry, M.: Thermal control of tunable photonic optical bandgaps in different cholesteric liquid crystals mixtures. J. Mol. Liq. **340**, 117179 (2021). https://doi.org/10.1016/j.molliq.2021.117179
54. Gollapelli, B., Potu, S., Rajaboina, R., Vallamkondu, J.: Optical anti-counterfeiting with cholesteric liquid crystal emulsions: preparation, properties, and applications. Mater. Adv. **5**(18), 7113–7129 (2024). https://doi.org/10.1039/D4MA00604F
55. Chilaya, G.S., Lisetski, L.N.: Cholesteric liquid crystals: physical properties and molecular-statistical theories. Mol. Cryst. Liq. Cryst. **140**(2–4), 243–286 (1986). https://doi.org/10.1080/00268948608080157
56. Chemingui, M., Donghao Yang, Y., Wang, H.A., Cai, W., Tang, A., Drevensek-Olenik, I., Zhang, X., Jingjun, X.: The influence of quantum dots on the optical properties of a room temperature cholesteric liquid crystal. Liq. Cryst. **49**(15), 2095–2107 (2022). https://doi.org/10.1080/02678292.2022.2103850
57. Sadigh, M.K., Naziri, P., Zakerhamidi, M.S., Ranjkesh, A., Yoon, T.-H.: Temperature dependent features of polymer stabilized cholesteric liquid crystals based on selected liquid crystal characteristics. Optik **230**, 166354 (2021). https://doi.org/10.1016/j.ijleo.2021.166354
58. Kocakülah, G.: Comparison of electro-optical and dielectric properties of cobalt (II) phthalocyanine doped cholesteric liquid crystals. Opt. Mater. **141**, 113949 (2023). https://doi.org/10.1016/j.optmat.2023.113949
59. Jung, J.Y., Shin, M.J., Kim, B., Pyo, U.J., Son, J.G., Yoon, D.K., Han, S.W.: Cholesteric liquid crystal-mediated chiral plasmonic films with strong chiroptical response, dynamic tunability, and reversible thermal reconfigurability. Adv. Opt. Mater. **12**(24), 2400760 (2024). https://doi.org/10.1002/adom.202400760
60. Zhang, L.-Y., Gao, Y.-Z., Song, P., Wu, X.-J., Yuan, X., He, B.-F., Chen, X.-W., et al.: Research progress of cholesteric liquid crystals with broadband reflection characteristics in application of intelligent optical modulation materials. Chin. Phys. B **25**(9), 096101 (2016). https://doi.org/10.1088/1674-1056/25/9/096101
61. Lee, H.-G., Munir, S., Park, S.-Y.: Cholesteric liquid crystal droplets for biosensors. ACS Appl. Mater. Interfaces **8**(39), 26407–26417 (2016). https://doi.org/10.1021/acsami.6b09624
62. Fuh, A.Y.-G., Ho, S.-J., Wu, S.-T., Li, M.-S.: Optical filter with tunable wavelength and bandwidth based on phototunable cholesteric liquid crystals. Appl. Opt. **53**(8), 1658–1662 (2014). https://doi.org/10.1364/AO.53.001658
63. Kim, J.-K., Joo, S.-H., Song, J.-K.: Reflective–emissive photoluminescent cholesteric liquid crystal display. Appl. Opt. **52**(34), 8280–8286 (2013). https://doi.org/10.1364/AO.52.008280
64. Xia, Y., Li, H., Vaskeviciute, M., Faccio, D., Karimullah, A.S., Heidari, H., Ghannam, R.: Temperature-tunable cholesteric liquid crystal optical combiners for extended reality applications. Adv. Intell. Syst. 2400411 (2024). https://doi.org/10.1002/aisy.202400411

65. Sun, C., Jiangang, L.: Optical filters based on cholesteric, blue and sphere mesophases. Polymers **14**(22), 4898 (2022). https://doi.org/10.3390/polym14224898
66. Chen, H.-Q., Wang, X.-Y., Bisoyi, H.K., Chen, L.-J., Li, Q.: Liquid crystals in curved confined geometries: Microfluidics bring new capabilities for photonic applications and beyond. Langmuir **37**(13), 3789–3807 (2021). https://doi.org/10.1021/acs.langmuir.1c00256
67. Chen, P., Wei, B.-Y., Hu, W., Lu, Y.-Q.: Liquid-crystal-mediated geometric phase: from transmissive to broadband reflective planar optics. Adv. Mater. **32**(27), 1903665 (2020). https://doi.org/10.1002/adma.201903665
68. Tran, L.: Kind of blue phases: controlling topological defects with geometry and anchoring energy in nematic and cholesteric liquid crystals. PhD diss., University of Pennsylvania (2018)
69. Lee, S.S., Kim, J.B., Kim, Y.H., Kim, S.-H.: Wavelength-tunable and shape-reconfigurable photonic capsule resonators containing cholesteric liquid crystals. Sci. Adv. **4**(6), eaat8276 (2018). https://doi.org/10.1126/sciadv.aat8276
70. Vernon, J.P., Hrozhyk, U.A., Serak, S.V., Tondiglia, V.P., White, T.J., Tabiryan, N.V., Bunning, T.J.: Optically reconfigurable reflective/scattering states enabled with photosensitive cholesteric liquid crystal cells. Adv. Opt. Mater. **1**(1), 84–91 (2013). https://doi.org/10.1002/adom.201200014
71. Esteves, C., Ramou, E., Porteira, A.R.P., Barbosa, A.J.M., Roque, A.C.A.: Seeing the unseen: the role of liquid crystals in gas-sensing technologies. Adv. Opt. Mater. **8**(11), 1902117 (2020). https://doi.org/10.1002/adom.201902117
72. Kocaman, C., Batir, O., Bukusoglu, E.: Optically responsive dry cholesteric liquid crystal marbles. J. Colloid Interface Sci. (2024). https://doi.org/10.1016/j.jcis.2024.05.194
73. Bae, S., Kim, M., Kang, G., Lee, H.-S., Kim, I.S., Choi, S.-W., Kang, J.G.: Efficient full-color daytime radiative cooling with diffuse-reflection-dominant cholesteric liquid crystals. Chem. Eng. J. **483**, 149245 (2024). https://doi.org/10.1016/j.cej.2024.149245
74. Lesot, P., Aroulanda, C., Berdagué, P., Meddour, A., Merlet, D., Farjon, J., Giraud, N., Lafon, O.: Multinuclear NMR in polypeptide liquid crystals: Three fertile decades of methodological developments and analytical challenges. Prog. Nucl. Mag. Reson. Spectrosc. **116**, 85–154 (2020). https://doi.org/10.1016/j.pnmrs.2019.10.001
75. Luh, H.-T., Hsu, C.-C., Chene, F.-L., Chang, H.-C., Hsieh, S.-T., Lai, D.-M., Hsiao, Y.-C.: Efficient colorimetric point-of-care detection and imaging of multiple biomolecules utilizing photonic liquid crystal composite on gold nanoisland thin films for label-free sensing. AIP Adv. **14**(7) (2024). https://doi.org/10.1063/5.0215262
76. Hsiao, Y.-C., Chen, F.-L., Hsu, C.-C., Chen, W.-H., Shang, L.-D.: Rapid colorimetric point-of-care detection and imaging of multiple biomolecules using liquid crystal composite on gold nanoisland thin films as a label-free optical biosensor (2023)
77. Zhai, F., Feng, Y., Zhou, K., Wang, L., Zheng, Z., Feng, W.: Graphene-based chiral liquid crystal materials for optical applications. J. Mater. Chem. C **7**(8), 2146–2171 (2019). https://doi.org/10.1039/C8TC04947E
78. Ranjkesh, A., Yoon, T.-H.: Thermal and electrical wavelength tuning of Bragg reflection with ultraviolet light absorbers in polymer-stabilized cholesteric liquid crystals. J. Mater. Chem. C **6**(45), 12377–12385 (2018). https://doi.org/10.1039/C8TC04563A
79. Li, S.-L., Chen, Z.-Y., Chen, P., Hu, W., Huang, C., Li, S.-S., Hu, X., Lu, Y.-Q., Chen, L.-J.: Geometric phase-encoded stimuli-responsive cholesteric liquid crystals for visualizing real-time remote monitoring: humidity sensing as a proof of concept. Light: Sci. Appl. **13**(1), 27 (2024). https://doi.org/10.1038/s41377-023-01360-7
80. Noh, K.-G., Park, S.-Y.: Biosensor array of interpenetrating polymer network with photonic film templated from reactive cholesteric liquid crystal and enzyme-immobilized hydrogel polymer. Adv. Funct. Mater. **28**(22), 1707562 (2018). https://doi.org/10.1002/adfm.201707562
81. Wang, I.-T., Lee, Y.-H., Chuang, E.-Y., Hsiao, Y.-C.: Sensitive, color-indicating and labeling-free multi-detection cholesteric liquid crystal biosensing chips for detecting albumin. Polymers **13**(9), 1463 (2021). https://www.mdpi.com/2073-4360/13/9/1463

82. Prakash, J., Khan, S., Chauhan, S., Biradar, A.M.: Metal oxide-nanoparticles and liquid crystal composites: a review of recent progress. J. Mol. Liq. **297**, 112052 (2020). https://doi.org/10.1016/j.molliq.2019.112052
83. Ye, Y., Guo, L., Zhong, T.: A review of developments in polymer stabilized liquid crystals. Polymers **15**(13), 2962 (2023). https://doi.org/10.3390/polym15132962
84. Zheng, W., Meng, Z., Murtaza, G., Zhang, N., Wu, L., Qiu, L.: Designing and tailoring optical properties: theory and simulation of photonic band gaps in photonic materials. Photonics **11**(2), 109 (2024). https://doi.org/10.3390/photonics11020109
85. Kao, T.-H., Hsu, H.-H., Chen, J.-J., Lee, L.-R., Chen, H.-Y., Chen, J.-T.: Fabrication of polymer/cholesteric liquid crystal films and fibers using the nonsolvent and phase separation method. Langmuir **40**(27), 14166–14172 (2024). https://doi.org/10.1021/acs.langmuir.4c01759
86. Ilchishin, I.P., Tikhonov, E.A.: Dye-doped cholesteric lasers: distributed feedback and photonic bandgap lasing models. Prog. Quantum Electron. **41**, 1–22 (2015). https://doi.org/10.1016/j.pquantelec.2015.02.001
87. Khoo, I.C.: Cholesteric and blue-phase liquid photonic crystals for nonlinear optics and ultrafast laser pulse modulations. Liq. Cryst. Rev. **6**(1), 53–77 (2018). https://doi.org/10.1080/21680396.2018.1509387
88. Liu, Y., Li, Y., Jiang, Y., Xiong, W., Wang, Q., Sun, J., Yang, H., Zhao, X.-Z.: Self-organized cholesteric liquid crystal polymer films with tunable photonic band gap as transparent and flexible back-reflectors for dye-sensitized solar cells. Nano Energy **26**, 648–656 (2016). https://doi.org/10.1016/j.nanoen.2016.06.021
89. Jiang, L., Shui, Q., Han, B., Huang, R., Deng, J., Hu, N., Yang, J.: Recent advances in liquid crystal droplet-based biosensors. TrAC Trends Anal. Chem. 117920 (2024). https://doi.org/10.1016/j.trac.2024.117920
90. Kudreyko, A., Chigrinov, V.: Optical biosensing of polarized light. Crystals **13**(12), 1680 (2023). https://www.mdpi.com/2073-4352/13/12/1680
91. Nesterkina, M., Kravchenko, I., Hirsch, A.K.H., Lehr, C.-M.: Thermotropic liquid crystals in drug delivery: a versatile carrier for controlled release. Eur. J. Pharm. Biopharm. 114343 (2024). https://doi.org/10.1016/j.ejpb.2024.114343
92. Chen, F., Zheng, J., Xing, C., Sang, J., Shen, T.: Applications of liquid crystal planer optical elements based on photoalignment technology in display and photonic devices. Displays 102632 (2024). https://doi.org/10.1016/j.displa.2023.102632
93. Mujahid, A., Stathopulos, H., Lieberzeit, P.A., Dickert, F.L.: Solvent vapour detection with cholesteric liquid crystals—optical and mass-sensitive evaluation of the sensor mechanism. Sensors **10**(5), 4887–4897 (2010). https://doi.org/10.3390/s100504887
94. Zhou, D., Sun, J., Wang, T., Zhang, S., Zhang, W., Zhou, N., Liu, Y.: Highly sensitive miniature optical fiber liquid crystal sensor for tumor temperature detection. IEEE Sens. J. (2024)
95. Yeh, T.-Y., Liu, M.-F., Lin, R.-D., Hwang, S.-J.: Alcohol selective optical sensor based on porous cholesteric liquid crystal polymer networks. Molecules **27**(3), 773 (2022). https://doi.org/10.3390/molecules27030773
96. Rajesh, L.K.G., Mishra, S.K., Choudhary, A., Biradar, A.M., Sumana, G.: Technological advancements in bio-recognition using liquid crystals: Techniques, applications, and performance. Luminescence **38**(7), 811–833 (2023). https://doi.org/10.1002/bio.4242
97. Oladepo, S.A.: Development and application of liquid crystals as stimuli-responsive sensors. Molecules **27**(4), 1453 (2022). https://doi.org/10.3390/molecules27041453
98. Zhang, Y.-S., Jiang, S.-A., Lin, J.-D., Lee, C.-R.: Bio-inspired design of active photo-mechanochemically dual-responsive photonic film based on cholesteric liquid crystal elastomers. J. Mater. Chem. C **8**(16), 5517–5524 (2020). https://doi.org/10.1039/C9TC05758G
99. Honaker, L.W., Chen, C., Dautzenberg, F.M.H., Brugman, S., Deshpande, S.: Designing biological microsensors with chiral nematic liquid crystal droplets. ACS Appl. Mater. Interfaces **14**(33), 37316–37329 (2022). https://doi.org/10.1021/acsami.2c06923
100. Wang, Y., Li, Q.: Light-driven chiral molecular switches or motors in liquid crystals. Adv. Mater. **24**(15), 1926–1945 (2012). https://doi.org/10.1002/adma.201200241

101. Agez, G., Bitar, R., Mitov, M.: Color selectivity lent to a cholesteric liquid crystal by monitoring interface-induced deformations. Soft Matter **7**(6), 2841–2847 (2011). https://doi.org/10.1039/C0SM00950D
102. Fan, Q., Li, Z., Jiang, K., Gao, J., Lin, S., Guo, J.: Tunable circular polarization room temperature phosphorescence with ultrahigh dissymmetric factor by cholesteric liquid crystal elastomers. Cell Rep. Phys. Sci. **4**(10) (2023)
103. Smalyukh, I.I., Lavrentovich, O.D.: Three-dimensional director structures of defects in Grandjean-Cano wedges of cholesteric liquid crystals studied by fluorescence confocal polarizing microscopy. Phys. Rev. E **66**(5), 051703 (2002). https://doi.org/10.1103/PhysRevE.66.051703
104. Belalia, M., Mitov, M., Bourgerette, C., Krallafa, A., Belhakem, M., Bormann, D.: Cholesteric liquid crystals with a helical pitch gradient: Spatial distribution of the concentration of chiral groups by Raman mapping in relation with the optical response and the microstructure. Phys. Rev. E—Stat. Nonlinear Soft Matter Phys. **74**(5), 051704 (2006). https://doi.org/10.1103/PhysRevE.74.051704
105. Shen, Y., Yan-Chao, X., Ge, Y.-H., Jiang, R.-G., Wang, X.-Z., Li, S.-S., Chen, L.-J.: Photoalignment of dye-doped cholesteric liquid crystals for electrically tunable patterns with fingerprint textures. Opt. Express **26**(2), 1422–1432 (2018). https://doi.org/10.1364/OE.26.001422
106. Kopp, V.I., Zhang, Z.-Q., Genack, A.Z.: Lasing in chiral photonic structures. Prog. Quantum Electron. **27**(6), 369–416 (2003). https://doi.org/10.1016/S0079-6727(03)00003-X
107. Chuang, Y.-T., Li, Y.-C., Feng, T.-M., Wang, C.-T.: Electrically switchable transmissive and reflective liquid crystal polarization gratings based on cholesteric liquid crystals. Adv. Mater. Technol. 2400726 (2025). https://doi.org/10.1002/admt.202400726
108. Xie, R., Li, N., Li, Z., Chen, J., Li, K., He, Q., Liu, L., Zhang, S.: Liquid crystal droplet-based biosensors: promising for point-of-care testing. Biosensors **12**(9), 758 (2022). https://doi.org/10.3390/bios12090758
109. Chuang, E.-Y., Lin, P.-Y., Wang, P.-F., Kuo, T.-R., Chen, C.-H., Manga, Y.B., Hsiao, Y.-C.: Label-free, smartphone-based, and sensitive nano-structural liquid crystal aligned by ceramic silicon compound–constructed dmoap-based biosensor for the detection of urine albumin. Int. J. Nanomed. 763–773 (2021). https://doi.org/10.2147/IJN.S285125
110. Moreddu, R., Elsherif, M., Butt, H., Vigolo, D., Yetisen, A.K.: Contact lenses for continuous corneal temperature monitoring. RSC Adv. **9**(20), 11433–11442 (2019). https://doi.org/10.1039/C9RA00601J
111. Araz, O.U., Kemiklioglu, E., Gurboga, B.: Modeling of the lyotropic cholesteric liquid crystal based toxic gas sensor using adaptive neuro-fuzzy inference systems. Exp. Syst. Appl. **240**, 122326 (2024). https://doi.org/10.1016/j.eswa.2023.122326
112. de Araujo, W.R., Lukas, H., Torres, M.D.T., Gao, W., de la Fuente-Nunez, C.: Low-cost biosensor technologies for rapid detection of COVID-19 and future pandemics. ACS Nano **18**(3), 1757–1777 (2024). https://doi.org/10.1021/acsnano.3c01629
113. Caligiuri, V., Leone, F., Annesi, F., Pane, A., Bartolino, R., De Luca, A.: Envisioning quantum electrodynamic frameworks based on bio-photonic cavities. Photonics **8**(11), 470 (2021). https://www.mdpi.com/2304-6732/8/11/470
114. Wei, X., Lin, T., Duan, M., Du, H., Yin, X.: Cellulose nanocrystal-based liquid crystal structures and the unique optical characteristics of cellulose nanocrystal films. Bioresources **16**(1) (2021). https://doi.org/10.15376/biores.16.1.Wei
115. Wang, R., Feng, Y., Li, D., Li, K., Yan, Y.: Towards the sustainable production of biomass-derived materials with smart functionality: a tutorial review. Green Chem. (2024). https://doi.org/10.1039/D4GC01771D
116. Dore, C.: Hydroxypropyl cellulose as enabling material for the nanofabrication of photonic architectures (2021). https://hdl.handle.net/10803/671940
117. Wang, Z., Tianhua, X., Noel, A., Chen, Y.-C., Liu, T.: Applications of liquid crystals in biosensing. Soft Matter **17**(18), 4675–4702 (2021). https://doi.org/10.1039/D0SM02088E

118. Hrozhyk, U.A., Serak, S.V., Tabiryan, N.V., Bunning, T.J.: Optical tuning of the reflection of cholesterics doped with azobenzene liquid crystals. Adv. Func. Mater. **17**(11), 1735–1742 (2007). https://doi.org/10.1002/adfm.200600776
119. Wang, Q., Zhang, Z., Wang, C., Yang, X., Fang, Z., Shang, L.: Bioinspired confined assembly of cellulosic cholesteric liquid crystal bubbles. Adv. Sci. **11**(11), 2308442 (2024). https://doi.org/10.1002/advs.202308442
120. Li, X., Chen, Y., Du, C., Liao, X., Yang, Y., Feng, W.: Cholesteric liquid crystal elastomer coatings with brilliant structural colors and mechanochromic response fabricated by spray deposition. Adv. Funct. Mater. 2412298 (2024). https://doi.org/10.1002/adfm.202412298
121. Akman, A., Bukusoglu, E.: Understanding directed assembly of concentrated nanoparticles at energetically heterogeneous interfaces of cholesteric liquid crystal droplets. J. Colloid Interface Sci. **649**, 772–784 (2023). https://doi.org/10.1016/j.jcis.2023.06.143
122. Zola, R.S., Li, Q.: Stimuli-directed helical axis switching in chiral liquid crystal nanostructures. Functional Organic and Hybrid Nanostructured Materials: Fabrication, Properties, and Applications, pp. 307–357 (2018). https://doi.org/10.1002/9783527807369.ch8
123. Zheng, R., Wei, Y., Zhang, Z.-C., Wang, Z.-Y., Ling-Ling Ma, Y., Wang, L.H., Yan-Qing, L.: Stimuli-responsive active materials for dynamic control of light field. Responsive Mater. **1**(2), e20230017 (2023). https://doi.org/10.1002/rpm.20230017
124. Saadaoui, L., Hamdi, R.: Electrically tunable cholesteric liquid crystal lines defects. Opt. Mater. **114**, 110960 (2021). https://doi.org/10.1016/j.optmat.2021.110960
125. Yu, C.-H., Po-Chang, W., Lee, W.: Electro-thermal formation of uniform lying helix alignment in a cholesteric liquid crystal cell. Crystals **9**(4), 183 (2019). https://doi.org/10.3390/cryst9040183
126. Lucchetti, L., Nava, G.: Nonlinear optical director reorientation in heliconical cholesteric liquid crystals: a brief review. Giant 100311 (2024). https://doi.org/10.1016/j.giant.2024.100311
127. Manna, S.K., George, T.F., Li, G.: Electrically driven self-organized chiral liquid-crystalline nanostructures: organic molecular photonic crystal with tunable bandgap. Functional Organic and Hybrid Nanostructured Materials: Fabrication, Properties, and Applications, pp. 359–381 (2018). https://doi.org/10.1002/9783527807369.ch9
128. Lan, R., Shen, W., Yao, W., Chen, J., Chen, X., Yang, H.: Bioinspired humidity-responsive liquid crystalline materials: from adaptive soft actuators to visualized sensors and detectors. Mater. Horiz. **10**(8), 2824–2844 (2023). https://doi.org/10.1039/D3MH00392B
129. Xie, M., Hisano, K., Zhu, M., Toyoshi, T., Pan, M., Okada, S., Tsutsumi, O., Kawamura, S., Bowen, C.: Flexible multifunctional sensors for wearable and robotic applications. Adv. Mater. Technol. **4**(3), 1800626 (2019). https://doi.org/10.1002/admt.201800626
130. Pyo, S., Lee, J., Bae, K., Sim, S., Kim, J.: Recent progress in flexible tactile sensors for human-interactive systems: from sensors to advanced applications. Adv. Mater. **33**(47), 2005902 (2021). https://doi.org/10.1002/adma.202005902
131. Li, W.-D., Ke, K., Jia, J., Pu, J.-H., Zhao, X., Bao, R.-Y., Liu, Z.-Y., et al.: Recent advances in multiresponsive flexible sensors towards E-skin: a delicate design for versatile sensing. Small **18**(7), 2103734 (2022). https://doi.org/10.1002/smll.202103734

Role of Chiral Dopants, Luminescence Applications of Cholesteric Liquid Crystals

Raghavendra P. Bakale, Sushant Kakati, Sheela A. Sangam, S. N. Mathad, and Gangadhar Bagihalli

Abstract Cholesteric liquid crystals (CLCs) are a fascinating class of materials that have garnered significant attention in recent years due to their unique optical properties. The incorporation of chiral dopants into CLCs has been shown to further enhance these properties, leading to potential applications in a wide range of fields, including displays, sensing, biological imaging, and photonic devices. CLCs are known for their ability to selectively reflect light, circular dichroism, and optical activity, making them ideal for use in various applications. The addition of chiral dopants to CLCs can modify their optical properties, leading to enhanced luminescence, tunable emission wavelengths, and polarized emission. This chapter aims to explore the role of chiral dopants in modulating the luminescence behavior of CLCs, with a focus on their applications in displays, sensing, biological imaging, and photonic devices. We will delve into the underlying mechanisms of chiral dopant-induced luminescence and discuss the potential benefits and challenges of using CLCs in these applications. Additionally, we will examine the current state of research in this field and discuss potential future directions, including the development of new CLC materials and the exploration of novel applications.

Keywords Cholesteric Liquid Crystals · Chiral Dopants · Luminescence · Optical Properties · Photonic Devices

R. P. Bakale (✉)
Department of Chemistry, Jain College of Engineering, Belagavi, Karnataka, India
e-mail: raghubakale@gmail.com

S. Kakati · S. N. Mathad (✉)
Department of Engineering Physics, KLE Institute of Technology, Airport Road, Gokul, Hubballi 580030, Karnataka, India
e-mail: physicssiddu@kleit.ac.in

S. A. Sangam
Department of Chemistry, K.L.S Gogte Institute of Technology, Belagavi, Karnataka, India
e-mail: sasangam@git.edu

G. Bagihalli
Department of Engineering Physics, KLE Institute of Technology, Airport Road, Gokul, Hubballi 580030, Karnataka, India

P. Kumar et al. (eds.), *Cholesteric Liquid Crystals*, Engineering Materials,
https://doi.org/10.1007/978-981-96-3821-5_4

1 Introduction to Cholesteric Liquid Crystals (CLCs)

Cholesteric liquid crystals (CLCs) are a subclass of liquid crystals that exhibit a helical arrangement of molecules [1–3], resulting in unique optical properties [4–6]. The helical structure of CLCs is characterized by a pitch [7–10], which determines their selective reflection and circular dichroism properties [11–14]. CLCs have been widely studied due to their potential applications in displays [15–18], optical devices [19–21], and biological imaging [22–25]. Figures 1 and 2 illustrates the three-dimensional distribution of the director components nx, ny, and nz in the vicinity of the homeotropic boundary of the cholesteric liquid crystal (CLC) layer. Notably, as the anchoring energy at the homeotropic boundary of the CLC layer decreases, the area exhibiting a screw-type texture diminishes, and the hybrid-cell texture gradually transitions towards the Grandjean structure.

In this section, we will provide an overview of the properties and characteristics of CLCs, different types of CLCs, Potential applications of CLCs (displays, optical devices, biological imaging) and Challenges associated with the use of CLCs (stability, scalability, toxicity).

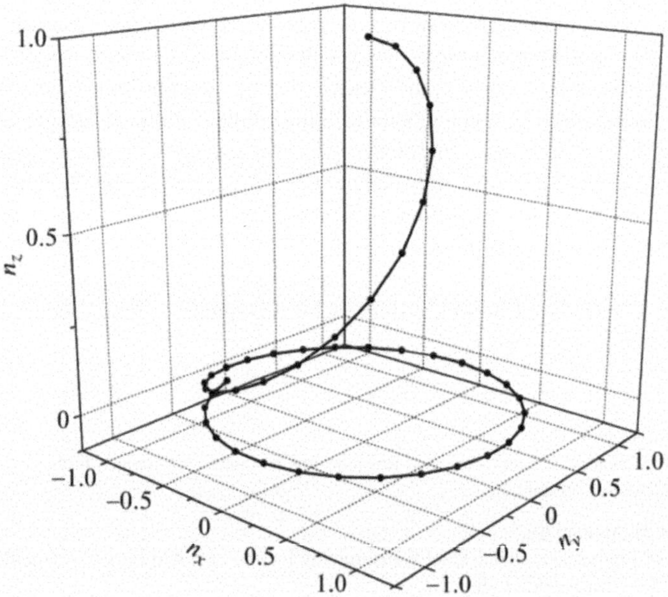

Fig. 1 A three-dimensional illustration of the screw-type texture near the homeotropic boundary. (Reproduced from Ref. [7], under the terms of the CC BY 4.0 license, http://creativecommons.org/licenses/by/4.0/.)

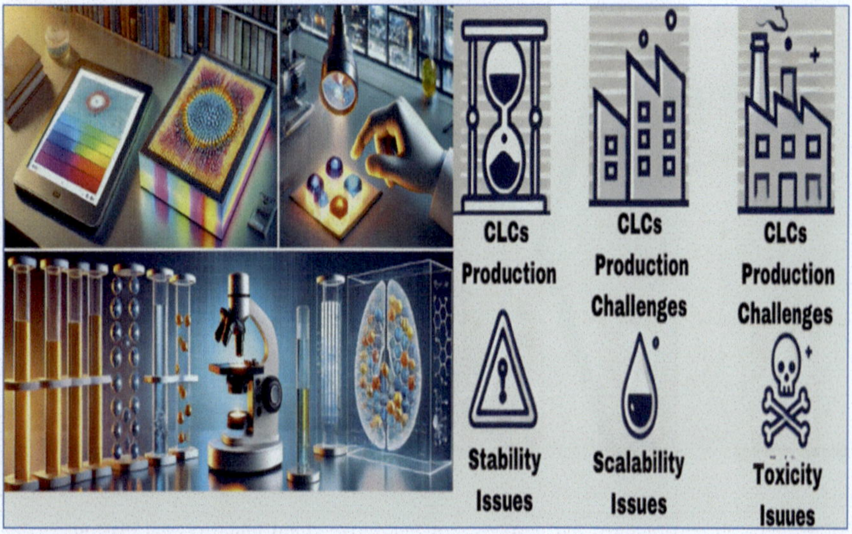

Fig. 2 Different types of CLCs and Challenges associated with the use of CLCs

1.1 Properties and Characteristics of CLCs

1.1.1 Selective Reflection

One of the most distinctive properties of CLCs is their ability to reflect light, selectively [26]. This phenomenon occurs due to the helical structure of the CLC molecules, which creates a periodic variation in the refractive index of the material [27]. Selective reflection in CLCs is characterized by the following features are shown in Fig. 3.

a. **Wavelength Selectivity**: CLCs selectively reflect light of a specific wavelength [28], which is determined by the pitch of the helical structure.
b. **Bandwidth**: The range of wavelengths that are selectively reflected is typically narrow, making CLCs useful for applications requiring precise wavelength control [29].
c. **Polarization**: The selectively reflected light is typically circularly polarized, with the handedness of the polarization depending on the handedness of the CLC helix [30].
d. **Angle of Incidence**: The selective reflection properties of CLCs are angle-dependent [31], with the reflectivity varying as a function of the angle of incidence.

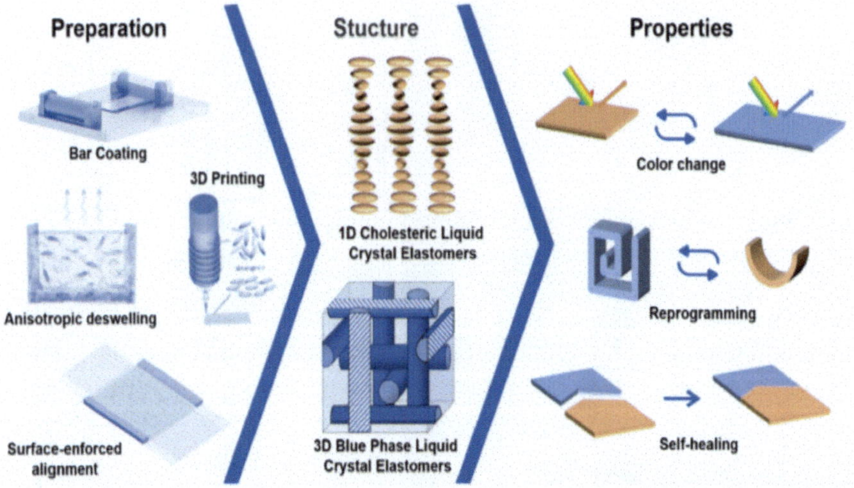

Fig. 3 Preparation, structure, and properties of innovative chiral liquid crystal elastomers that exhibit: 1. color change, 2. self-healing capabilities, 3. re-programmability. (Reproduced from Ref. [19], under the terms of the CC BY 4.0 license, http://creativecommons.org/licenses/by/4.0/.)

1.2 Potential Applications of CLCs

The selective reflection properties of CLCs make them useful for a range of applications, including:

a. **Displays**: CLCs can be used to create displays with high color purity and contrast [32].
b. **Optical Filters**: CLCs can be used as optical filters to selectively transmit or reflect specific wavelengths [33].
c. **Sensing**: CLCs can be used as sensors to detect changes in their environment, such as temperature [34] or concentration changes [35].
d. **Circular Dichroism**: CLCs exhibit circular dichroism [36], work by Zhao et al. successfully demonstrates the development of a highly sensitive and stable temperature sensor by integrating a CLC polymer film onto the end face of an optical fiber. The proposed sensor exhibits a wide temperature sensing range of 30–270 °C, improved sensitivity compared to existing non-polymerized LC-filled optical fiber temperature sensors. Its compact, lightweight, and easy-to-manufacture design make it an ideal solution for real-time temperature monitoring in various applications, particularly in micron-scale spaces. The influence of CLCP film thickness on reflected light wavelength is investigated, as shown in Fig. 4a. The optimal thickness range for experimentation is 20–40 μm. The temperature-sensitive polymerized CLC is prepared using a Michael mercaptan-ene addition reaction. Figure 4b illustrates the thermal expansion principle of CLCP film, where the pitch elongation and thickness change occur upon heating. And where the absorption of left- and right-handed circularly polarized light

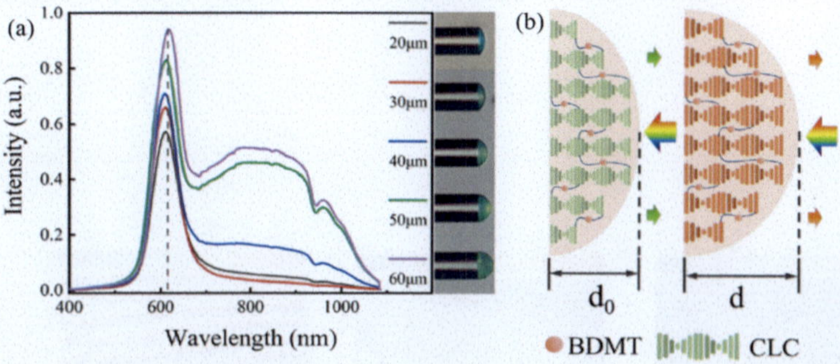

Fig. 4 Thermal expansion characteristics of CLCP film. **a** Thickness-dependent wavelength shift: the influence of CLCP film thickness on the reflected light wavelength. **b** Internal structure and thermal expansion: a detailed illustration of the internal network structure and thermal expansion mechanism of CLCP film. (Reproduced from Ref. [36], under the terms of the CC BY 4.0 license, http://creativecommons.org/licenses/by/4.0/.)

differs, leading to potential applications in biological imaging [37] and sensing [38].

e. **Optical Activity**: CLCs exhibit optical activity [39], where the rotation of polarized light occurs, makes them useful for applications such as optical filtersas depicted in Fig. 4 the application of an external magnetic field significantly enhances the accumulation of light energy at the edge modes. The polarization of the total wave-field changes from linear to elliptical with the presence of an external magnetic field. The suppression of absorption is observed near the short-wavelength boundary of the photonic bandgap (PBG) for certain values of the magnetic field [40] and modulators [41].
f. **Helical Structure**: The helical structure of CLCs is characterized by a pitch, which can be modified by the addition of chiral dopants [42], leading to changes in their optical properties [43] (Figs. 5 and 6).

2 Types of CLCs

- **Thermotropic CLCs**: These CLCs exhibit a helical structure in response to changes in temperature. Examples: Cholesteryl oleyl carbonate (COC) [44], Cholesteryl acetate [45], cholesteryl benzoate [46]
- **Lyotropic CLCs**: These CLCs exhibit a helical structure in response to changes in concentration. Examples: Mixtures of cholesterol and bile salts [47, 48], Mixtures of cholesterol and water [49], mixtures of liquid crystals and surfactants [50]
- **Polymer CLCs**: These CLCs exhibit a helical structure due to the presence of polymer chains. Example: Polymer-stabilized CLCs (PSCLCs) [51], Polymer networks doped with chiral dopants [52], polymer-dispersed liquid crystals (PDLCs) [53]. Various types of CLCs are shown in Fig. 7.

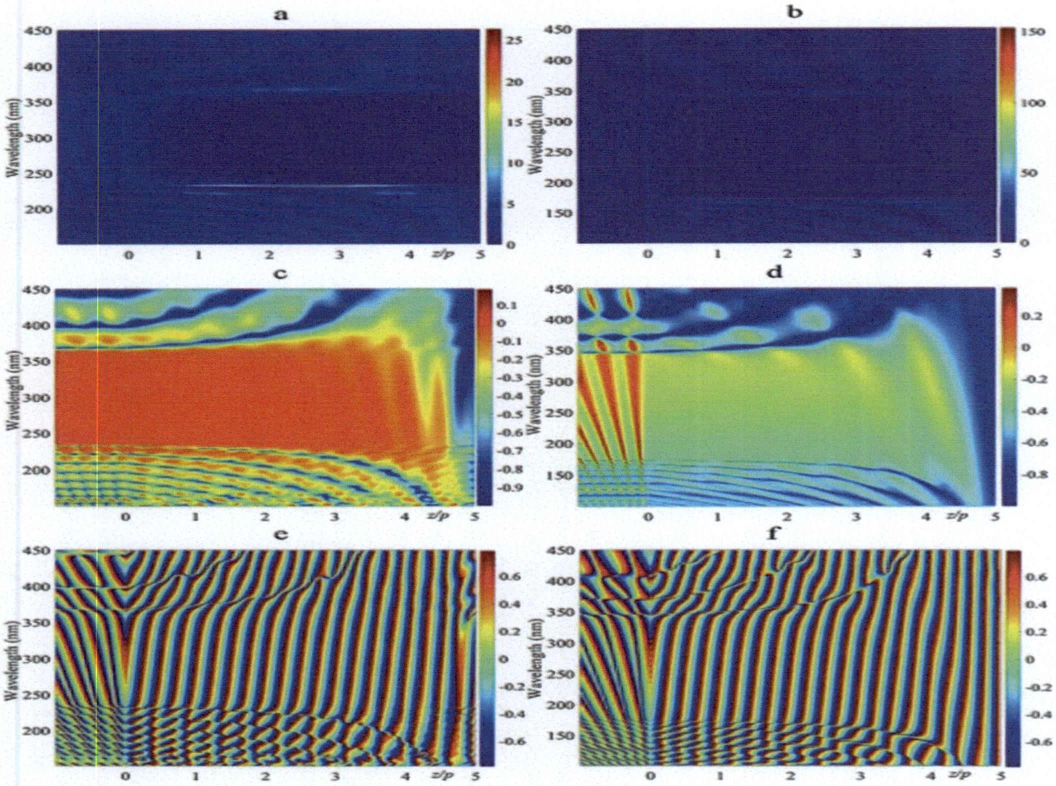

Fig. 5 The density plots of the spectra (first row), the total wave ellipticity (second row) and the azimuth (third row) as functions of z = p for the diffracting EP in the absence of absorption and at the absence of external magnetic field (the left column) and at g ¼ 0:45 (the right column). (Reproduced from Ref. [41], under the terms of the CC BY 4.0 license, http://creativecommons.org/licenses/by/4.0/.)

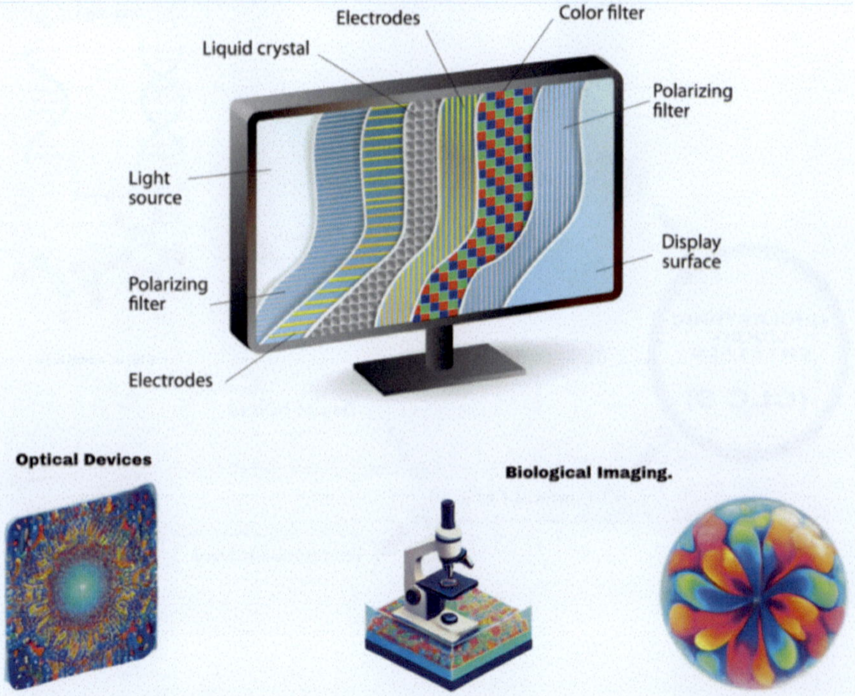

Fig. 6 CLCs various range of applications

3 Challenges Associated with the Use of CLCs

CLCs offer unique optical properties, but their practical applications are often hindered by certain challenges shown in Fig. 8.

3.1 *Stability*

a. **Temperature Sensitivity**: CLCs are sensitive to temperature fluctuations [54]. Changes in temperature can alter the pitch of the helical structure, affecting the reflected wavelength and potentially leading to color shifts [34] or degradation of optical properties [55].
b. **Photodegradation**: Exposure to UV light or other forms of radiation can cause photodegradation of the CLC material, leading to a loss of optical activity and color [56].
c. **Chemical Degradation**: CLCs can be susceptible to chemical degradation, especially in the presence of certain solvents or contaminants [57].

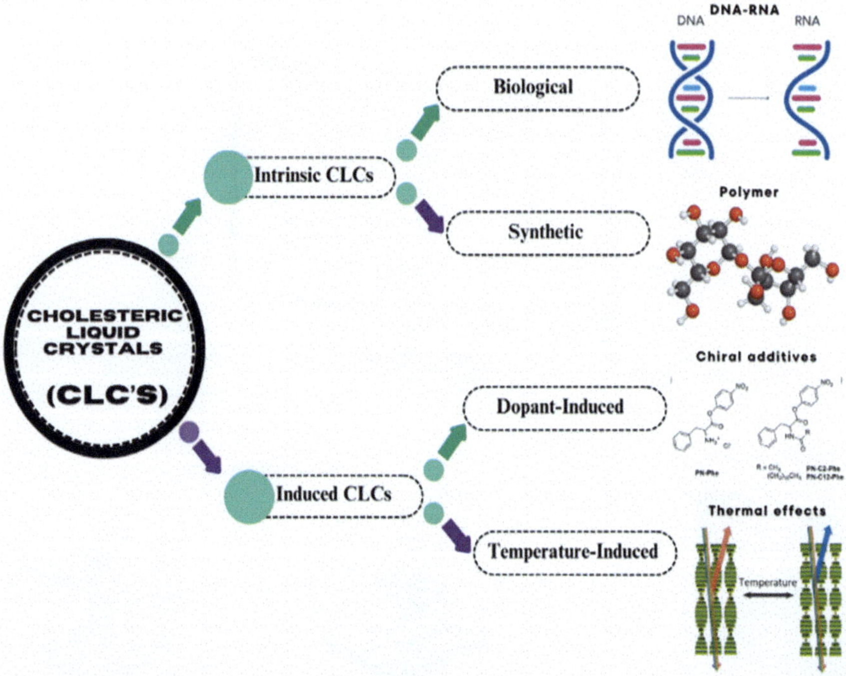

Fig. 7 Various types of CLCs

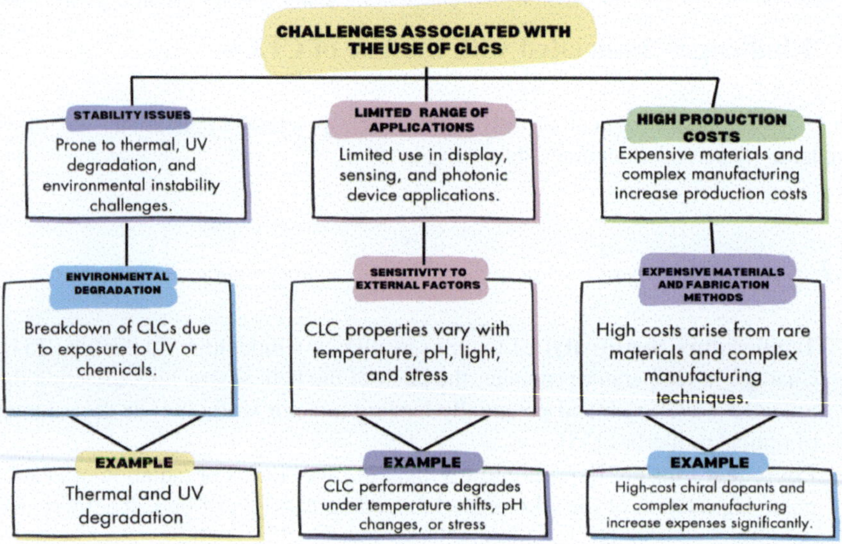

Fig. 4.8 Challenges associated with the use of CLCs

3.2 *Scalability*

a. **Manufacturing Challenges:** Large-scale production of high-quality CLC materials can be challenging due to factors such as uniformity [58], reproducibility [59], and cost.
b. **Device Fabrication:** Integrating CLCs into devices like displays or sensors can require complex fabrication processes [60], which may increase costs and reduce yields.

3.3 *Toxicity*

a. **Health Concerns**: While most CLC materials are considered non-toxic, some components may have potential health risks [61]. It is essential to assess the toxicity of CLC materials and ensure safe handling and disposal.
b. **Environmental Impact**: The environmental impact of CLC materials and their manufacturing processes should be considered [62]. Efforts should be made to minimize waste and pollution.

The challenges associated with CLCs significantly impact their applications in various fields. For instance, in displays, temperature sensitivity and photodegradation limit their use in outdoor settings or applications requiring long-term stability. In sensors, high manufacturing costs and integration complexity hinder commercialization. Furthermore, potential toxicity concerns restrict CLC use in biological imaging, particularly when in contact with living organisms. Addressing these challenges is vital for the widespread adoption of CLCs. Consequently, researchers and engineers are tirelessly working to develop more stable, scalable, and environmentally friendly CLC materials and devices, paving the way for innovative applications in displays, sensing, biological imaging, and beyond.

4 Luminescence Applications of CLCs

CLCs have been explored for various luminescence applications [63], including optical displays [64], sensing [19] and detection [65], biological imaging [66], and photonic devices [67]. Some of the potential benefits and challenges of using CLCs in these applications is as follows. In displays, CLCs are used in reflective displays for e-readers and digital signage, offering advantages like wide viewing angles and low power consumption [68]. However, temperature sensitivity and limited viewing angles can pose challenges [69]. In optical sensing, CLCs are employed as temperature sensors based on color changes, but they may be sensitive to external factors like humidity and light intensity [70]. Biosensors utilizing CLCs can detect biomolecules like proteins [71] or nucleic acids [33], but interference from other biological components and sensitivity to environmental conditions are potential limitations [72]. CLCs also have applications in photonic devices, such as tunable lasers [73] and optical filters [74], where precise control over optical properties and compatibility with other components are crucial. Optical data storage using CLCs offers high data density and readout efficiency [75], but long-term stability of the stored information remains a challenge [76]. In security features, CLCs are used in anti-counterfeiting labels [77] and authentication tags [78], but reproducibility and security against counterfeiting techniques are essential [79]. CLCs have also found artistic applications in interactive installations [80] and decorative objects [81], although cost, durability, and maintenance can be factors. Finally, in medical imaging, CLCs are used as contrast agents for techniques like polarized light microscopy [82], but biocompatibility and safety are critical considerations for in vivo use [83].

Current research CLC-based luminescence applications is focused on addressing the limitations [59] and exploring new opportunities [84]. Researchers are actively developing novel chiral dopants to enhance the tunability and stability of CLCs [85]. A full-color-tunable circularly polarized luminescence (CPL) device was developed using a CLC matrix doped with luminescent dyes. The device exhibited a dynamic reconfiguration of the CLC structure upon UV light irradiation, leading to a redshift of the photonic bandgap. The relative location of the photoluminescence peak and the photonic bandgap was found to be crucial in determining the CPL intensity. The device demonstrated potential applications in optical labelling and encryption shown in Fig. 9.

Flexible CLC devices are being investigated for applications in wearable technology [86] and foldable displays [87] Liu et al. provides a comprehensive overview of the recent advances in applying chiral liquid crystal elastomers (CLCEs) in various fields. The design and self-assembly strategies for CLCEs and their applications in biomimetic camouflage, information encryption, and shape-programmable materials. Also, highlights the challenges and future directions of CLCEs, aiming to inspire multidisciplinary research and development of these materials for diverse applications, including soft robotics, information encryption, and artificial intelligent systems. Additionally, integrating CLCs with other materials like quantum dots [88] and graphene [89] is leading to hybrid devices with improved properties. The exploration of new applications, such as optical data storage [90], security features, and environmental monitoring [91], is also a significant area of research.

Looking ahead, CLCs have the potential to revolutionize various fields. In the biomedical sector, they are being explored for biosensing [92], drug delivery [93], and tissue engineering [94]. Energy applications include their potential use in solar cells [95] and energy storage devices [96]. Optical communications can benefit from CLCs as tunable filters [97] and wavelength selectors [98]. Even consumer electronics like smartphones [99] and televisions [100] could see the integration of CLCs. However, challenges related to cost, durability, and manufacturing scalability need to be addressed to fully realize these potential applications [101]. As research progresses, we can anticipate even more innovative and exciting applications of CLCs in the near future.

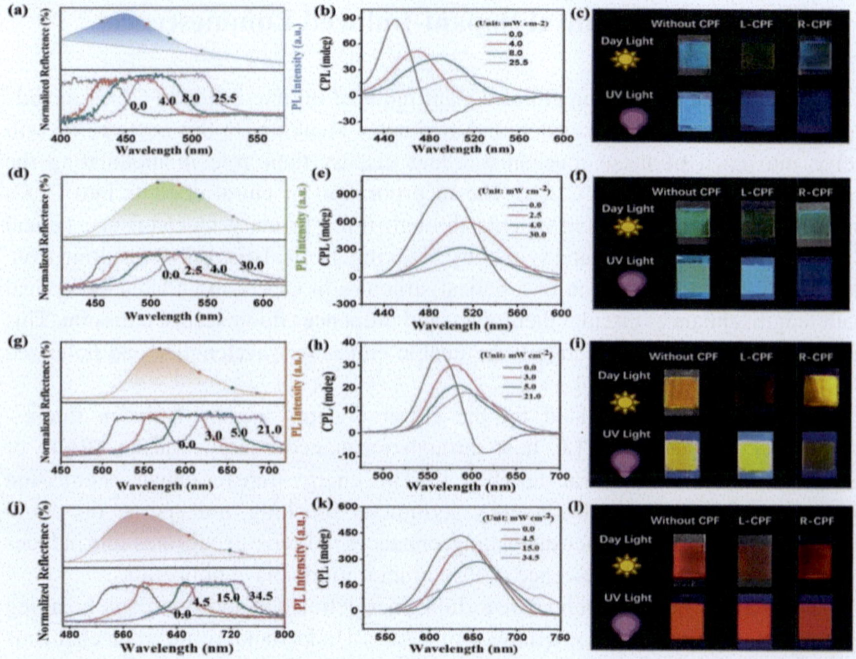

Fig. 9 Light-dependent characteristics of selective reflection bands and CPL in various films. **a** Photoluminescence (PL) and reflection spectra of a right-handed N*-LC cell from BHA. **b** CPL spectra of the BHA film under different irradiation intensities. **c** Photographs of CPL photonic films reflecting blue light, taken with and without circularly polarized filters (CPF) under natural and UV light. Additional figures (**d–l**) present similar characterization spectra and photographs for DAC, N-R, and R-6G films, respectively. [92] (Reproduced from Ref., under the terms of the CC BY 4.0 license, http://creativecommons.org/licenses/by/4.0/.)

5 Role of Chiral Dopants on Luminescence Applications of CLCs

Chiral dopants play a crucial role in modulating the luminescence behavior of CLCs. The incorporation of chiral dopants into CLCs has revolutionized their luminescence applications, offering unparalleled benefits. Chiral dopants induce a helical structure in CLCs, influencing their optical properties. Some exemplary benefits include enhanced emission intensity and polarization of fluorescent dyes, CPL emission for display applications, improved display brightness and contrast, enhanced sensitivity of CLC-based biosensors, tunable emission wavelengths, polarized electroluminescence control, induction of lasing, optical switching capabilities, chiral recognition in CLC-based sensors, and increased luminescence efficiency of quantum dots.

6 Mechanisms of Chiral Dopant-Induced Luminescence

The mechanisms underlying chiral dopant-induced luminescence in CLCs include helical distortion, energy transfer, and symmetry breaking. In this section, we will delve into each of these mechanisms and discuss their role in modulating the luminescence behavior of CLCs. The incorporation of chiral dopants into CLCs induces luminescence enhancements through three primary mechanisms: helical distortion, energy transfer, and symmetry breaking. Helical distortion, resulting from the chiral dopant's induction of a helical structure in CLCs, modulates the optical path length, enhances circular dichroism, and influences fluorescence emission. This distortion enables chiral recognition, tunable emission wavelengths, and polarized electroluminescence control.

Energy transfer, facilitated by the chiral dopants, occurs between fluorescent molecules and the CLC host through resonance energy transfer (RET) or Förster resonance energy transfer (FRET). This energy transfer enhances emission intensity, polarization, and efficiency. Symmetry breaking, induced by the chiral dopants, reduces fluorescence quenching, enhances radiative decay rates, and induces circularly polarized luminescence (CPL), crucial for display applications.

The interplay between helical distortion, energy transfer, and symmetry breaking modulates the luminescence behavior of CLCs. Understanding these mechanisms is crucial for designing optimized CLC systems. Key factors influencing chiral dopant-induced luminescence include dopant concentration and structure, CLC host material properties, fluorescent molecule characteristics, temperature, and environmental conditions is shown in Fig. 10. Theoretical models, such as density functional theory (DFT) and time-dependent density functional theory (TDDFT), and molecular dynamics simulations help elucidate the mechanisms underlying chiral dopant-induced luminescence. Experimental techniques, including spectroscopy and microscopy, provide valuable insights into the luminescence behavior of CLCs.

7 Potential Benefits, Challenges and Overview of the Current State of Research

In this section, we will discuss the potential benefits and challenges of using chiral dopants in CLCs. We will also provide an overview of the current state of research in this area and discuss potential future directions. The incorporation of chiral dopants into CLCs presents numerous benefits, including enhanced luminescence efficiency, tunable emission wavelengths, improved display performance, biosensing applications, quantum dot enhancement, circularly polarized luminescence, optical switching, and chiral recognition. Chiral dopants increase emission intensity and polarization, enable adjustable emission wavelengths, and enhance brightness, contrast, and viewing angle. Additionally, they facilitate label-free detection and chiral recognition in biosensing applications.

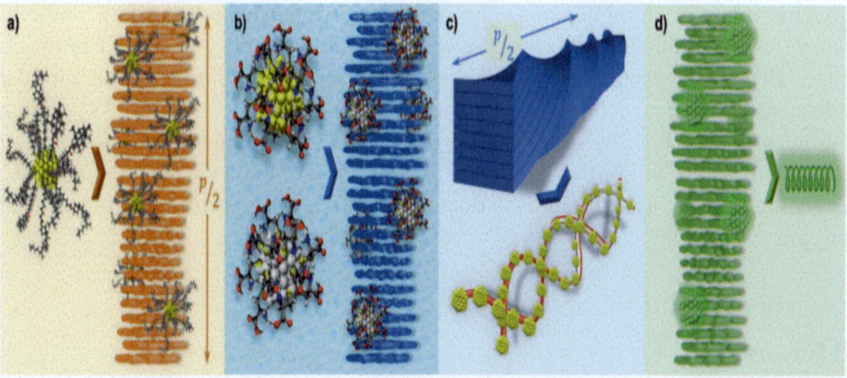

Fig. 10 Illustration of the four approaches (**a**). Quantifying nanoparticle chirality using induced chiral liquid crystal phases, where the helical pitch and twisting power are measured. **b** A similar approach using induced lyotropic chiral liquid crystal phases. **c** Helical assembly of achiral nanoparticles using a chiral organic nanoscale template, such as bent-core molecules forming helical nanofilaments. **d** Chiral assembly of emissive nanomaterials using chiral liquid crystal phases to achieve circularly polarized luminescence (CPL). Reproduced with permission from Ref. (Wang and Urbas 2020)

However, using chiral dopants in CLCs also poses challenges. These include optimizing dopant concentration, ensuring stability and durability, scaling up production while minimizing costs, ensuring material compatibility, and developing efficient synthesis methods. Balancing dopant concentration and CLC properties is crucial, as is ensuring long-term stability and durability. Currently, research focuses on designing novel chiral dopants, investigating dopant-CLC interactions, exploring new applications, developing scalable synthesis methods, and improving CLC device fabrication techniques. Future directions include developing high-performance CLC displays, biosensing and diagnostics, quantum dot-based devices, chiral recognition and separation, and theoretical modeling. Emerging trends in this field include integrating CLCs with nanomaterials for enhanced performance, developing stimuli-responsive CLCs, and exploring biomedical applications. Addressing the challenges and exploring new research directions will pave the way for breakthroughs in display technology, biosensing, and beyond.

8 Conclusion

In conclusion, the incorporation of chiral dopants into CLCs has revolutionized their optical properties, unlocking vast potential for innovative applications in displays, sensing, biological imaging, and photonic devices. By modulating the luminescence behavior of CLCs, chiral dopants enhance emission intensity, polarization, and efficiency, enabling tunable emission wavelengths and circularly polarized luminescence. The unique properties of CLCs, including selective light reflection, circular dichroism, and optical activity, make them ideal for various applications. As research continues to advance, the development of new CLC materials and exploration of novel applications will further expand the possibilities of these fascinating materials, paving the way for breakthroughs in display technology, biosensing, and biomedical imaging.

References

1. Chilaya, G.S., Lisetski, L.N.: Cholesteric liquid crystals: physical properties and molecular-statistical theories. Mol. Cryst. Liq. Cryst. **140**(2–4), 243–286 (1986). https://doi.org/10.1080/00268948608080157
2. Adimule, V., Yallur, B.C., Keri, R., Bathula, C., Batakurki, S.: Microstructure, photoluminescence and electrical properties of SmxGd (1 x)SrO hybrid nano-materials synthesized via facile coprecipitation method. Electron. Mater. Lett. **19**(3), 278–297 (2023). https://doi.org/10.1007/s13391-022-00394-0
3. Adimule, V., Keri, R., Rajendrachari, S., Kendrekar, P., Kulkarni, C.V.: Highly photoluminescence and wide band gap insulating metal hybrid nanoparticles array of samarium-doped SrO: CoO: synthesis, characterizations and sensor characteristics. J. Mater. Sci.: Mater. Electron. **34**(5), 442 (2023). https://doi.org/10.1007/s10854-023-09899-2
4. Blinov, L.M.: Structure and Properties of Liquid Crystals, vol. 123. Springer Science and Business Media (2010). https://doi.org/10.1007/978-90-481-8829-1
5. Challa, M., Yallur, B.C., Ambika, M.R., Adimule, V.: Influence of nano particles on optical properties of Cu-MOFs. Adv. Mater. Res. **1173**, 23–33 (2022). https://doi.org/10.4028/p-vn4hd4
6. Nandi, S.S., Adimule, V., Kadapure, S.A., Kerur, S.S.: Rare earth based nanocomposite materials for prominent performance supercapacitor: a review. Appl. Mech. Mater. **908**, 3–18 (2022). https://doi.org/10.4028/p-rff302
7. Shtykov, N.M., Palto, S.P., Umanskii, B.A., Rybakov, D.O., Simdyankin, I.V.: Director distribution in a CLC hybrid cell with a small helicoidal pitch. Crystallogr. Rep. **68**(1), 88–96 (2023). https://doi.org/10.1134/S1063774522060219
8. Manhas, N., Adimule, V., Lakshminarayana, P., Keri, R., Kumar, L.S., Bathula, C.: Utilization of CO^{2+} doped Eu_2O_3 nanostructures and their effect on the dielectric and photoisomerization of liquid crystals. Phys. Fluids **35**(10) (2023). https://doi.org/10.1063/5.0171578
9. Nagalakshmi, G., Nandeesh, I.M., Yallur, B.C., Adimule, V., Batakurki, S.: Synthesis and optical properties of copper terephthalate metal organic frame works. Eng. Chem. **2**, 3–11 (2023). https://doi.org/10.4028/p-fdqs03
10. Adimule, V., Yallur, B.C., Gowda, A.H.: Advanced sensors based on carbon nanomaterials. In: Carbon nanomaterials-based sensors, pp. 259–268. Elsevier (2022). https://doi.org/10.1016/B978-0-323-91174-0.00004-4
11. Mitov, M.: Cholesteric liquid crystals with a broad light reflection band. Adv. Mater. **24**(47), 6260–6276 (2012). https://doi.org/10.1002/adma.201202913
12. Sharma, V., Kumar, P., Chinky, C., Gahrotra, R., Raina, K.K., Malik, P.: Effect of nano particles on electro optic properties of polymer dispersed liquid crystal in normal mode. In: AIP Conference Proceedings, vol. 2142, no. 1. AIP Publishing (2019). https://doi.org/10.3390/polym15163420

13. Adimule, V., Sharma, V., Kumar, P., Keri, R., Sharma, K.: Perylene-3, 4, 9, 10-tetracarboxylic dianhydride dye intercalated liquid crystal molecules: synthesis, spectroscopic characterizations and photoluminescence dynamics. J. Mol. Liq. 125990 (2024). https://doi.org/10.1016/j.molliq.2024.125990
14. Dogra, A.R., Kumar, P., Sharma, V., Malik, P., Khanra, P., Kapoor, M.: Deposition of nanoparticles layer on ITO substrate and its application for alignment control of liquid crystal. In: AIP Conference Proceedings, vol. 2357. AIP Publishing (2022). https://doi.org/10.1063/5.0080664
15. Bisoyi, H.K., Li, Q.: Light-directing chiral liquid crystal nanostructures: from 1D to 3D. Acc. Chem. Res. **47**(10), 3184–3195 (2014). https://doi.org/10.1021/ar500249k
16. Sharma, V., Kumar, P., Chinky, C., Malik, P., Raina, K.K.: Morphological and electro optic studies of polymer dispersed liquid crystal in reverse mode. In: AIP Conference Proceedings, vol. 1953. AIP Publishing (2018). https://doi.org/10.1063/1.5032969
17. Gahrotra, R., Sharma, V., Dogra, A.R., Malik, P., Kumar, P.: Temperature dependent morphological and electro-optical characteristics of dye doped cholesteric liquid crystal. ECS Trans. **107**(1), 5481 (2022). https://doi.org/10.1149/10701.5481ecst
18. Adimule, V., Nandi, S., Sharma, K., Keri, R.: Microstructure, zeta potential and enhanced photocatalytic dye degradation performances of mesoporous flower like $Nb_2O_5@CeO_2$ bimetallic oxide nanomaterials decorated with reduced graphene oxide. Inorg. Chem. Commun. 112592 (2024). https://doi.org/10.1016/j.inoche.2024.112592
19. Furumi, S.: Recent progress in chiral photonic band-gap liquid crystals for laser applications. Chem. Rec. **10**(6), 394–408 (2010). https://doi.org/10.1002/tcr.201000013
20. Dogra, A.R., Sharma, V., Gahrotra, R., Khanra, P., Dogra, A.: Thickness dependent electro-optical behaviour of homeotropically aligned nematic liquid crystal. ECS Trans. **107**(1), 5021 (2022). https://doi.org/10.1149/10701.5021ecst
21. Adimule, V., Bathula, C., Sharma, K., Keri, R.: Study of structural, optical, dielectric properties poly-3-butyl thiophene reinforced with CO_3O_4 nanoparticles. Nano-Struct. Nano-Objects **38**, 101150 (2024). https://doi.org/10.1016/j.nanoso.2024.101150
22. Stauber, T., Weinert, S., Jentsch, T.J.: Cell biology and physiology of CLC chloride channels and transporters. Compr. Physiol. **2**(3), 1701–1744 (2012). https://doi.org/10.1002/cphy.c110038
23. Dogra, A.R., Sharma,V., Malik, P., Kumar, P.: In-situ homeotropic alignment of dye doped liquid crystal molecules on multi-layered self assembled nanoparticles in confined cells for next generation display devices. Opt. Mater. **154**, 115731 (2024). https://doi.org/10.1002/adma.201300730. https://doi.org/10.1016/j.molliq.2021.118030
24. Sharma, V., Kumar, P., Sharma, A., Raina, K.K.: Droplet configuration control with orange azo dichroic dye in polymer dispersed liquid crystal for advanced electro-optic characteristics. J. Mol. Liq. **233**, 122–130 (2017). https://doi.org/10.1016/j.molliq.2017.03.012
25. Mitov, M.: Cholesteric liquid crystals in living matter. Soft Matter **13**(23), 4176–4209 (2017). https://doi.org/10.1039/C7SM00384F
26. Ahn, S., Ko, M.O., Kim, J.H., Chen, Z., Jeon, M.Y.: Characterization of second-order reflection bands from a cholesteric liquid crystal cell based on a wavelength-swept laser. Sensors **20**(16), 4643 (2020). https://doi.org/10.3390/s20164643
27. Balenko, N., Shibaev, V., Bobrovsky, A.: Polymer dispersed cholesteric liquid crystals with combined photo-and mechanochromic response. J. Mol. Liq. **401**, 124637 (2024). https://doi.org/10.1016/j.molliq.2024.124637
28. Ranjkesh, A., Yoon, T.H.: Thermal and electrical wavelength tuning of Bragg reflection with ultraviolet light absorbers in polymer-stabilized cholesteric liquid crystals. J. Mater. Chem. C **6**(45), 12377–12385 (2018). https://doi.org/10.1039/C8TC04563A
29. Balamurugan, R., Liu, J.H.: A review of the fabrication of photonic band gap materials based on cholesteric liquid crystals. React. Funct. Polym. **105**, 9–34 (2016). https://doi.org/10.1016/j.reactfunctpolym.2016.04.012
30. Palto, S.P., Barnik, M.I., Geivandov, A.R., Kasyanova, I.V., Palto, V.S.: Spectral and polarization structure of field-induced photonic bands in cholesteric liquid crystals. Phys. Rev. E **92**(3), 032502 (2015). https://doi.org/10.1103/PhysRevE.92.032502

31. Lee, K.M., Rumi, M., Mills, M.S., Reshetnyak, V., Evans, D.R., Bunning, T.J., McConney, M.E.: A different perspective on cholesteric liquid crystals reveals unique color and polarization changes. ACS Appl. Mater. Interfaces **12**(33), 37400–37408 (2020). https://doi.org/10.1021/acsami.0c09845
32. Hsiang, E.L., Li, Y., He, Z., Zhan, T., Zhang, C., Lan, Y.F., Wu, S.T.: Enhancing the efficiency of color conversion micro-LED display with a patterned cholesteric liquid crystal polymer film. Nanomaterials **10**(12), 2430 (2020). https://doi.org/10.3390/nano10122430
33. Fuh, A.Y.G., Ho, S.J., Wu, S.T., Li, M.S.: Optical filter with tunable wavelength and bandwidth based on phototunable cholesteric liquid crystals. Appl. Opt. **53**(8), 1658–1662 (2014). https://doi.org/10.1364/AO.53.001658
34. Ahn, S., Lee, G.H., Lee, J.Y., Kim, Y., Kim, M.S., Pagidi, S., Jeon, M.Y.: Fiber-optic temperature sensor using cholesteric liquid crystals on the optical fiber ferrules. Sensors **22**(15), 5752 (2022). https://doi.org/10.3390/s22155752
35. Hao, N., Li, X., Chen, Y., Zhou, D., Huang, Y., Liu, Y.: A compact, highly sensitive optical fiber temperature sensor based on a cholesteric liquid crystal polymer film. Opt. Commun. 131241 (2024). https://doi.org/10.1016/j.optcom.2024.131241
36. Umanskii, B.A., Simdyankin, I.V.: Circular dichroism in cholesteric liquid crystals. Crystallogr. Rep. **64**, 437–442 (2019). https://doi.org/10.1134/S1063774519030283
37. Qiao, J., He, Y., Lin, S., Fan, Q., Guo, J.: A photoswitchable circularly polarized luminescent cholesteric superstructure: direct visualization and dynamic modulation of the amplified luminescence dissymmetry factor. J. Mater. Chem. C **10**(18), 7311–7318 (2022). https://doi.org/10.1039/D2TC00424K
38. Zhang, X., Xu, Y., Valenzuela, C., Zhang, X., Wang, L., Feng, W., Li, Q.: Liquid crystal-templated chiral nanomaterials: from chiral plasmonics to circularly polarized luminescence. Light: Sci. Appl. **11**(1), 223 (2022). https://doi.org/10.1038/s41377-022-00913-6
39. Gevorgyan, A.H.: Light absorption suppression in cholesteric liquid crystals with magneto-optical activity. J. Mol. Liq. **335**, 116289 (2021). https://doi.org/10.1016/j.molliq.2021.116289
40. Golovina, T.G., Konstantinova, A.F., Timofeev, V.I.: Features of optical activity in inorganic and organic materials. Crystallogr. Rep. **65**, 653–677 (2020). https://doi.org/10.1134/S1063774520050077
41. Vetrov, S.Y., Timofeev, I.V., Shabanov, V.F.: Localized modes in chiral photonic structures. Phys. Usp. **63**(1), 33 (2020). https://doi.org/10.1134/S1063774520050077
42. Kidowaki, M., Moriyama, M., Wada, M., Tamaoki, N.: Molecular mechanism of anomalous increase in the helical pitch of cholesteric liquid crystals induced by achiral dopants. J. Phys. Chem. B **107**(44), 12054–12061 (2003). https://doi.org/10.1021/jp034916e
43. Zhai, F., Feng, Y., Zhou, K., Wang, L., Zheng, Z., Feng, W.: Graphene- based chiral liquid crystal materials for optical applications. J. Mater. Chem. C **7**(8), 2146–2171 (2019). https://doi.org/10.1039/C8TC04947E
44. Singh, S.: Chiral nematic liquid crystals. In: Handbook of Liquid Crystals—Volume II: Advanced Aspects and Applications, pp. 1–69. Springer Nature Switzerland, Cham (2024). https://doi.org/10.1007/978-3-031-52621-3
45. Almeida, A.P.C.: Cellulose Filaments for Responsive and Functional Materials. Doctoral dissertation, Universidade NOVA de Lisboa (Portugal) (2021)
46. Radka, B.P.: Fundamental Examinations of the Ion-Mediated Electro-Optical Response of Polymer Stabilized Cholesteric Liquid Crystals. Doctoral dissertation, University of Colorado at Boulder (2023)
47. Wang, C.: Physiological Relevance of Novel Lysosomal Ion Channels. Doctoral dissertation (2021)
101. Hung, Y.H., Liu, C.Y., Chen, W.C., Liu, J.H.: Thermal and optical properties of amphitropic liquid crystals derived from cholesterol and cinnamic acid. Liq. Cryst. **48**(4), 542–554 (2021). https://doi.org/10.7302/1150
48. Hoffmann, H.: From micellar solutions to liquid crystalline phases. Berichte der Bunsengesellschaft fu¨r physikalischeChemie **88**(11), 1078–1093 (1984). https://doi.org/10.1002/bbpc.198400012

49. Ye, Y., Guo, L., Zhong, T.: A review of developments in polymer stabilized liquid crystals. Polymers **15**(13), 2962 (2023). https://doi.org/10.3390/polym15132962
50. Mani, S., Mani, S., Khosla, S., Sarawade, P.: (2022) Tuning of thermal and electrical properties of polymer dispersed cholesteric liquid crystal. In: Materials Science Forum, vol. 1070, pp. 33–43. https://doi.org/10.4028/p-w7lky2
51. Ltd, T.T.P., Balenko, N., Shibaev, V., Bobrovsky, A.: Mechanosensitive polymer-dispersed cholesteric liquid crystal composites based on various polymer matrices. Polymer **281**, 126119 (2023). https://doi.org/10.1016/j.polymer.2023.126119
52. Zhang, W., Froyen, A.A., Schenning, A.P., Zhou, G., Debije, M.G., de Haan, L.T.: Temperature-responsive photonic devices based on cholesteric liquid crystals. Adv. Photonics Res. **2**(7), 2100016 (2021). https://doi.org/10.1002/adpr.202100016
53. Zhou, H., Wang, H., He, W., Yang, Z., Cao, H., Wang, D., Li, Y.: Research progress of cholesteric liquid crystals with broadband reflection. Molecules **27**(14), 4427 (2022). https://doi.org/10.3390/molecules27144427
54. McConney, M.E., Rumi, M., Godman, N.P., Tohgha, U.N., Bunning, T.J.: Photoresponsive structural color in liquid crystalline materials. Adv. Opt. Mater. **7**(16), 1900429 (2019). https://doi.org/10.1002/adom.201900429
55. Alalwan, H.A., Alminshid, A.H.: CO_2 capturing methods: chemical looping combustion (CLC) as a promising technique. Sci. Total Environ. **788**, 147850 (2021). https://doi.org/10.1016/j.scitotenv.2021.147850
56. de Souza Carvalho, T.A., Magalhaes, L.F., do Livramento Santos, C.I., de Freitas, T.A.Z., Carvalho Vale, B.R., Vale da Fonseca, A.F., Schiavon, M.A.: Lead-free metal halide perovskite nanocrystals: from fundamentals to applications. Chem. A Eur. J. **29**(4), e202202518 (2023). https://doi.org/10.1002/chem.202202518
57. de Araujo, W.R., Lukas, H., Torres, M.D., Gao, W., de la Fuente-Nunez, C.: Low-cost biosensor technologies for rapid detection of COVID-19 and future pandemics. ACS Nano **18**(3), 1757–1777 (2024). https://doi.org/10.1021/acsnano.3c01629
58. Xia, Y.: Reconfigurable Liquid Crystals for Wearable Applications. Doctoral dissertation, University of Glasgow (2024). https://doi.org/10.1002/adem.202301165
59. Idziak, K., Czakiert, T., Krzywanski, J., Zylka, A., Kozlowska, M., Nowak, W.: Safety and environmental reasons for the use of Ni-, Co-, Cu-, Mn-and Fe- based oxygen carriers in CLC/CLOU applications: an overview. Fuel **268**, 117245 (2020). https://doi.org/10.1016/j.fuel.2020.117245
60. Pathak, G., Hegde, G., Prasad, V.: Octadecylamine-capped CdSe/ZnS quantum dot dispersed cholesteric liquid crystal for potential display application: Investigation on photoluminescence and UV absorbance. Liq. Cryst. **48**(4), 579–587 (2021). https://doi.org/10.1080/02678292.2020.1799085
61. Kumar, A., Singh, G.: Recent advances and future perspectives of photoluminescent liquid crystals and their nanocomposites for emissive displays and other tunable photonic devices. J. Mol. Liq. (2023). https://doi.org/10.1016/j.molliq.2023.122607
62. Setia, S., Sidiq, S., De, J., Pani, I., Pal, S.K.: Applications of liquid crystals in biosensing and organic light-emitting devices: future aspects. Liq. Cryst. **43**(13–15), 2009–2050 (2016). https://doi.org/10.1080/02678292.2016.1213002
63. Su, Y., Deng, D., Zhang, L., Song, H., Lv, Y.: Strategies in liquid-phase chemiluminescence and their applications in bioassay. TrAC, Trends Anal. Chem. **82**, 394–411 (2016). https://doi.org/10.1016/j.trac.2016.07.002
64. Bai, P.F., Hayes, R.A., Jin, M., Shui, L., Yi, Z.C., Wang, L., Zhou, G.: Review of paper-like display technologies (invited review). Prog. Electromag. Res. **147**, 95–116 (2014). https://doi.org/10.2528/PIER13120405
65. Yin, K., Hsiang, E.L., Zou, J., Li, Y., Yang, Z., Yang, Q., Wu, S.T.: Advanced liquid crystal devices for augmented reality and virtual reality displays: principles and applications. Light: Sci. Appl. **11**(1), 161 (2022). https://www.nature.com/articles/s41377-022-00851-3.pdf

66. Li, S.L., Chen, Z.Y., Chen, P., Hu, W., Huang, C., Li, S.S., Chen, L.J.:. Geometric phase-encoded stimuli-responsive cholesteric liquid crystals for visualizing real-time remote monitoring: humidity sensing as a proof of concept. Light: Sci. Appl. **13**(1), 27 (2024). https://www.nature.com/articles/s41377-023-01360-7.pdf
67. Sidiq, S., Pal, S.K.: Liquid crystal biosensors: new approaches. In: Proc. Indian Natl. Sci. Acad., vol. 82, pp. 75–98 (2016). https://doi.org/10.16943/ptinsa/2016/v82i1/48445
68. Qu, R., Li, G.: Overview of liquid crystal biosensors: from basic theory to advanced applications. Biosensors **12**(4), 205 (2022). https://doi.org/10.3390/bios12040205
69. Leal-Junior, A., Soares, M.S., de Almeida, P.M., Marques, C.: Cholesteric liquid crystals sensors based on nanocellulose derivatives for improvement of quality of human life: a review. Adv. Sens. Res. **2**(10), 2300022 (2023). https://doi.org/10.1002/adsr.202300022
70. Labeeb, A.M., Ward, Y.A., Fikry, M.: Thermal control of tunable photonic optical bandgaps in different cholesteric liquid crystals mixtures. J. Mol. Liq. **340**, 117179 (2021). https://doi.org/10.1016/j.molliq.2021.117179
71. Zhan, X., Zhou, Z., Zhou, W., Yan, Y., Yao, J., Zhao, Y.S.: Wavelength tunable circularly polarized laser arrays for multidimensional information encryption. Adv. Opt. Mater. **11**(13), 2200872 (2023). https://doi.org/10.1002/adom.202200872
72. Chen, F., Zheng, J., Xing, C., Sang, J., Shen, T.: Applications of liquid crystal planer optical elements based on photoalignment technology in display and photonic devices. Displays (2024). https://doi.org/10.1016/j.displa.2023.102632
73. Wang, Q., Zhou, D., Gong, K., Lv, X., Li, P., Liu, Y.: A multidimensional anti-counterfeiting dynamic flexible label based on light/thermal responsive CLC microspheres. J. Mater. Chem. C **12**(37), 15092–15100 (2024). https://doi.org/10.1039/D4TC02630F
74. Nayak, P., Swapna, G.: Security issues in IoT applications using certificateless aggregate signcryption schemes: an overview. Internet of Things **21**, 100641 (2023). https://doi.org/10.1016/j.iot.2022.100641
75. Qu, G., Zhang, X., Li, S., Lu, L., Gao, J., Yu, B., Hu, Z.: Liquid crystal random lasers. Phys. Chem. Chem. Phys. **25**(1), 48–63 (2023). https://doi.org/10.1039/D2CP02859J
76. Subiyantoro, S., Prabowo, T.: The learning of arts and cultures in the farm field school for the illegal Indonesian migrant workers' children: a case study of from Community Learning Center (CLC) at Sabah, Malaysia. In International Conference on Teacher Training and Education 2017 (ICTTE 2017), pp. 223–231. Atlantis Press (2017). https://doi.org/10.2991/ictte-17.2017.25
77. Bravo Mlina, L.M.: A CLIL Didactic Proposal of Arts and Crafts for Primary Education (2022)
78. Lee, K.M., Rumi, M., Mills, M.S., Reshetnyak, V., Evans, D.R., Bunning, T.J., McConney, M.E.: A different perspective on cholesteric liquid crystals reveals unique color and polarization changes. ACS Appl. Mater. Interfaces **12**(33), 37400–37408 (2020). https://doi.org/10.1021/acsami.0c09845
79. Bassi da Silva, J., Ferreira, S.B.D.S., de Freitas, O., Bruschi, M.L.: A critical review about methodologies for the analysis of mucoadhesive properties of drug delivery systems. Drug Dev. Ind. Pharm. **43**(7), 1053–1070 (2017). https://doi.org/10.1080/03639045.2017.1294600
80. Ortega, J., Folcia, C.L., Etxebarria, J.: Upgrading the performance of cholesteric liquid crystal lasers: improvement margins and limitations. Materials **11**(1), 5 (2017). https://doi.org/10.3390/ma11010005
81. Jiang, L., Shui, Q., Han, B., Huang, R., Deng, J., Hu, N., Yang, J.: Recent advances in liquid crystal droplet-based biosensors. TrAC, Trends Anal. Chem. (2024). https://doi.org/10.1016/j.trac.2024.117920
82. Bao, J., Lan, R., Shen, C., Huang, R., Wang, Z., Hu, W., Yang, H.: Modulation of chirality and intensity of circularly polarized luminescence emitting from cholesteric liquid crystals triggered by photoresponsive molecular motor. Adv. Opt. Mater. **10**(3), 2101910 (2022). https://doi.org/10.1002/adom.202101910
83. Liu, Y., Ma, J., Yang, Y., Valenzuela, C., Zhang, X., Wang, L., Feng, W.: Smart chiral liquid crystal elastomers: design, properties and application. Smart Mol. **2**(1), e20230025 (2024). https://doi.org/10.1002/smo.20230025

84. Singh, G.: Recent advances on cadmium free quantum dots-liquid crystal nanocomposites. Appl. Mater. Today **21**, 100840 (2020). https://doi.org/10.1016/j.apmt.2020.100840
85. Wu, Q., Lv, H., Zhao, L.: Applications of carbon nanomaterials in chiral separation. TrAC, Trends Anal. Chem. **129**, 115941 (2020). https://doi.org/10.1016/j.trac.2020.115941
86. Zhang, R., Zhang, Z., Han, J., Yang, L., Li, J., Song, Z., Zhu, J.: Advanced liquid crystal-based switchable optical devices for light protection applications: principles and strategies. Light: Sci. Appl. **12**(1), 11 (2023). https://doi.org/10.1038/s41377-022-01032-y
87. Schwartz, M., Lenzini, G., Geng, Y., Rønne, P.B., Ryan, P.Y., Lagerwall, J.P.: Cholesteric liquid crystal shells as enabling material for information-rich design and architecture. Adv. Mater. **30**(30), 1707382 (2018). https://doi.org/10.1002/adma.201707382
88. Bhatia, M.S., Jakhar, S.K., Mangla, S.K., Gangwani, K.K.: Critical factors to environment management in a closed loop supply chain. J. Clean. Prod. **255**, 120239 (2020). https://doi.org/10.1016/j.jclepro.2020.120239
89. Cezar, S.V.S., Santos, A.B., Santos, A.M., Brito, J.R.L.R., dos Passos Menezes, P., Serafini, M.R.: Patents on the move: the therapeutic future of liquid crystals in cancer. J. Drug Deliv. Sci. Technol. (2024). https://doi.org/10.1016/j.jddst.2024.105822
90. Zhang, Z., Yang, X., Zhao, Y., Ye, F., Shang, L.: Liquid crystal materials for biomedical applications. Adv. Mater. **35**(36), 2300220 (2023). https://doi.org/10.1002/adma.202300220
91. Liu, Y., Yu, L., Jiang, Y., Xiong, W., Wang, Q., Sun, J., Zhao, X.Z.: Self-organized cholesteric liquid crystal polymer films with tunable photonic band gap as transparent and flexible back-reflectors for dye-sensitized solar cells. Nano Energy **26**, 648–656 (2016). https://doi.org/10.1016/j.nanoen.2016.06.021
92. Prakash, J., Chandran, A., Biradar, A.M.: Scientific developments of liquid crystal-based optical memory: a review. Rep. Prog. Phys. **80**(1), 016601 (2016). https://doi.org/10.1088/0034-4885/80/1/016601
93. Ryabchun, A., Bobrovsky, A.: Cholesteric liquid crystal materials for tunable diffractive optics. Adv. Opt. Mater. **6**(15), 1800335 (2018). https://doi.org/10.1002/adom.201800335
94. Tocnaye, J.: Engineering liquid crystals for optimal uses in optical communication systems. Liq. Cryst. **31**(2), 241–269 (2004). https://doi.org/10.1080/02678290410001648570
95. Mathew, C., Pathinti, R.S., Pradhan, S.R., Gollapelli, B., Chithari, K., Ghosh, M., Vallamkondu, J.: Enhanced security through fluorescent DyeDoped cholesteric liquid crystal shells for anti-counterfeiting. Opt. Mater. 115614 (2024)
96. Li, Y.: Reflective planar optics with cholesteric liquid crystal for near-eye displays (2023). https://doi.org/10.1016/j.optmat.2024.115614
97. Bisoyi, H.K., Li, Q.: Liquid crystals: versatile self-organized smartsoft materials. Chem. Rev. **122**(5), 4887–4926 (2021). https://doi.org/10.1021/acs.chemrev.1c0076
98. Liu, J., Song, Z.P., Sun, L.Y., Li, B.X., Lu, Y.Q., Li, Q.: Circularly polarized luminescence in chiral orientationally ordered soft matter systems. Responsive Mater. **1**(1), e20230005 (2023). https://doi.org/10.1002/rpm.20230005
99. Gollapelli, B., Potu, S., Rajaboina, R., Vallamkondu, J.: Optical anticounterfeiting with cholesteric liquid crystal emulsions: preparation, properties, and applications. Mater. Adv. **5**(18), 7113–7129 (2024). https://doi.org/10.1039/D4MA00604F
100. Liu, J., Zhou, X., Tang, X., Tang, Y., Wu, J., Song, Z., Li, Q.: Circularly polarized organic ultra-long room-temperature phosphorescence: generation, enhancement, and application. Adv. Funct. Mater. 2414086 (2024). https://doi.org/10.1002/adfm.202414086

The Chemistry of Synthesis of Cholesteric Liquid Crystals: Their Self-Assembly and Colour Interferences

Santosh Nandi, Savita Hanaji, Vinayak Adimule, Praveen Barmavatu, S. S. Nesargi, and Madhavaprasad Dasari

Abstract Cholesteric liquid crystals (CLCs) are a unique class of liquid crystalline materials known for their helical structure and selective reflection of light, resulting in vibrant color interference patterns making them attractive candidates for a wide range of photonic applications. Designing, creating, and working with these optical microstructures in soft matter systems, such liquid crystals (LCs), would provide new avenues for both academic study and real-world applications. This book chapter explores the chemistry involved in creating CLCs, emphasising the molecular architecture, the function of chiral dopants, and the variables affecting the self-assembly of these crystals. The helical pitch and optical properties of CLCs can be accurately controlled through careful manipulation of chemical interactions, temperature, and concentration of chiral agents, enabling a wide range of applications, from display technologies to sensors. Selective light reflection and the capacity to generate structural colour are two of the distinctive optical characteristics of CLCs that result from

S. Nandi (✉)
Department of Chemistry, KLE Technological University, Dr. M. S. Sheshgiri Campus, Udyambag, Belagavi, Karnataka, India
e-mail: santosh.nandi603@gmail.com

S. Hanaji
Department of Engineering Science and Humanities, KLE Technological University, Dr. M. S. Sheshgiri Campus, Udyambag, Belagavi, Karnataka, India

V. Adimule
Department of Chemistry, Angadi Institute of Technology and Management (AITM), Belagavi, Karnataka, India

P. Barmavatu
Department of Mechanical Engineering, Faculty of Engineering, Universidad Tecnológica Metropolitana, Santiago, Chile
e-mail: pbarmavatu@utem.cl

S. S. Nesargi
Chemistry Section, Department of Chemistry, KLE Technological University, Dr. M. S. Sheshgiri Campus, Udyambag, Belagavi, Karnataka, India

M. Dasari
Department of Physics, GITAM Institute of Science, GITAM (Deemed to be University), Visakhapatnam, India

P. Kumar et al. (eds.), *Cholesteric Liquid Crystals*, Engineering Materials,
https://doi.org/10.1007/978-981-96-3821-5_5

their inherent ability to self-assemble into ordered, periodic structures. Additionally, we investigate the connection between the observed colour interferences and molecular structure, emphasising the recent progress in the design and functionalisation of CLC materials, which highlights the potential of CLCs to generate dynamic and responsive systems for the next generation of photonic and optoelectronic devices.

Keywords Cholesteric liquid crystals · Synthesis · Chiral Dopants · Bragg Reflection · Color Interference · Pitch Modulation

1 Introduction

Smart buildings with intelligent control and low energy consumption have attracted a lot of attention as urbanisation and modernisation progress, and they are anticipated to offer a productive, warm, and pleasant environment for people to live and work in [1, 2]. Windows are a key component of smart buildings because they allow heat radiation from the sun and outside light to enter the interior of the structure. A promising option for light control in contemporary intelligent buildings are "smart windows," which are windows that have the ability to selectively alter colour and light transmittance on demand [3, 4]. Among the smart window technologies that are accessible are mechanochromism, magnetochromism, humidity-triggered chromism, photochromism, electrochromism, and thermochromism [5–8]. Over the past few decades, a number of materials, including phase change materials (PCMs), hydrogels, organic and inorganic electrochromic materials, perovskites, liquid crystals (LCs), and hydrogels, have been created as essential components for smart windows. Because of their optical anisotropy and molecule orientation dependency of optical characteristics. LCs represent a unique state of matter, displaying properties typical of both liquids and crystalline solids [9, 10]. They have anisotropic qualities, meaning that the direction of measurement affects their physical attributes. According to their molecular structure, LCs can be divided into a number of categories, such as cholesteric, smectic, and nematic phases (Fig. 1a. A helical molecule arrangement distinguishes cholesteric liquid crystals (CLCs), also referred to as chiral nematic LCs, as a subclass of LCs [11]. CLCs have special optical qualities due to their helical form, most notably the ability to selectively reflect circularly polarised light (CPL). CLCs can exhibit vivid structural colours that change in response to external stimuli like temperature, pressure, or electric fields, depending on the helix's pitch. Rod-like molecules organise themselves into layers in CLCs, and the molecules aligned parallel to each other in a certain direction are known as alignment directors. By rotating at a specific angle, the molecules between two adjacent layers create a periodic helical nanostructure. A pattern resembling a spiral is produced along the helix's axis by this rotation. Pitch of CLCs is defined by the pattern's repetition along the helical axis. Pitch is the distance between layers in a full cycle of the helix (Fig. 1b), when alignment directors spin 360° [12]. A subclass of chiral nematic LCs, CLCs have special optical and structural characteristics because of their helical molecular structure. These materials have attracted a lot of interest in both scholarly studies and real-world applications, particularly in the domains of sensors, tuneable optical devices, and display technologies. One of CLCs' most intriguing properties is their wavelength-selective light reflection, which produces interference colours. This chapter examines the chemistry involved in the creation of CLCs, the mechanisms of self-assembly that produce their helical structures, and the colour interference phenomena that ensue. Comprehending these procedures is essential for modifying CLCs for upcoming technological advancements.

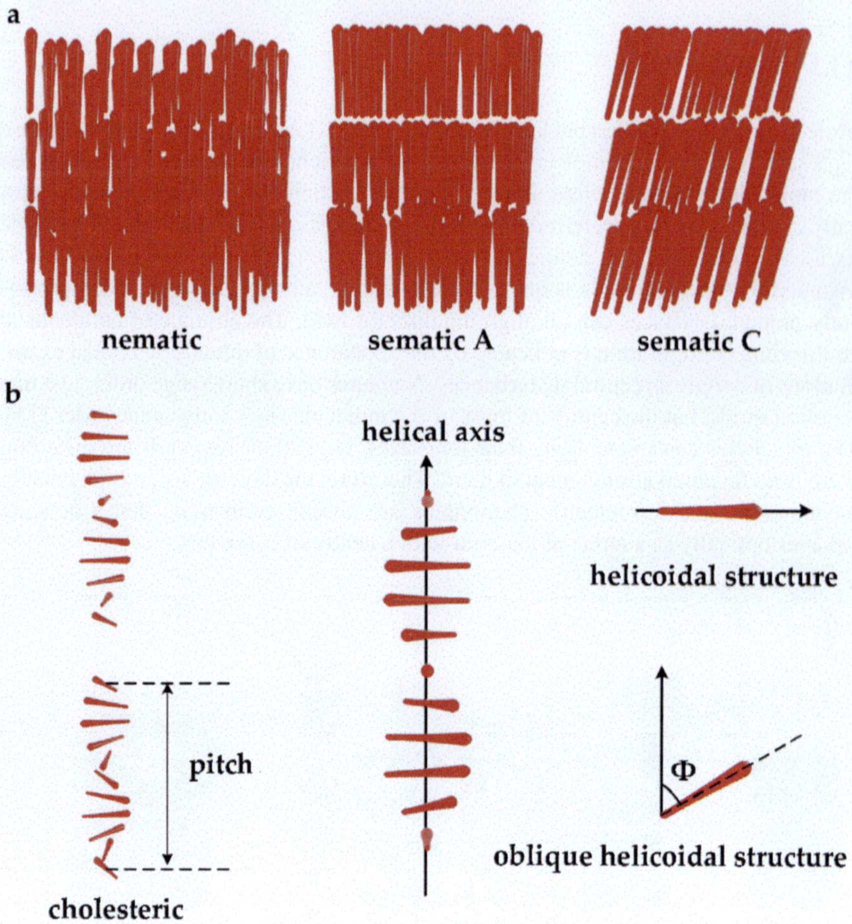

Fig. 1 Illustration of LC molecules, alignment, and chiral pitch. **a** LC molecular arrangement of nematic, sematic A, and sematic C. **b** The arrangement of CLC molecules (where one molecule represents the molecular arrangement direction of the entire layer) and the pitch of CLCs are shown here as two cycles. To visually display the number of cycles, a virtual interval space is added. And two types of CLC nanostructures: helicoidal structure and oblique helicoidal structure. (Adapted from Ref. [12], Copyright, 2024 MDPI)

2 Chemical Synthesis of Cholesteric Liquid Crystals (CLCs)

2.1 Basics of Liquid Crystal (LC) Phases

LCs are substances that have characteristics intermediate between those of solid crystals and typical liquids [13]. LCs are divided into three main classes: (a) nematic, (b) cholesteric, and (c) smectic phases (Fig. 2).

2.1.1 Nematic Phase

Molecules align in parallel but lack positional order. The ellipses in the nematic phase represent the long planar molecules. A long-range orientational order is present when the molecules long axes align along a chosen direction. The strained nematic typically exhibits a locally preferred direction that changes over the medium. To provide its local orientation, one defines a vector field n(r), called the director [14]. It is assumed that its magnitude is one. An electric field, a magnetic field, and appropriately prepared surfaces can all align the director field. The absence of an idealised equilibrium configuration is indicated by the appearance of threads on optical examinations of certain structural disturbances. Nematics have short-range order, just like regular liquids, but the centres of mass of the molecules lack long-range order [15]. The molecules can rotate along their long axes, but if their two ends are different, there is no favoured arrangement of them. Therefore, the director's sign is physically meaningless, and ferroelectric phenomena are absent, confirming that a nematic behaves optically as a uniaxial material with a centre of symmetry.

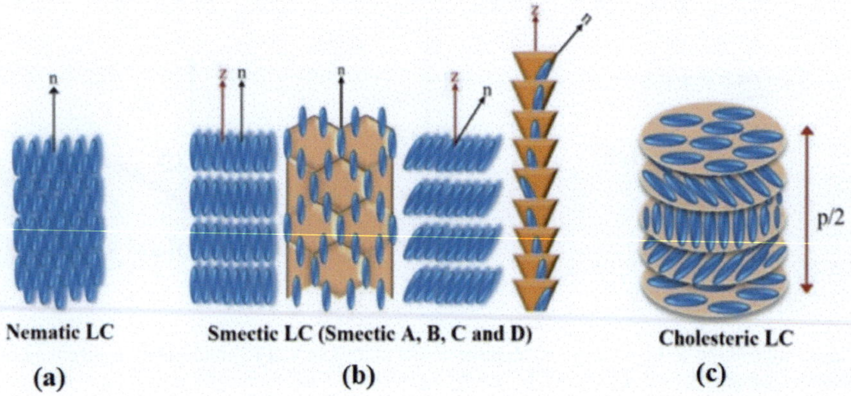

Fig. 2 Molecular arrangement **a** Nematic, **b** Smectic (A, B, C and D) and **c** Cholesteric liquid crystal. (Adapted from Ref. [13], Intech Open 2020). https://creativecommons.org/licenses/by/3.0/

2.1.2 Smectic Phase

Molecules maintain both positional and orientational order. In the smectic phase, the molecules are layered and show orientational ordering in addition to orientational layering [16]. The following is a brief summary of the various classes of smectic LC.

Smectic A: There is no long-range crystalline organisation within a layer in the smectic A phase, yet the molecules are orientated perpendicular to the layers [17]. The layers are able to easily slide over one other.

Smectic B: Within the layers of Smectic B, there is a hexagonal crystalline arrangement. The layers are unable to rotate on one another, but they can slip on one another.

Smectic C: This phase possesses biaxial symmetry because the preferred axis is not perpendicular to the layers.

Smectic D: The X-ray patterns are compatible with a cubic packing, and optically, the D phase seems to have a cubic structure.

Smectic E: A layered structure and a high degree of order arrangement within the planes are seen in the X-ray patterns derived from the smectic E phase.

2.1.3 Cholesteric (Chiral) Nematic Phase

Molecules are arranged in a helical structure, a feature that distinguishes CLCs from other LC phases [18]. Unlike the nematic phase, the cholesteric phase is characterised by a regular variation in the direction of the director **n** in the medium. A cholesteric configuration would be produced by twisting a nematic that was originally aligned along the y-axis twisted about the x-axis. One moves along the twist axis as the Fresnel ellipsoid and the director spin. In any plane perpendicular to the twist axis, the molecules' long axes have a tendency to align along a single favoured direction. Since this specific kind of liquid crystalline structure was first seen in cholesterol esters, the term "cholesteric" has a historical origin [19]. Selective reflection and other special optical characteristics of the phase are caused by the helical organisation in the cholesteric state. The plane of polarisation of this chiral macroscopic structure rotates in the direction of the helix when plane-polarized light interacts with it. When the chiral mesophase's pitch matches a wavelength in the visible spectrum (about 400–800 nm), colour results [20]. It is important to note that the reflected wavelength and the helix's pitch are both temperature-sensitive. An electric field can also be used to unwind helical structures, which causes the molecule axis to rotate along the field direction.

3 Synthesis of CLCs

There are three main kinds of these phases that are frequently seen: the cholesteric, smectic, and nematic phases. The layered structure that LC molecules self-assemble into during the smectic phase is evocative of the structure found in solid crystals. Unlike the nematic phase, which is primarily characterised by orientational ordering, the smectic phase exhibits both positional and orientational ordering. During this stage, LC molecules align themselves into layers with a certain orientation, producing a configuration that is well-organised and structured. On the other hand, the nematic phase is characterised by exclusive orientational ordering, in which LC molecules align along a certain direction without any corresponding positional ordering. This results in a condition that is less structured and more fluid than the structured [21]. The preferred direction of molecular alignment is often referred to as the director (n). Nematic phase molecules exhibit their orientational order when they align in a particular direction in the absence of external effects or stimuli. Chirality has a big impact on LCs' physical characteristics. When chirality is introduced into LCs, new mesophases with interesting physical properties appear. The cholesteric phase is produced when a chiral molecule is added to an achiral nematic matrix. Since CLCs have a helical shape and function as a Bragg reflector with a reflection peak that depends on both the cholesteric pitch and the average refractive index, they are desirable materials for photonic surfaces. The distance that a full revolution of the molecular director is made is known as the cholesteric pitch. There are a number of methods for creating emissive CLCs, such as combining chiral and fluorescent dopants to create nematic LCs and grafting chiral and luminous moieties onto LC molecules. Creating emissive CLCs seems to be easiest and most convenient when chiral fluorescent dopants, which integrate fluorescence and chirality into one entity, are added to nematic LCs. Chiral fluorescent dopants that are frequently utilised include compounds of binaphthyl and cholesterol [22]. Doped emissive CLC of this kind has a number of benefits. In particular, light-responsive compounds like azobenzene and spiropyran diarylethene are of more interest to the researchers. Temperature, radiation, mechanical stress, and electromagnetic fields are examples of external stimuli that can cause these stimuli-responsive materials to dynamically change their fluorescence or reflective colour over time.

4 Preparation of CLC Emulsions

Emulsions, which can be single, double, or multiple emulsions, are a type of colloidal dispersion made up of two or more immiscible liquids. With one phase dispersed in the other, single emulsions the most prevalent kind of emulsions—are made up of two immiscible phases. In addition to offering unique properties and applications, they can have more complex structures with multiple phases. The fabrication techniques used to create LC emulsions are frequently comparable to those used to create isotropic emulsions. Stirring or vertexing is one of the techniques now employed to create emulsions. An LC emulsion is a specific type of emulsion in which at least one of the phases is an LC. Compared to alternative techniques, stirring and vertexing are quick and inexpensive ways to create LC emulsions. The bulk LC is mechanically divided into tiny droplets using these methods, which makes it difficult to control the size and achieve homogeneity (Fig. 3a) [23]. Another popular method for creating LC emulsions is ultrasonication, which creates a high oscillating shear flow when pressure waves are applied at ultrasonic frequencies (Fig. 3b) [23]. This technique can create droplets with a somewhat smaller size range and smaller than the micron scale in the form of a miniemulsion. The production of monodisperse droplets is not achieved even by ultrasonication, which strives for size homogeneity. Microfluidic emulsification is one method that has grown in popularity because it gives significant control over the size, shape, and compartments of emulsions. It entails processing and making an effort to control fluids moving via tens to hundreds of micrometer-sized channels. Monodisperse emulsions, as seen in Fig. 3c–g [23], are created in their most basic form by forcing the dispersed phase into the continuous phase through microscopic channels. In order to keep the dispersed phase from sticking to the walls, oil in water emulsions require hydrophilic pathways. Using glass capillaries on a glass slide, the David A. Weitz group presented a completely new method for creating microfluidic devices in 2007 [24]. The co-flow setup (Fig. 3d) is the most basic axisymmetric geometry in capillary microfluidics. In this configuration, LC flows within the inner capillary while water flows through the outer capillary in the same direction, creating individual monodisperse LC droplets at the inner capillary's tip. As illustrated in Fig. 3e, unlike co-flow capillary devices, a second symmetrical capillary is inserted in opposition to the initial inner capillary to produce a flow-focusing effect that helps to form stable droplets. Droplets of LC double emulsion, in which the LC phase acts as the emulsifier, can be created by combining co-flow and flow-focusing geometries in coaxial flow-focusing glass microcapillary devices. Via a cylindrical injection capillary, an aqueous solution is injected as the inner fluid. Between the square and cylindrical injection capillaries, the middle fluid, LC, is injected. As the LCs separate the inner aqueous phase, capillary instability causes water-in-(W/LC) emulsions to form at the tip of the left inner capillary [25]. These droplets merge to form W/LC/W double emulsion droplets, where LCs occupy the middle layer, as they come into contact with the aqueous continuous phase, which is travelling in the opposite direction. Three coaxially aligned capillary device can be used to create triple emulsions using microfluidics. Other ways to create LC droplets with smaller sizes (<10 μm) are provided by classical polymerisation procedures, as shown in Fig. 3h. The production of almost monodisperse LC polymer droplets with sizes of roughly ≈1 μm is made possible by precipitation polymerisation, which is particularly interesting among these methods. Polymerisation is the process that starts with a homogeneous solution of reactive LC monomers. As a result, freshly synthesised polymer chains precipitate, forming beginning particles that later develop into the final LC particles. On the other hand, suspension and mini-emulsion polymerisation involve using bulk emulsification to disperse a mixture of LC monomers in the continuous phase, then proceeding with the polymerisation process [26]. Both techniques

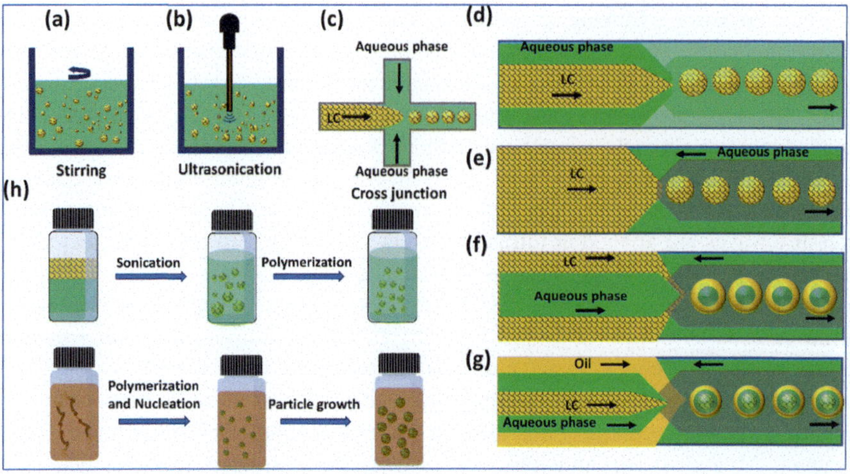

Fig. 3 Schematics depicting the production process of LC emulsions: emulsification through **a** stirring, **b** ultrasonication represent bulk emulsification strategies. Microfluidic emulsification using **c** PDMS cross-junction microchannel, while **d**–**g** involve microfluidic emulsification using a tip-to-tip microchannel, enabling the fabrication of single emulsion (**d**) and (**e**), double emulsion (**f**), and triple emulsion (**g**), respectively. **h** Suspension (top) and precipitate polymerization (bottom). (Adapted from Ref. [23]. Copyright, 2024 Royal Society of Chemistry)

offer flexibility in creating LC polymer particles with a variety of size distributions since they may be applied to a broad range of LC phases. While mini-emulsion polymerisation achieves average diameters below 1 μm, suspension polymerisation produces average diameters greater than 1 μm.

5 Synthesis of Polymer/CLC

LC materials and polymers have been the subject of greater research in recent years. These materials include LC elastomers, polymer-dispersed LCs, and most importantly LC fibres. Even though the wide range of uses is essential for contemporary technology, incorporating LCs into polymer matrices still presents difficulties. Among the problems are LCs leaking light and the liquid crystal phase being destroyed.

Variety of methods for producing polymer/LC fibres have been proposed in earlier research, with CLC being particularly useful due to their many colour transitions and ease of manipulation. For instance, Lagerwall et al. [27] combined CLCs with polyvinylpyrrolidone (PVP) in a solvent to create fibres via coaxial electrospinning. Due to the intense electric fields used in this procedure, the solution was forced to accrue charges and eject from the needle tip, forming PVP/LC fibres with a diameter of around 7 μm. However, this method necessitates meticulous control over the electrospinning parameters. Tang et al. [28]. used a different coaxial electrospinning method to produce thermochromic CLC fibres. They used solvents for the electrospinning process to combine CLCs with polystyrene utilising a miscibility approach. These fibres had an average diameter of roughly 23 μm, but their ability to display colour was compromised since the solvent's breakdown upset the ordered structures of CLC. Fu et al. [29] showed how to implant CLC microcapsules with intelligent functional fibres to stop LCs from leaking. The CLCs were encapsulated using the emulsion polymerisation process. Following the PVP-microcapsule blend, the electrospinning technique was used to produce fibres with a diameter of around 10 μm. Although this technique successfully stopped LC leaks, it was difficult to fabricate and its potential for scaling up was yet unknown. Chen et al. [30]. and colleagues offer a straightforward and effective nonsolvent and phase separation technique that uses polar solvents to create polymer/CLC composite films and fibres in order to resolve the aforementioned problems. By combining a commercial nematic LC DLC-111 with a left-handed chiral dopant called 4′-(1,3-dimethyl-3-chloro) propoxy-4-cyanobiphenyl (NYCL), a CLC is created that forms a blue phase at a higher temperature (Fig. 4). Due of its great transparency and ease of handling, poly (vinyl alcohol) (PVA), a water-soluble polymer, is used as the polymer matrix. The aqueous PVA solution can stabilise CLC microspheres by utilising polar solvents to capitalise on the hydrophobic characteristics of CLC. The CLC microspheres can be inserted in the PVA matrix to create PVA/CLC films and fibres using spin-coating and wet spinning processes. When preparing the fibres, the wet spinning process offers a larger window of opportunity than the electrospinning approach. It is common knowledge that chiral units are found in LC molecules or that chiral chemicals can be added to nematic LCs to produce CLCs. However, because tiny molecule CLCs have a low viscosity, their pitch is often consistent, single, and narrow. Consequently, the broadband reflection of a CLC is typically obtained in a polymer/CLC combination. The formation of microstructures with varying length pitch distributions may be the result of the stable network structure and the high viscosity of the polymer, which restrict diffusion between various length pitch structures. Choleric liquid crystalline polymers (with 100% monomer), small-molecule CLCs (without monomer), and polymer-stabilized CLCs are the three categories into which the polymer/CLC system can be classified based on the monomer content.

5.1 Synthesis and Direct Ink Writing of the Water-Responsive Cholesteric Ink

The race to produce responsive inks has now begun due to the success of additive manufacturing (AM) in both academic and industry contexts [31, 32]. "4D printing" refers to the combination of additive manufacturing and responsive inks, where the fourth dimension is a one- or two-way change in the material following 3D printing. Direct ink writing (DIW) is a popular 4D printing method because it can handle a

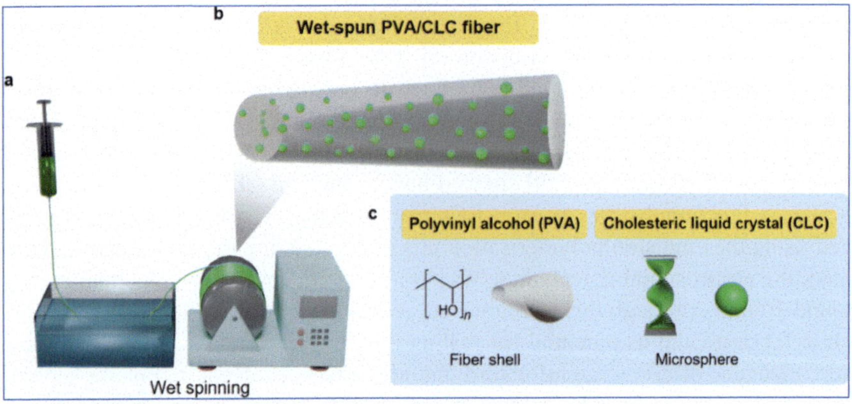

Fig. 5.4 Schematic illustration of the wet-spinning technique (**a**), PVA/CLC fibers (**b**), and the constituent materials PVA and CLC (**c**). (Adapted from Ref. [30], Copyright 2024, American Chemical Society)

wide variety of material types and provides flexibility in terms of print path designs, printing speeds, and ambient temperatures during the print process [33]. Colloidal photonic crystals, block copolymers, hydrogels with photonic arrays, and CLCs both lyotropic cellulose derivatives and LC oligomer inks are among the materials that have been reported thus far for printing structurally coloured designs using DIW. The inclusion of a reversible macroscopic alteration following printing is what the research community currently requires in order for these materials to move closer to "four dimensionality."

Because CLCs may be designed to react to mechanical forces, water, and (chemical) analytes, they hold promise for creating stimuli-responsive, architecturally coloured objects [34]. Debije et al. [35] used the ink, which is a water-sensitive precursor of CLC elastomer (CLCE), is created via a simple amine–acrylate "aza-Michael" addition reaction. The commercially accessible diacrylate mesogens "RM82," 1,4-di(4-(6-acryloyloxyhexyloxy) benzoyloxy)-2-methylbenzene (**1**), and "RM257," 1,4-di(4-(3-acryloyloxypropyloxy)benzoyloxy)-2-methylbenzene (**2**), are the source of the LC behaviour (see Fig. 5a). To get a cholesteric mesophase out of the DIW ink, a reactive chiral dopant is required. To do this, the reaction mixture is supplemented with a specially prepared reactive chiral dopant (3R,3aR,6S,6aR)- hexahydrofuro[3,2-b]furan-3,6-diyl bis(4-((6-(acryloyloxy) hexyl)oxy)benzoate) (**3**). The creation of an ionic species in the LC is necessary to maximise the material's reaction to water. This can be accomplished by employing a chain extender with a pendant tertiary amine group, such as *N*, *N*-dimethylpropane-1, 3-diamine (4), which can be positively charged following the ink's processing into an object, as shown for DIW 4D-printed soft actuators. Carbonate-primary amine reactions have also been extensively documented; as explained in the "Experimental Section," chiral dopant **3** is used in an amine–acrylate addition reaction procedure to synthesise 4-(6-acryloyloxyhexyloxy) benzoic acid (3a) and (3R,3aR,6S,6aR)-hexahydrofuro[3,2-b]furan-3,6-diol (3b). In nematic LC "E7," this chiral dopant has a helical twisting power (HTP) of 38.8 μm^{-1}, which is marginally less than that of the previously stated "LC 756". The two-day oligomerisation procedure begins by refluxing the reaction mixture in tetrahydrofuran for an entire night at 60 °C in an argon environment. Next, the reaction mixture is supplemented with the antioxidant 2,6-ditert-butyl-4-methylphenol (5). The reaction takes place in bulk at a high temperature on the second night (see Fig. 5b).

As seen in Fig. 6, a novel method for creating broadband reflective CLC films using non-stick technology and inkjet printing was recently proposed [36]. Wanli He and his colleagues used software (Acro-rip) to manage the inkjet volume while filling the distinct C, M, and Y channels with LCs, polymerizable monomers, and chiral compounds, respectively. We individually printed the precursor LC layer with various chiral modifications on each substrate to create a broadband reflective film by stacking layers. Following the initial LC layer's UV polymerisation, the second LC layer was moved to the first LC layer to undergo polymerisation. A third, fourth, or even more layers were then piled in the same manner, and polyimide films were added to adjust the thickness of each PSCLC film. The simplicity of the preparation procedure and the controllability of each film's wavelength and reflection range are two benefits of the single-pitch multilayer stacking method. Furthermore, the previously mentioned approach produces a comparatively thick film without dynamic adjustment, which lessens its practicability.

Using a wash-out/filling technique, Guo employed CLCs with varying pitches in CLCs/polymer template composites, allowing for simultaneous red, green, and blue reflections as well as broadband reflections in monolayer CLC films (as illustrated in Fig. 7) [37]. Furthermore, thermally sensitive hydrogen-bonded self-assembled CLCs were used to fill the polymer template, resulting in a temperature-tunable photonic bandgap of CLC films. Two reflection bands of the aforementioned CLC mixtures 5 and 6 superimpose to allow the bandwidth of cell E to reach 340 nm (450–790 nm), which includes red, green, and blue colours in the visible light spectrum.

6 Color Interference in CLCs

6.1 *Principles of Selective Reflection of Light*

The cholesteric phase is one of the several LC phase states that possesses specific light reflection properties and a distinctive self-assembled helical structure. Since it is considered a specific state of the nematic phase, the cholesteric phase is frequently called the chiral nematic phase [38]. The formation of the cholesteric phase occurs through two main methods. First, the molecule contains chiral carbon atoms or other asymmetric structures; second, chiral compounds are integrated into smectic or nematic LCs. Nowadays, the great majority of CLCs are made using the second method. The application of CLCs has been made possible by their selective reflection

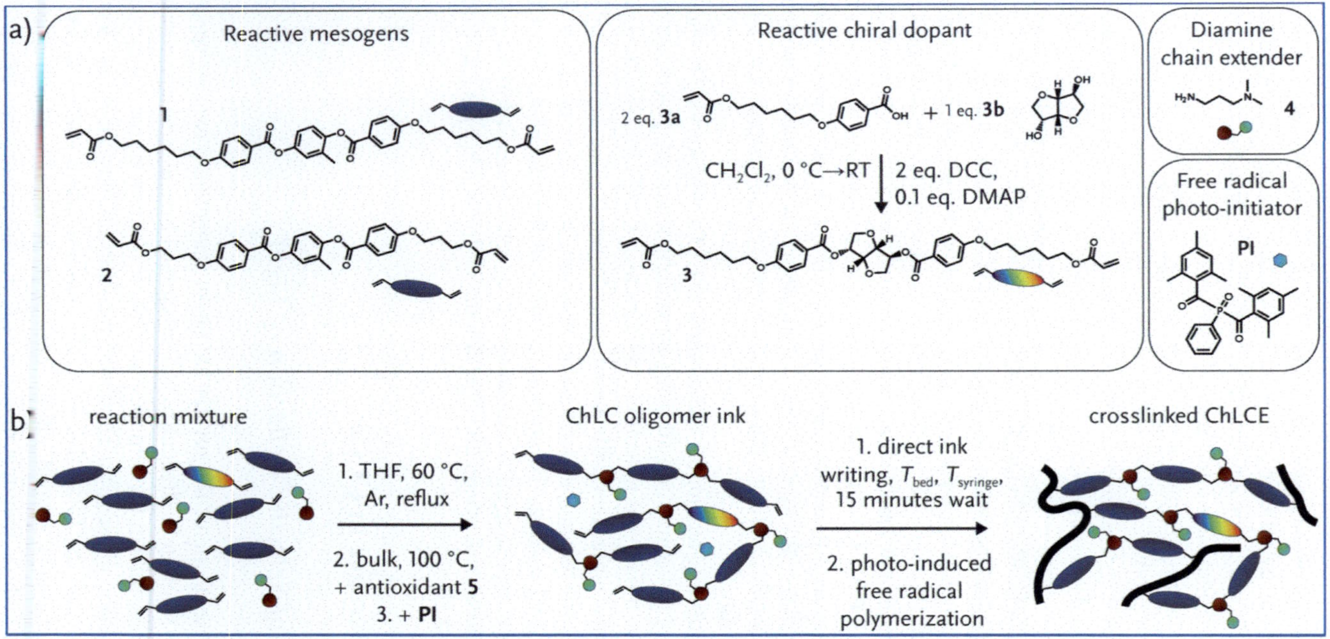

Fig. 5 **a** Components used for synthesizing the cholesteric liquid crystal (CLC) oligomer ink—left to right: reactive mesogens **1** and **2**, reactive chiral dopant **3**, diamine chain extender **4**, and free radical photoinitiator **PI**. **b** Schematic drawing of the molecular composition of the CLC mix before the chain extension reaction, after oligomerization, and after acrylate crosslinking. Also given are the reaction conditions for both steps. (Adapted from Ref. [35], Copyright 2022 Wiley)

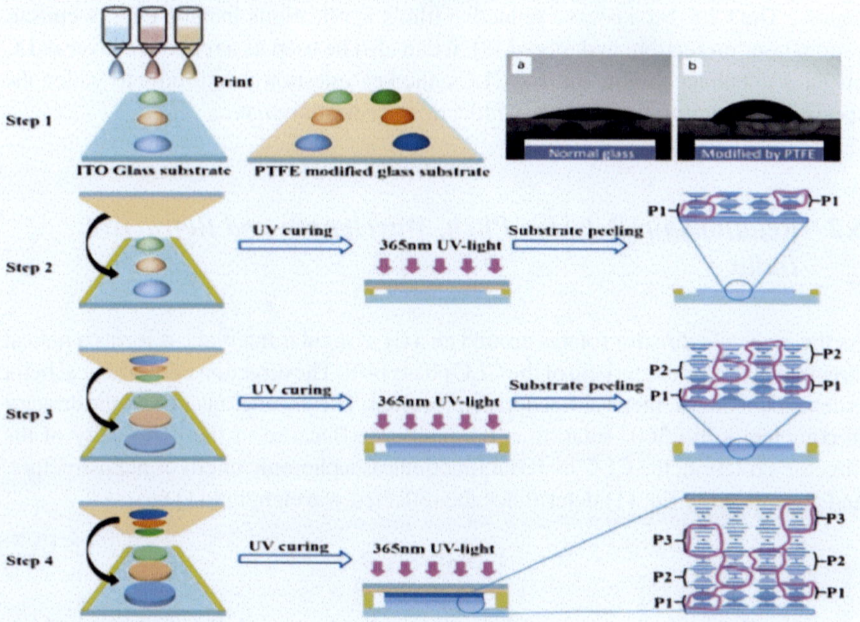

Fig. 6 The scheme of a broadband reflective film prepared by inkjet printing. (Adapted from Ref. [36], Copyright 2022 MDPI)

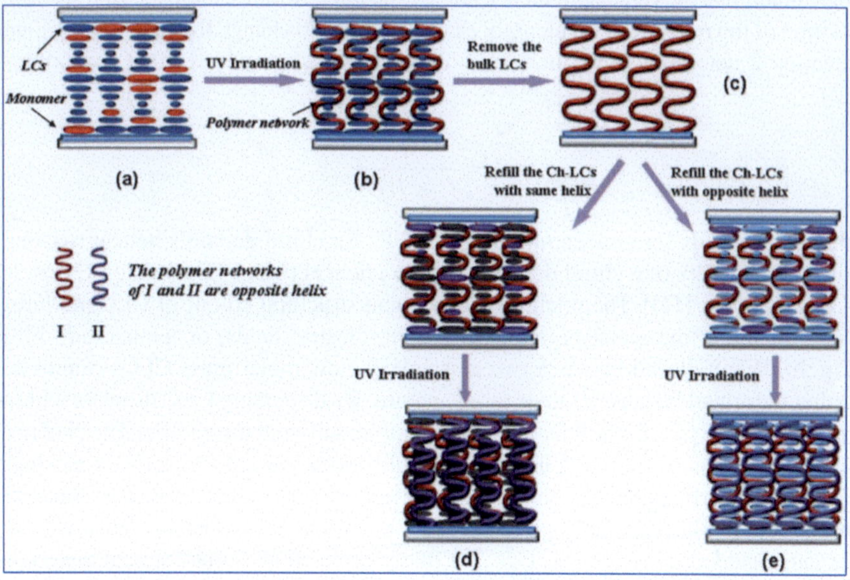

Fig. 7 Illustration outlining the procedure for preparing the Ch-LCs film. (Adapted from Ref. [37], Copyright 2010, Royal Society of Chemistry)

feature. The CLC broad-wave reflective film's applications include energy conservation, laser protection, and more [39]. It can also be used as a reflective layer in LC displays to enhance light. But for CLCs, the key question is still how to widen the spectral band within the necessary reflection spectrum range.

6.2 Relationship Between Pitch, Wavelength and Refractive Index

As the molecule director rotates around an axis at regular intervals, it forms a helical structure that is characteristic of the CLC phase [40]. The direction of this helical helix is dependent on whether a chiral dopant is present. The distance needed by the director to complete a full 360° rotation is the pitch (P). Because of the periodicity of the director's rotation, the CLC material functions as a photonic micro or nanostructure, and the following Eq. (1) determines the reflected wavelength [41].

$$\lambda = \mathrm{nP}\cos\theta \tag{1}$$

where θ is the incident angle, n is the average refractive index, P is the pitch, and λ is the wavelength that is reflected. With regard to the incident light, CLCs show some degree of angular dependence, which appears as a blue shift when viewed from an angle other than the typical incidence angle. The pitch, which establishes the spectral position of the reflection wavelength, can be changed to change the structural colour. Equation 2 determines the pitch when chiral dopants are added to CLC mixtures [42].

$$P = \frac{1}{C \times HTP} \tag{2}$$

where "C" stands for concentration and "HTP" for chiral dopant's helical twisting power. The ability of a chiral dopant to cause a helical twist in a nematic LC host is measured by its HTP. The pitch and consequent structural colour of CLC mixtures with low molar mass can be altered by light, electric fields, or temperature [43] (Fig. 8). When placed between two glass plates, low molar mass CLCs cannot be further deformed because of their fluidic nature. By dispersing CLC droplets within a polymer matrix, it is possible to create a flexible, self-supporting film. The choleric LCs' helical architecture enables them to split iridescent light into two opposing-handed circularly polarised photons. However, they can simultaneously transmit another circularly polarised light and reflect the circularly polarised lights that are the same handedness as themselves. For many applications, smart optical materials are especially attractive due to their ability to dynamically change their microstructures and optical characteristics in response to external stimuli. Achieving high levels of anti-counterfeiting performance requires creating materials with multi-mode properties, such as reflecting colour, fluorescence, and circularly polarised luminescence

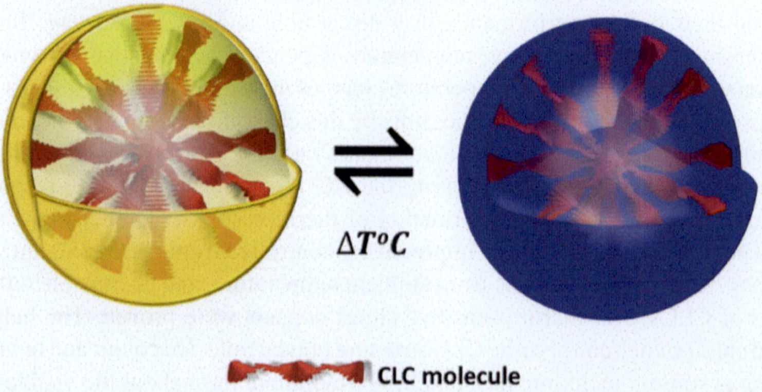

Fig. 8 Principle of anticounterfeiting using temperature as stimuli. (Adapted from Ref. [43] with permission. Copyright 2022, American Chemical Society)

(CPL) signal, because each mode has multiple states and allows for the reversible conversion of those states [44]. CPL-active materials with very high activity can be created using CLCs. Two main categories can be used to classify the mechanisms that produce CPL in CLCs. In the first, the chirality amplification effect is realised by the helical assembling superstructure of CLCs through optical rotation. Here, the polarised direction of the resultant CPL follows the director twist and it is transmitted down the helical axis of the CLC [45]. The second mechanism, which is used to produce strong CPL with high glum, is the selective reflection of circularly polarised light. The centre of the CLCs selective reflection band is where the emission peak is situated in this instance. CLCs reflect CPL that has the same helical handedness as the CLCs, whereas the planar CLCs cell transmits CPL that has the opposite helical handedness.

6.3 *Thermochromism and Structural Coloration in CLCs*

CLCs reflect particular light wavelengths because of their helicoidal molecular architecture, which results in unusual optical phenomena. They are useful for applications including sensors, thermochromic devices, and smart displays because of these qualities. A special class of materials known as thermochromic LCs responds to temperature changes by changing their texturing or optical characteristics [46]. The phase-transition properties of thermochromic LCs determine their behaviour. varied LC phases, such as the nematic phase, smectic phase, cholesteric phase, blue phase, etc., are caused by the varied molecular configurations that these materials display at different temperatures. Reversible nature is one of the main characteristics of thermochromic LCs. The LCs undergo phase transitions as the temperature rises or falls, changing their optical performance in a discernible and reversible way. Temperature sensing, thermal mapping, temperature-dependent coatings, smart windows, and even visual indicators for temperature changes in different systems are just a few of the uses for them that are made possible by this characteristic. Functional dopants and temperature programming have an impact on the LC structures in addition to temperature [47]. Furthermore, doping the LC host with temperature-responsive chiral dopants can result in the formation of thermoresponsive helical superstructures. The centre wavelength was impressively controlled from 780 to 380 nm with a small 3 °C temperature increase from ambient temperature, and the tunable reflected colours of CLCs with thermo-sensitive chiral dopants were proven. The halogen-bonded chiral switch can give the CLC structure optical reflected colour and tunability for various stimuli, including heat and light irradiation, throughout the visible spectrum [48]. Self-assembled asymmetric LC microlenses with polarisation selectivity and multifocal capability were created by Ling-Ling Ma et al. to demonstrate four-dimensional visual imaging (Fig. 9a). Thermoresponsive LC architectures can be tuned to achieve both linear and nonlinear optical effects for chiral LCs. Ma et al. [49] developed programmable self-propelling actuators using a molecular assembly in a photoresponsive cholesteric medium (Fig. 9b). Logically presetting alignments were used to illustrate the parallel transport of microspheres in customised trajectories, such as orbital revolution, convergence, divergence, and gathering. Wang et al. [50] showed that CLCs doped with chiral molecular switches containing terminal bromo atoms exhibit reversible handedness inversion when exposed to visible light of various wavelengths (Fig. 9c). Additionally, they initially showed that handedness inversion in CLCs was achieved by attaching the binaphthyl to the ortho-positions of azobenzene and shorter substituents to the 6,6′-positions of binaphthyl. The light-driven handedness inversion of CLCs was carried out at 530 nm light, which resulted in trans-to-cis photoisomerization and a photo stationary state. At 440 nm light, reverse isomerisation was accomplished.

6.4 Light and Electrically-Driven CLCs

Light stimuli on LC structures have clear advantages over heat stimulus, including great spatial and temporal precision even for digital and remote control. The tuning of the LC structure and optical performance would be affected differently by several aspects of the light, including wavelength, intensity, polarisation, etc. [51]. When doped with photoresponsive matter that exhibits photochemical or photothermal effects, LCs are given the ability to be structurally tuned by light. A specific light wavelength can cause photoisomerization in azo-chiral dopants, azo-dyes, and other photoresponsive materials. Because of their dielectric anisotropy, LC molecules would be reoriented in an electric field. In 1962, Richard Williams made the initial discovery of an intriguing electro-optic property: the reorientation of NLCs produced a regular pattern under the external electric field [52]. Additionally, LCs that were feasible for display devices—a precursor to the LCD—were exhibited. Subsequently,

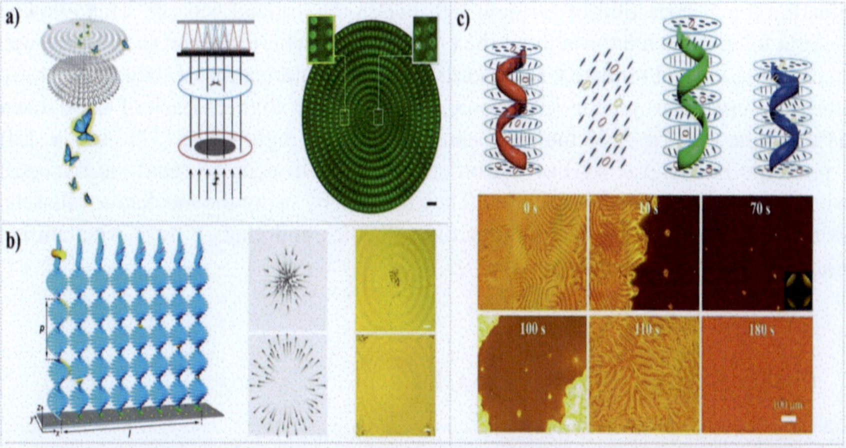

Fig. 9 Performances of thermal and light stimuli-responsive LCs. **a** Depth/polarization information demultiplexing of microlens. Scale bar: 20 μm. (Adapted from Ref. [48] Copyright 2019, American Chemical Society). **b** Semi-free CLC film with helical configuration for converging and diverging the particles. Scale bar: 50 μm. (Adapted from Ref. [49] Copyright 2021, The American Association for the Advancement of Science). **c** Visible-light-driven handedness inversion of CLC mixture. Scale bar: 100 μm. (Adapted from Ref. [50] Copyright 2023, John Wiley and Sons)

it was shown that the applied electric field's frequency and intensity both showed significant influence on LC realignment, suggesting a potential route to binary responsiveness. Chen et al. [53] developed ultra-broadband (over 1000 nm) adjustable planar optical components with light stimulation by continually adjusting the pitch and handedness of CLCs. They utilised an opposing chiral dopant and a photoresponsive azobenzene chiral molecule. Enhancing the stability of the helical structure and optical performance is highly desired for photosensitive CLCs with orientation disorder, phase defects, fatigue resistance, and thermal instability. It is difficult to simultaneously adjust the spectrum from the NIR area to the visible spectrum due to the bistable constraint of azobenzene-based chiral switches.

Helicoidal CLCs can undergo two different electrical modifications: realigning the helicoid axis and adjusting the helicoid pitch [54]. Similar to NLCs, when CLCs have a negative dielectric, the pitch remains constant since there is no torque and the LC molecules align vertically with the electrical field. Realigning the helical axis perpendicular to the electrical field is possible with positive dielectric anisotropy. Additionally, recent studies have shown that the heliconical CLCs system is where the frequency-dependent photonic bandgap (PBG) occurs over a broad-spectrum range. Liu et al. [55] mixed a positive dielectric LC combination of E7 with the dimer CB7CB to study a dynamic helicoid architecture. A heliconical structure was created by the interaction and rivalry between the dielectric torque and elastic effects. The reflection spectrum can undergo a reversible change from 460 nm to about 1000 nm by adjusting the frequency (Fig. 10a). A hybrid reflective display was proposed by Franklin et al. [56]. In their study, the optimised LC layer was responsible for adjusting the polarisation of incident light when the voltage was off. This allowed the light to be reflected throughout the entire device, whereas when the voltage was on, the light was not reflected. In order to adjust the polarisation of light, CLCs have three different states: planar, focal conic, and vertically aligned. Each of these states has a unique electric field intensity and frequency. Lingling Ma [57] used a shift of photonic bandgap (PBG) of the order of 100 ns to achieve rapid electro-optic switching in soft photonic structures. They did this by using the mechanism that the electric fields quench the fluctuations of molecules and change their orientational order, not the orientation (Fig. 10b).

7 Other Stimuli-Responsive CLCs

Typically seen in polymerised LC systems, the structure of the LC system can be tuned to alter its optical performance by applying mechanical stress and humidity. By manipulating the helical pitch through mechanical stress and humidity, the CLC polymer can also be used for active optical designs and applications, including information encryption and structural colour [58]. CLC polymers can reversibly adjust their structure under humid stimuli following specific chemical reactions, combining the optical characteristics of CLCs with the flexibility of polymers. To make it react to humidity, Sol et al. [59] applied acid to the CLCE. The CLCs inflated and changed volume when exposed to water; hence, the colour that was reflected also altered. In both dry and wet air, the bright reflected spectra changed across the visible spectrum (Fig. 11a). The mechanical stimulation can be used to modify the soft LC structure. Kim et al. used the considerable elasticity anisotropy and Poisson's ratio of Chloestric LC to create a pixelated structural colour with a spectrum colour shift from NIR to UV light. The CLC, which was implanted with air passages and mounted on poly(dimethylsiloxane) (PDMS), experienced a shift in spectrum due to a change in aspect ratio brought on by variations in the inflating region (Fig. 11b) [60]. Controlling multicolour switching enhances the use of information encryption and concealing by means of mechanical stretching.

8 Conclusion

A special kind of LC materials called CLCs are characterised by their helical structure and selective light reflection, which produces vivid colour interference patterns and makes them desirable candidates for a variety of photonic applications. These optical microstructures could be designed, made, and used in soft matter systems like LCs to open up new research and practical applications. In this book chapter, the chemistry of CLC formation is examined, with a focus on the molecular structure, the role of chiral dopants, and the factors influencing the crystals' self-assembly. By carefully controlling the temperature, concentration of chiral agents, and chemical interactions,

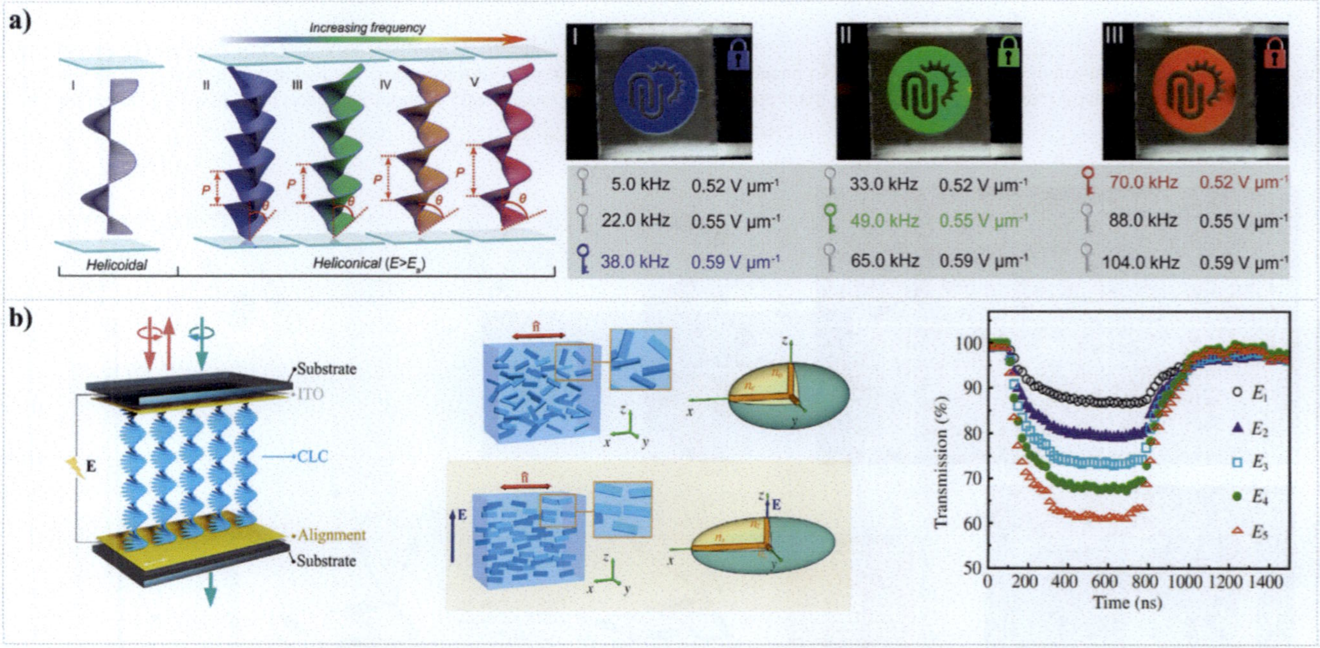

Fig. 10 **a** Frequency-actuated heliconical soft architecture and active information encryption. (Adapted from Ref. [55]. Copyright 2022, Springer Nature). **b** Sub microsecond electro-optic switch of CLCs. (Adapted from Ref. [57]. Copyright 2022, Chinese Laser Press)

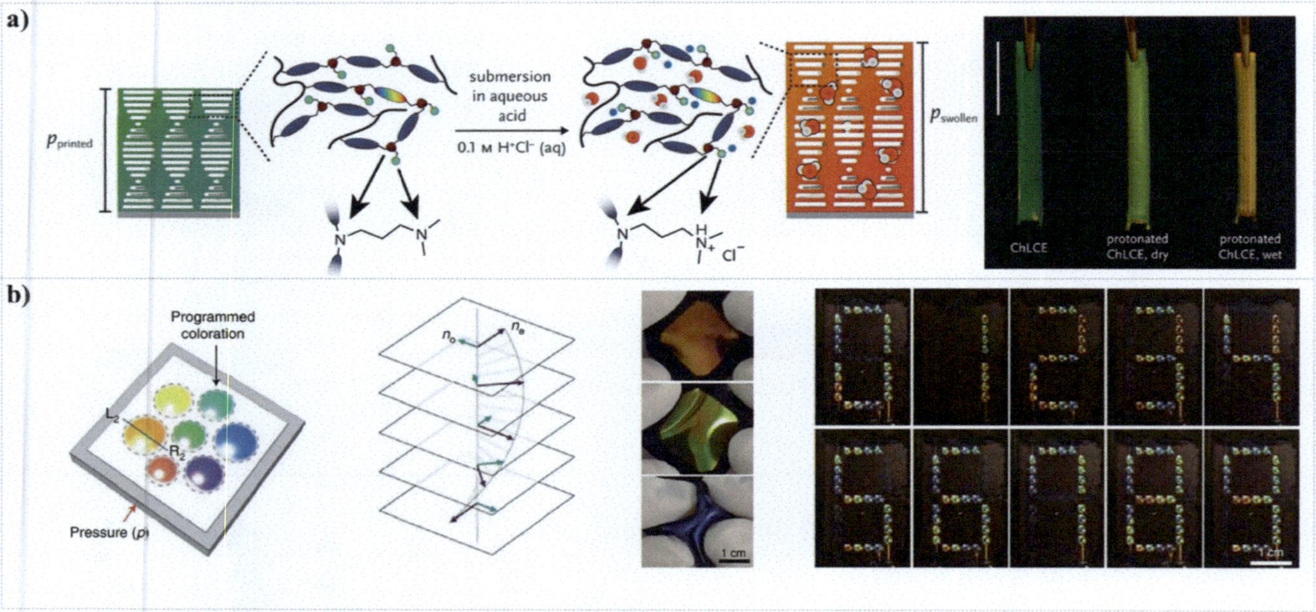

Fig. 11 **a** Cholesteric LCs with acid treatment induced concurrent change in the cholesteric pitch length. Scale bar: 1 cm. (Adapted from Ref. [59]. Copyright 2022, John Wiley and Sons). **b** The active structural coloration of a CLCE membrane induced by in-plane strain upon an arbitrary out-of-plane deformation. Scale bar: 1 cm. (Adapted from Ref. [60]. Copyright 2022, Springer Nature)

it is possible to precisely regulate the helical pitch and optical properties of CLCs, opening up a variety of applications, including sensors and display technologies. CLCs' intrinsic ability to self-assemble into ordered, periodic structures give rise to a number of unique optical properties, including selective light reflection and the potential to produce structural colour. Furthermore, we examine the relationship between the observed colour interferences and molecular structure, highlighting the recent advancements in CLC material design and functionalisation that demonstrate the potential of CLCs to produce dynamic and responsive systems for the upcoming generation of photonic and optoelectronic devices.

Acknowledgements All authors are thankful to Chitkara University Research and Innovation Network, Chitkara University, Patiala, Punjab, India. Authors are also thankful to KLE technological University, Dr. M. S. Sheshgiri Campus, Udhyambag, Belagavi, Karnataka, India for constant encouragement and support for the research.

Funding Authors declare that they have not received any funding from any source or institutions or organizations.

Conflict of Interest All authors declare that they do not have any conflict of interest.

References

1. Liang, X., Guo, S., Chen, M., Li, C., Wang, Q., Zou, C., Zhang, C., Zhang, L., Guo, S., Yang, H.A.: Temperature and electric field-responsive flexible smart film with full broadband optical modulation. Mater. Horiz. **4**, 878–884 (2017). https://doi.org/10.1039/C7MH00224F
2. La, T.G., Li, X., Kumar, A., Fu, Y., Yang, S., Chung, H.J.: Highly flexible, multipixelated thermosensitive smart windows made of tough hydrogels. ACS Appl. Mater. Interfaces **9**, 33100–33106 (2017). https://doi.org/10.1021/acsami.7b08907
3. Zhang, Y., Yang, X., Zhan, Y., Zhang, Y., He, J., Lv, P., Yuan, D., Hu, X., Liu, D., Broer, D.J.: Electroconvection in zwitterion-doped nematic liquid crystals and application as smart windows. Adv. Opt. Mater. **9**, 2001465 (2021). https://doi.org/10.1002/adom.202001465
4. Hu, X., Zeng, W., Zhang, X., Wang, K., Liao, X., Jiang, X., Jiang, X.F., Jin, M., Shui, L., Zhou, G.: Pitch gradation by ion-dragging effect in polymer-stabilized cholesteric liquid crystal reflector device. Polymers **12**, 96 (2020). https://doi.org/10.3390/polym12010096
5. Zhang, Y.S., Emelyanenko, A.V., Liu, J.H.: Fabrication and optical characterization of imprinted broad-band photonic films via multiple gradient UV photopolymerization. J. Polym. Sci. Part B Polym. Phys. **55**, 1427–1435 (2017). https://doi.org/10.1002/polb.24392
6. Timuçin, H., de Pieter, W.: The interaction between humans and buildings for energy efficiency: a critical review. Energy Res. Soc. Sci. **71**, 101828 (2021). https://doi.org/10.1016/j.erss.2020.101828
7. Xinle, L., Cai, S., Sun, B., Yang, C., Zhang, J., Liu, Y.: Chemically robust covalent organic frameworks: progress and perspective. Matter **3**(5), 1507–1540 (2020). https://doi.org/10.1038/s41467-018-05462-4
8. Nandi, S.S., Adimule, V., Kadapure, S.A., Kerur, S.S.: Rare earth-based nanocomposite materials for prominent performance supercapacitor: a review. Appl. Mech. Mater. **908**, 3–18 (2022). https://doi.org/10.4028/p-rff302

9. Zhou, M., Shang, L., Li, B., Huang, L., Dong, S.: Highly ordered mesoporous carbons as electrode material for the construction of electrochemical dehydrogenase-and oxidase-based biosensors. Biosen. Bioelectron. **24**, 442–447 (2008). https://doi.org/10.1016/j.bios.2008.04.025
10. Abuzeid, H.R., El-Mahdy, A.F., Kuo, S.W.: Covalent organic frameworks: Design principles, synthetic strategies, and diverse applications. Giant **6**, 100054 (2021). https://doi.org/10.1016/J.GIANT.2021.100054
11. Shen, W., Li, G.: Recent progress in liquid crystal-based smart windows: materials, structures, and design. Laser Photonics Rev. **17**, 2200207 (2022). https://doi.org/10.1002/lpor.202200207
12. Zhang, J., Zhang, Y., Yang, J., Wang, X.: Beyond color boundaries: pioneering developments in cholesteric liquid crystal photonic actuators. Micromachines **15**(6), 808 (2024). https://doi.org/10.3390/mi15060808
13. Anuja, K.J., Rajendra, R.D.: Edited by Morteza Sasani Ghamsari and Irina Carlescu, pp. 1–68 (2021). https://doi.org/10.5772/intechopen.91889
14. Yang, Q., Luo, M., Liu, K., Cao, H., Yan, H.: Covalent organic frameworks for photocatalytic applications. Appl. Catal. B: Environ. **276**, 119174 (2020). https://doi.org/10.3389/fenrg.2021.800382
15. de Laurens, T.H., Julien, M.N.V., Dirk, J.B., Cees, W.M.B., Albertus, P.H.J.S.: Humidity-responsive liquid crystalline polymer actuators with an asymmetry in the molecular trigger that bend, fold, and curl. J. Am. Chem. Soc. **136**(30), 10585–10588 (2014). https://doi.org/10.1021/ja505475x
16. Zhang, R.C., Zhang, Z.B., Han, J.C., Yang, L., Li, J.J., Song, Z.C., Wang, T.Y., Zhu, J.Q.: Advanced liquid crystal-based switchable optical devices for light protection applications: principles and strategies. Light-Sci. Appl. **12**, 30 (2023). https://doi.org/10.1038/s41377-022-01032-y
17. Belmonte, A., Bus, T., Broer, D.J., Schenning, A.: Patterned full-color reflective coatings based on photonic cholesteric liquid-crystalline particles. ACS Appl. Mater. Interfaces **11**, 14376–14382 (2019). https://doi.org/10.1021/acsami.9b02680
18. Zheng, X.X., Jia, Y.N., Chen, A.H.: Azobenzene-containing liquid crystalline composites for robust ultraviolet detectors based on conversion of illuminance-mechanical stress-electric signals. Nat. Commun. **12**, 8 (2021). https://doi.org/10.1038/s41467-021-25178-2
19. Xie, X., Huang, P., Xu, M., Ding, Y., Qiu, L., Lu, H.: Tri-state switching of a high-order parameter, double-layered guest-host liquid-crystal shutter, doped with the mesogenic molecule 4HPB. Liq. Cryst. 1–6 (2021). https://doi.org/10.1080/02678292.2021.1885759
20. Davydova, T.E., Turchenko, E.A., Usachev, A.M.: Social and technological aspects of building materials production in the concept of a smart city. IOP Conf. Ser.: Mater. Sci. Eng. **775**, 012110 (2020). https://doi.org/10.1088/1757-899X/775/1/012110
21. Vinayak, A., Parashuram, L., Santosh, N., Rangappa, K.: Crystal structures, morphology, optical, supercapacitor properties of reduced graphene oxide doped cesium oxide@cobalt oxide bimetallic oxide nanostructures for visible light photocatalysis of Alizarin Red S dye. J. Energy Storage **104**, 114429 (2024). https://doi.org/10.1016/j.est.2024.114429
22. Ube, T., Yoshida, M., Kurihara, S., Ikeda, T.: Sunlight-driven smart windows with a wide temperature range of optical switching based on chiral nematic liquid crystals. ACS Appl. Mater. Interfaces **16**(22), 28638–28644 (2024). https://doi.org/10.1021/acsami.4c04370
23. Buchaiah, G., Supraja, P., Rakeshkumar, R., Jayalakshmi, V.: Optical anti-counterfeiting with cholesteric liquid crystal emulsions: preparation, properties, and applications. Mater. Adv. **5**, 7113–7129 (2024). https://doi.org/10.1039/D4MA00604F
24. Qing, X., Luming, M., Tingchao, H., Guangxi, H., Bing, S.L., Ben, Z.T.: Direct visualization of chiral amplification of chiral aggregation induced emission molecules in nematic liquid crystals. ACS Nano **15**(3), 4956–4966 (2021). https://doi.org/10.1021/acsnano.0c09802
25. Vinayak, A., Rangappa, K., Santosh, N., Parashuram, L.: Fullerene C60 functionalized Sm_2O_3@ Eu_2O_3 bimetallic oxides core shell nanocomposites for zeta potential and energy storage applications. J. Electroanal. Chem. **973**, 118663 (2024). https://doi.org/10.1016/j.jelechem.2024.118663

26. Vinayak, A., Kalpana, S., Vandna, S., Pankaj, K., Rangappa, K., Rajeev, J., Santosh, N.: Influence of core fluorination on the phase properties of fan-like azobenzene based supramolecules, their cis-trans photoisomerization and photoluminescence dynamics. Mater. Chem. Phys. **329**, 130140 (2024). https://doi.org/10.1016/j.matchemphys.2024.130140
27. Lagerwall, J.P.F., Mccann, J.T., Formo, E.: Coaxial electrospinning of microfibres with liquid crystal in the core. Chem. Commun. **2008**, 5420–5422 (2008). https://doi.org/10.1039/b810450f
28. Melissa, W.W., James, A.W., Ratib, M.S., Jimmy, N., Paola, A.D., Christina, T.: Temperature-responsive structurally coloured fibers via blend electrospinning. ACS Appl. Polym. Mater. **5**(4), 3065–3078 (2023). https://doi.org/10.1021/acsapm.3c00222
29. Guan, Y., Zhang, L.P., Wang, D., West, J.L., Fu, S.H.: Preparation of thermochromic liquid crystal microcapsules for intelligent functional fiber. Mater. Des. **147**, 28–34 (2018). https://doi.org/10.1016/j.matdes.2018.03.030
30. Tzu-Hsun, K., Hsun-Hao, H., Jui-Juin, C., Lin-Ruei, L., Hui-Yu, C., Jiun-Tai, C.: Fabrication of polymer/cholesteric liquid crystal films and fibers using the nonsolvent and phase separation method. Langmuir **40**(27), 14166–14172 (2024). https://doi.org/10.1021/acs.langmuir.4c01759
31. Saboor, S., Kirankumar, G., Venkata, R., Shantiswaroop, M., Kishor, S.K.: Thermal and cost assessment of various polymer-dispersed liquid crystal film smart windows for energy efficient buildings. Constr. Build. Mater. **263**, 120155 (2020). https://doi.org/10.1016/j.conbuildmat.2020.120155
32. Hari, K.B., Quan, L.: Light-driven liquid crystalline materials: from photo-induced phase transitions and property modulations to applications. Chem. Rev. **116**, 15089–15166 (2016). https://doi.org/10.1021/acs.chemrev.6b00415
33. Zheng-Wei, L., Victor, Y.Z., Wei, L.: Adjustable active–passive dual-frequency liquid crystal smart window based on thermally responsive chiral molecules. J. Mol. Liq. **415**, 126383 (2024). https://doi.org/10.1016/j.molliq.2024.126383
34. Vinayak, A., Revaiah, R.G., Santosh, S.N., Adarsha, H.J.: Synthesis, characterization of Cr Doped TeO_2 nanostructures and its application as EGFET pH sensor. Electroanalysis **33**, 579 (2021). https://doi.org/10.1002/elan.202060329
35. Sol, J.A.H.P., Smits, L.G., Schenning, A.P.H.J., Debije, M.G.: Direct ink writing of 4D structural colors. Adv. Funct. Mater. **32**, 2201766 (2022). https://doi.org/10.1002/adfm.202201766
36. He, W., Yao, D., Luo, S., Xiong, R., Yuan, X.: Broadband reflective liquid crystal films prepared by rapid inkjet printing and superposition polymerization. Crystals **12**(4), 473 (2022). https://doi.org/10.3390/cryst12040473
37. Jinbao, G., Hao, W., Fengjin, C., Lipei, Z., Wanli, H., Huai, Y., Jie, W.: Fabrication of multi-pitched photonic structure in cholesteric liquid crystals based on a polymer template with helical structure. J. Mater. Chem. **20**, 4094–4102 (2010). https://doi.org/10.1039/B926670D
38. Adimule, V., Suryavanshi, A., Yallur, B.C., Santosh, S.N.: A facile synthesis of poly(3-octyl thiophene):Ni0.4$Sr_{0.6}TiO_3$ hybrid nanocomposites for solar cell applications. Macromol. Symp. **392**, 2000001 (2020). https://doi.org/10.1002/masy.202000001
39. Meng, W., Gao, Y., Hu, X., Tan, L., Li, L., Zhou, G., Yang, H., Wang, J., Jiang, L.: Photothermal dual passively driven liquid crystal smart window. ACS Appl. Mater. Interfaces **14**, 28301–28309 (2022). https://doi.org/10.1021/acsami.2c06428
40. Vinayak, A., Santosh, N., Kalpana, S., Rangappa, K.: Microstructure, zeta potential and enhanced photocatalytic dye degradation performances of mesoporous flower like Nb_2O_5@CeO_2 bimetallic oxide nanomaterials decorated with reduced graphene oxide. Inorg. Chem. Commun. **166**, 112592 (2024). https://doi.org/10.1016/j.inoche.2023.112592
41. Chen, C., Brigeman, A., Ho, T., Khoo, I.: Normally transparent smart window based on electrically induced instability in dielectrically negative cholesteric liquid crystal. Opt. Mater. Express **8**, 691–697 (2018). https://doi.org/10.1364/OME.8.000691
42. Guo, S.M., Liang, X., Zhang, C.H., Chen, M., Shen, C., Zhang, L.Y., Yuan, X., He, B.F., Yang, H.: Preparation of a thermally light-transmittance-controllable film from a coexistent system

of polymer-dispersed and polymer-stabilized liquid crystals. ACS Appl. Mater. Interfaces **9**, 2942–2947 (2017). https://doi.org/10.1021/acsami.6b14967
43. Yang, T., Yuan, D., Liu, W., Zhang, Z., Wang, K., You, Y., Ye, H., De Haan, L.T., Zhang, Z., Zhou, G.: Thermochromic cholesteric liquid crystal microcapsules with cellulose nanocrystals and a melamine resin hybrid shell. ACS Appl. Mater. Interfaces **14**, 4588–4597 (2022). https://doi.org/10.1021/acsami.1c23101
44. Xianyu, M., Siyang, L., Shuo, C., Xuanzhe, S., Dekang, G., Jinbao, G.: Recent advances in smart windows based on photo-responsive liquid crystals featuring phase transition. Chem. Plus. Chem. **89**, 202300700 (2024). https://doi.org/10.1002/cplu.202300700
45. Seung-Won, O., Jong-Min, B., Sang-Hyeok, K., Tae-Hoon, Y.: Optical and electrical switching of cholesteric liquid crystals containing azo dye. RSC Adv. **7**, 19497 (2017). https://doi.org/10.1039/C7RA02294D
46. Javed, R.T., Hung-Yuan, L., Shin-Tson, W.: Photo- and electrical-responsive liquid crystal smart dimmer for augmented reality displays. Opt. Express **27**(13), 18169 (2019). https://doi.org/10.1364/OE.27.018169
47. Adimule, V., Keri, R., Rajendrachari, S., Kendrekar, P., Kulkarni, C.V.: Highly photoluminescence and wide band gap insulating metal hybrid nanoparticles array of samarium-doped SrO: CoO: synthesis, characterizations and sensor characteristics. J. Mater. Sci.: Mater. Electron. **34**(5), 442 (2023). https://doi.org/10.1007/s10854-023-09899-2
48. Ling-Ling, M., Sai-Bo, W., Wei, H., Chao, L., Peng, C., Hao, Q., Yandong, W., Lifeng, C., Yan-qing, L.: Self-assembled asymmetric microlenses for four-dimensional visual imaging. ACS Nano **13**(12), 13709–13715 (2019). https://doi.org/10.1021/acsnano.9b07104
49. Ling-Ling, M., Chao, L., Sai-Bo, W., Peng, C., Quan-Ming, C., Jia-Xin, Q., Shi-Jun, G., Yuan-Hang, W., Wei, H., Yan-Qing, L.: Programmable self-propelling actuators enabled by a dynamic helical medium. Sci. Adv. **7**, eabh3505 (2021). https://doi.org/10.1126/sciadv.abh3505
50. Wang, H., Tang, Y., Krishna Bisoyi, H., Li, Q.: Reversible handedness inversion and circularly polarized light reflection tuning in self-organized helical superstructures using visible-light-driven macrocyclic chiral switches. Angew. Chem. Int. Ed. **62**, e202216600 (2023). https://doi.org/10.1002/anie.202216600
51. Adimule, V., Yallur, B.C., Keri, R., Bathula, C., Batakurki, S.: Microstructure, photoluminescence and electrical properties of SmxGd(1−x): SrO hybrid nanomaterials synthesized via facile coprecipitation method. Electron. Mater. Lett. **19**(3), 278–297 (2023). https://doi.org/10.1007/s13391-022-00394-0
52. Zhang, H., Liu, J., Zhao, X., Gao, J., Ma, C., Zhao, Y., Yao, R., Miao, Z., Shen, W.: Electrically induced coloration of polymer-stabilized cholesteric liquid crystal films with broadband reflection capability for smart windows. Dye. Pigment. **203**, 110316 (2022). https://doi.org/10.1016/j.dyepig.2022.110316
53. Chen, P., Ma, L.L., Hu, W.: Chirality invertible superstructure mediated active planar optics. Nat. Commun. **10**, 2518 (2019). https://doi.org/10.1038/s41467-019-10538-w
54. Adimule, V., Sharma, V., Kumar, P., Keri, R., Sharma, K.: Perylene-3, 4, 9, 10-tetracarboxylic dianhydride dye intercalated liquid crystal molecules: synthesis, spectroscopic characterizations, and photoluminescence dynamics. J. Mol. Liq. 125990 (2024). https://doi.org/10.1016/j.molliq.2023.125990
55. Binghui, L., Cong-Long, Y., Hong-Long, H., Hao, W., Yu-Wen, Z., Pei-Zhi, S., Zhi-Ying, L., Zhi-Gang, Z., Quan, L.: Dynamically actuated soft heliconical architecture via frequency of electric fields. Nat. Commun. **13**, 2712 (2022). https://doi.org/10.1038/s41467-022-30486-2
56. Daniel, F., Ziqian, H., Pamela, M.O., Alireza, S., Pablo, C., Shin-Tson, W., Debashis, C.: Self-assembled plasmonics for angle-independent structural color displays with actively addressed black states. https://doi.org/10.1073/pnas.2001435117
57. Lingling, M., Chaoyi, L., Luyao, S., Zhenpeng, S., Yanqing, L., Bingxiang, L.: Submicrosecond electro-optical switching of one-dimensional soft photonic crystals. Photon. Res. **10**, 786–792 (2022). https://doi.org/10.1364/PRJ.449204
58. Adimule, V., Bathula, C., Sharma, K., Keri, R.: Study of structural, optical, dielectric properties poly-3-butyl thiophene reinforced with CO_3O_4 nanoparticles. Nano-Struct. Nano-Objects **38**, 101150 (2024). https://doi.org/10.1016/j.nanoso.2023.101150

59. del Marc, P., Jeroen, S., Albert, S., Michael, G.D.: 4D printing of liquid crystals: what's right for me? Adv. Mater. **34**, 2104390 (2022). https://doi.org/10.1002/adma.202104390
60. Kim, S.U., Lee, Y.J., Liu, J.: Broadband and pixelated camouflage in inflating chiral nematic liquid crystalline elastomers. Nat. Mater. **21**, 41–46 (2022). https://doi.org/10.1038/s41563-021-01075-3

Recent Advances in the Smart Window Applications of Cholesteric Liquid Crystals

Santosh Nandi, S. S. Nesargi, Vinayak Adimule, Savita Hanaji, Praveen Barmavatu, and Madhavaprasad Dasari

Abstract Windows are essential components of building environments because they allow light and heat penetration into interior spaces and serve as an interface for communication between a building's interior and exterior. The development of smart windows has placed a great deal of emphasis on cholesteric liquid crystals (CLCs) because of their exceptional capacity to selectively reflect certain light wavelengths through Bragg reflection. They are perfect candidates for energy-efficient window technologies because of this feature as well as their dynamic tunability in reaction to external stimuli including light, electric fields, and temperature. Current developments have concentrated on improving the thermal and optical characteristics of CLC-based smart windows, which has improved their capacity to manage heat transmission, control solar radiation, and offer privacy without sacrificing visibility. The performance and durability of these systems have been further enhanced by advances in material design, such as the addition of polymer-stabilized CLCs. Furthermore,

S. Nandi (✉)
Department of Chemistry, KLE Technological University, Dr. M. S. Sheshgiri Campus, Udyambag, Belagavi, Karnataka, India
e-mail: santosh.nandi603@gmail.com

S. S. Nesargi
Chemistry Section, Department of Chemistry, KLE Technological University's Dr. M. S. Sheshgiri Campus, Udyambag, Belagavi, Karnataka, India

V. Adimule
Department of Chemistry, Angadi Institute of Technology and Management (AITM), Belagavi, Karnataka, India

S. Hanaji
Department of Engineering Science and Humanities, KLE Technological University, Dr. M. S. Sheshgiri Campus, Udyambag, Belagavi, Karnataka, India

P. Barmavatu
Department of Mechanical Engineering, Faculty of Engineering, Universidad Tecnológica Metropolitana, Santiago, Chile
e-mail: pbarmavatu@utem.cl

M. Dasari
Department of Physics, GITAM Institute of Science, GITAM (Deemed to be University), Visakhapatnam, Visakhapatnam, India

P. Kumar et al. (eds.), *Cholesteric Liquid Crystals*, Engineering Materials,
https://doi.org/10.1007/978-981-96-3821-5_6

large-scale production and the integration of CLC-based windows in architectural applications have been made possible by improvements in fabrication techniques. In particular, their importance in energy-saving solutions for sustainable building design is discussed, along with the most recent advancements in CLC materials and their prospects for the future.

Keywords Smart Window · Cholesteric liquid crystal · Energy efficiency · Sustainable architecture · Tunable optical properties · Dynamic light modulation

1 Introduction

Smart buildings that have intelligent control and low energy usage have drawn a lot of attention as urbanisation and modernisation progress, and it is anticipated that these buildings will offer people an efficient, comfortable, and convenient place to live and work [1, 2]. Windows are vital to smart buildings because they serve as a vital conduit for heat radiation from the sun and outside light to reach the interior of the structures. In the context of contemporary intelligent building light management, smart windows with the ability to selectively regulate light transmittance and colour on demand have emerged as a viable option [3]. They can be used to give both inside lighting and an outside view, or to control light transmission to safeguard users privacy [4]. A variety of smart window technologies are available, including humidity-triggered chromism [5], photochromism [6], electrochromism [7], mechanochromism [8], and magnetochromism [9]. Liquid crystals (LCs), hydrogels, perovskites, organic and inorganic electrochromic materials, phase change materials (PCMs), and many more materials have been developed during the past few decades as vital elements in smart windows [10, 11]. Applications including switchable privacy glass [12], solar heat rejection [13], and switchable colour [14] have been demonstrated using LCs, which provide an excellent material system for constructing smart windows. To develop smart windows for controlling interior solar irradiation, particularly for visible light and infrared (IR) light, a variety of LC technologies based on light scattering or reflection have been investigated. While infrared light controls interior temperature, visible light maintains interior illumination levels [15]. By carefully regulating their chemical configuration, it is possible to modify the optical states of the LC materials [16]. Furthermore, a variety of external stimuli, including humidity, temperature, magnetic field, light, and electrical field, can be used to influence the arrangement of mesogenic molecules. Cholesteric LC materials show an LC organisation in which the successive pneumatic LC layers are spaced apart from one another by a little rotation in the molecular direction [17]. The cholesteric LC with planar textures can display Bragg reflection because of their helical molecule architectures. A certain wavelength of light can cause the light-active dopants in cholesteric LC materials to reorganise, resulting in a variation in the materials pitch and colour of reflection before and after exposure. Utilising smectic/cholesteric liquid crystal (CLC) phases, smart windows have been designed that allow the window to be switched reversibly

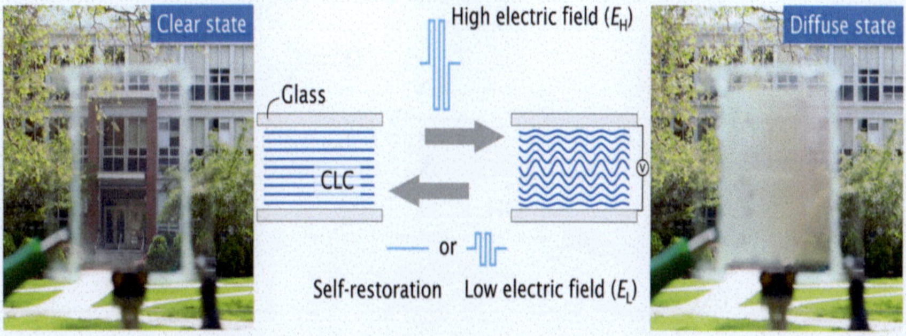

Fig. 1 Schematic and photographs of the proposed CLC smart window ($d \approx 45$ μm). (Adapted from Ref. [28], Optical Materials Express 2018)

between transparent (planar and homeotropic) and scattering (focal conic) states [18]. Currently, a variety of materials are accessible for the production of smart windows, such as perovskites [19], hydrogel [20], suspended particles [21], phase-changing materials [22], electrochromic materials [23], and thermochromic materials [24]. One of the most appealing materials for smart windows is LC, which has responsive qualities by nature [25]. Several LC smart windows are now on the market, and research on LC smart windows has advanced significantly in the last few decades [26, 27]. Specifically, LC smart windows that alternate between transparent and opaque states can be utilised for energy conservation in buildings as well as a useful tool in greenhouses, cars sunroofs, aeroplane portholes, private spaces, and augmented reality displays (Fig. 1) [28].

Recent developments in LC-based smart window technology are reviewed in this chapter. The latest developments in dual-responsive LC smart windows, such as thermal-/electrical-responsive and thermal-/photo-responsive smart windows, the control mechanisms of LC-based smart windows are also compiled. Electrochromism, thermochromism, photochromism, humidity-triggered chromosomes, mechanical self-powered techniques, and dynamic infrared regulating mechanisms are among the various sophisticated LC-based smart window processes that are covered. The uses of LC-based smart windows in thermal regulation for intelligent buildings, human–machine interface, and indoor lighting/privacy protection switching are emphasised. Additionally, the difficulties and potential paths for LC-based smart window development are also covered.

1.1 Photo-Responsive Liquid Crystal (LC) Smart Windows

LC is a stimuli-responsive (external stimuli, such as light, temperature, magnetic fields, or electricity) material. A photo-responsive liquid crystal (PRLC) smart

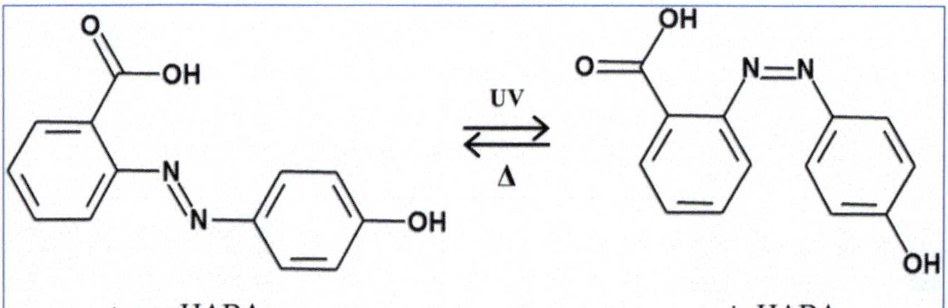

Fig. 2 *Trans–cis* isomerization of push–pull azo dye HABA. (Adapted from Ref. [31], Copyright RSC, 2017)

window is typically created by doping LC mixtures with photo-responsive materials [29]. Given their straightforward designs, passive light modulations, and zero-energy input features, PRLC smart windows are likewise regarded as cost-effective, stimuli-rational, and energy-efficient smart windows. One of the most popular photo-responsive materials for creating photo-responsive LC smart windows is azobenzene derivatives [30]. In the first state, the window scatters incident light through the focal conic (FC) state while the azobenzene chiral dopant causes LCs to produce CLCs. Photoisomerization of azobenzene lowers the order parameter of LCs, the window transitions from FC state to isotropic state, and the material turns transparent when exposed to ultraviolet (UV) light within a particular switching temperature range. The advantage of push–pull azobenzene (2-(4-hydroxyphenylazo)benzoic) acid (HABA) is its rapid thermal relaxation (Fig. 2), which allows the window to instantly return to its opaque condition after the UV light is removed [31]. As seen in Fig. 3 [32], Talukder et al. developed a method to use an azobenzene dimmer PSC-01, whose helical twist powers (HTP) of trans and cis states are 64.18 μm^{-1} and 16.31 μm^{-1}, respectively, to create an innovative transparent photo-responsive LC smart windows. The azobenzene undergoes *trans–cis* photoisomerization and the helical twist power varies when the cell is exposed to UV light. As a result, the cell transforms into a dark, opaque state and left-handed CLCs are produced and enter the FC state. Additionally, the LCs will need to be reversed from the FC state to the homeotropic state using heating or visible light. In addition to photo-isomerized materials, a useful tactic for creating PRLC smart windows is the use of photothermal materials that alter the LC phase [33].

1.2 Electrical-Responsive LC Smart Windows

A component that consists of the LC mixture sandwiched between two transparent conductive electrodes and controls the optical characteristics by varying the presence or absence of an electric field is known as an electrical-responsive liquid crystal

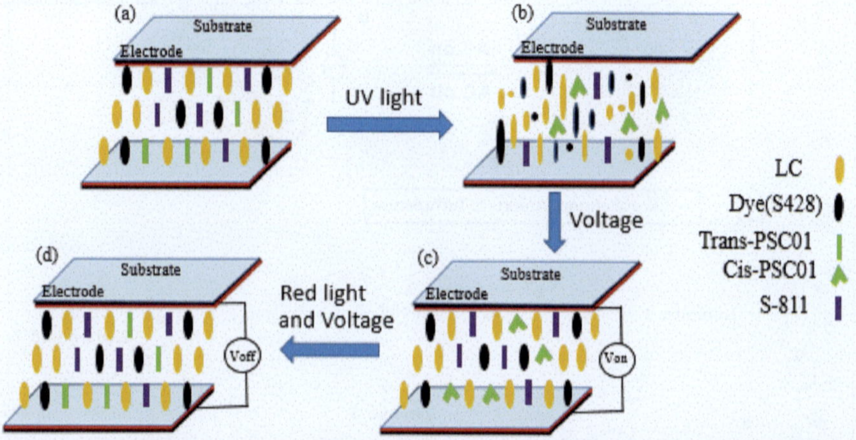

Fig. 3 Operation mechanisms of the proposed device: **a** initial state, **b** after UV exposure, **c** with a voltage, and **d** removing the voltage and red light. (Adapted from Ref. [32], Copyright Optics Express 2019)

(ERLC) smart window [34]. By electrically controlling the reorientation of the LC director, ERLC smart windows typically transition between transparent and opaque states through light scattering [35]. As a result, the ERLC smart window receives greater attention and is quickly commercialised, and the response time is in milli second short [36]. One extremely useful method for achieving the dynamic scattering phenomena is to dope the nematic LCs (NLCs) with the appropriate species of ions [37]. LC mixes align in the same direction under the alignment layer's operation, revealing the pattern underlying the apparatus. When the electric field is applied, the ions' motion creates a vortex and the LCs director disorder, which causes intense light scattering. Charge-neutral zwitterion molecules with many ionic groups have become a popular ionic material in recent years [38]. It has been demonstrated that zwitterion molecules can be used to further enhance the electrohydrodynamic instability smart windows long-term stability (Fig. 4). Apart from NLCs, doping ion materials into cholesteric or smectic LCs can be another useful method for creating bistable smart windows. This method has garnered a lot of interest because of its low energy consumption. The window changes from planar to FC texture when a low-frequency voltage is supplied, allowing ions to move freely [39, 40]. This causes incident light to scatter in random directions, creating the scattering state. Without applied fields, CLCs will continue to retain the metastable FC state. The window returns to its original transparent condition when the high-frequency voltage is applied since the ions inside hardly move. As a result, the negative dielectric CLCs react to the electric field and self-assemble to a planar state.

The composite polymer/LCs system is the basis of another well-liked ERLC smart window. polymer dispersed liquid crystal (PDLC), is the most well-known example [41]. Because low-molar-mass LCs are typically insoluble in the polymer network during PDLC synthesis, phase separation occurs, resulting in the embedding

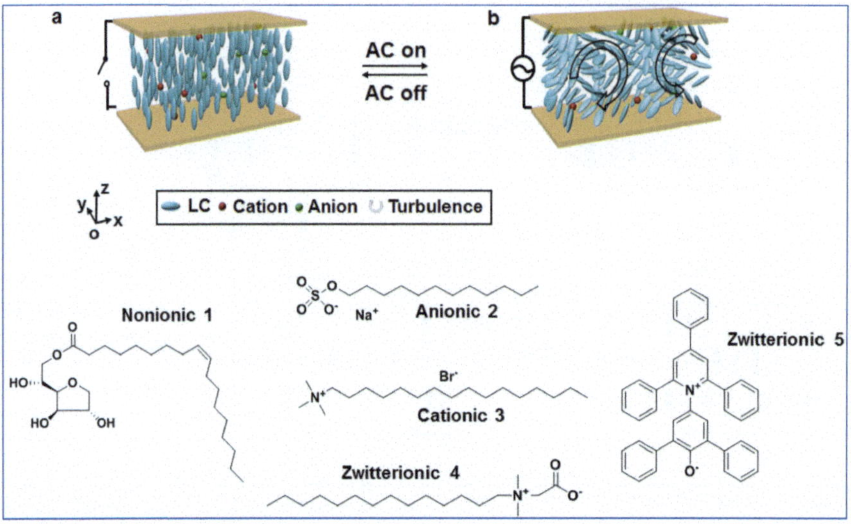

Fig. 4 Schematic representation of electrohydrodynamic instability (EHDI) **a** at 0 V and **b** under the application of an alternating electric field. **c** Chemical structures of four different species of ions: the non-ionic 1, the anionic 2, the cationic 3 and the zwitterionic 4, 5. (Adapted from Ref. [38], Copyright 2019, Taylor and Francis)

of nano- or micrometer-sized LC droplets in a continuous polymer matrix. Because the LCs and polymer in the initial state have different refractive indices, PDLC displays a milky scattering condition. The LC director aligns with the field when a high enough voltage is provided, which eliminates the refractive index mismatch and permits incident light to flow through (Fig. 5) [42]. Since the polymer matrix lacks LC orientation control, PDLC typically exhibits a fuzzy state when the electric field is not applied, making it inappropriate for conventional buildings. Additionally, a higher working voltage (~100 V) is needed to reach the transparent state, and the polymer content is greater than 40 weight percent. In the OFF state, the polymer network and LCs are aligned in the same direction because of the alignment layer on both sides of the substrate. Here, PSLC displays a clear state and the LC effective index is equal to that of the polymer matrix. Numerous studies have concentrated on methods to further improve the capabilities of polymer/LCs composite ERLC smart windows, including doping nanomaterials [43, 44], LC type (CLCs, SmLCs, dual-frequency LCs) [45], monomer chemical structure [46], polymerisation type [47], etc. Doping certain metal nanoparticles (ZnO, CuO, Fe_3O_4, TiO_2, et.) in PDLC might lower the driving voltage of PDLC because they can alter the LCs' dielectric constant or produce local field effects. Ag nanoparticles in particular, whose surface plasmon excitations at metal–LC interfaces raise local electric fields and cause the PDLC's driving voltage to decrease from 77 to 40 V. Additionally, it is important to note that Yang's group has addressed the issue of high working voltage of PDLC by developing a novel coexisting system of polymer dispersed and polymer stabilised liquid crystal

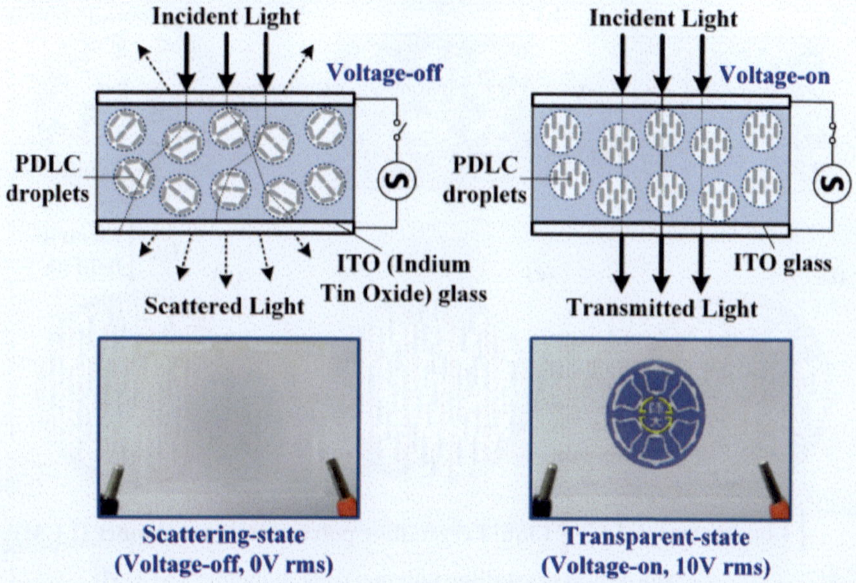

Fig. 5 Illustration of the operating principles of PDLC in the voltage-off and voltage-on states, respectively. (Adapted from Ref. [42], Copyright Elsevier, 2019)

(PD&SLC) [48]. They created a novel system in which the homeotropically polymer network structure of PSLC is randomly distributed in each cavity of the porous polymer matrix of PDLC, taking advantage of the difference in polymerisation rates of liquid crystalline monomers and non-liquid crystalline monomers (Fig. 6).

1.3 Thermal-Responsive LC Smart Windows

The most popular technique for creating thermal responsive LC smart windows involves taking use of the reversible phase transition characteristics of LCs, particularly the reversible phase transition between the chiral nematic phases and the smectic A. The LC materials used in thermal-responsive LC smart windows undergo a change in characteristics when the temperature exceeds a crucial value of transition temperature [49]. Since the windows may adaptively change the light transmittance based on the temperature outside, the thermal-responsive LC smart window mechanism is typically thought of as a passive method that does not require an extra control system. TRLC smart windows are therefore among the most alluring choices for energy-efficient structures. The anchoring of the homeotropically aligned polymer network at room temperature (25 °C) causes the Smectic A LC molecule to be orientated perpendicular to the substrate surface, and the film exhibits a clear condition. LC molecules rearrange to create the FC texture when the film is heated to a higher

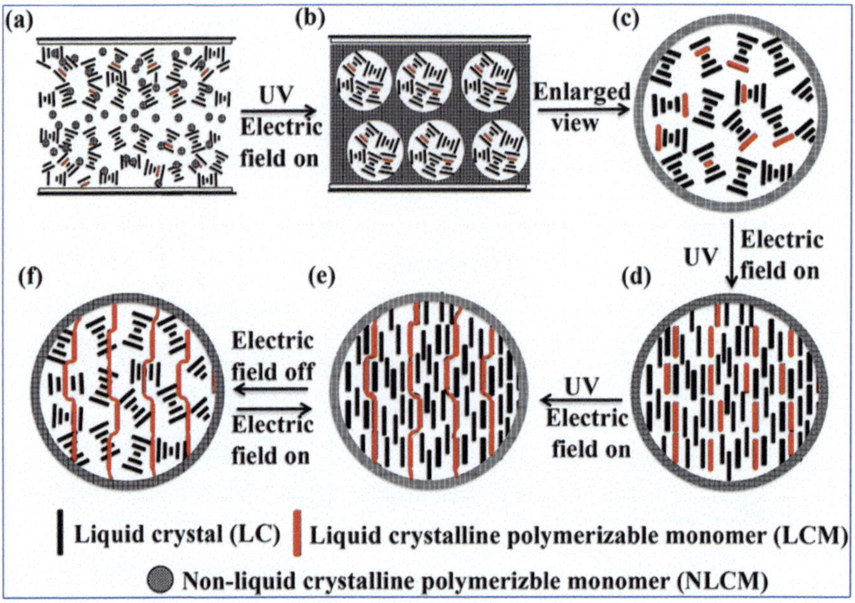

Fig. 6 The schematic illustration for preparing the PD&SLC. (Adapted from Ref. [48], Copyright Elsevier, 2019)

temperature into the chiral nematic phase, and the film appears opaque (Fig. 7) [50]. When the temperature reaches 31 °C, LCs move into the chiral nematic phase and create the FC domain. Additionally, the absorption is improved by the combination of LC molecules and black dyes. The film exhibits a high-haze opaque condition, with a transmittance of only 4.3%, and simultaneously produces the scattering state and deep black colour. Further raising the temperature to the isotropic phase causes the light scattering to vanish, the transmittance to slightly increase (19.4%), and the film to appear in an opaque state free of haze. Yang et al. [51] prepared coexistent system of PD&SLCs, which combines the advantages of PDLC and PSLC systems. Moreover, a unique thermally light-transmittance controlled (TLTC) film was prepared from the PD&SLC system utilising an LC with a smectic A (SmA)-chiral nematic (N*) phase transition (Fig. 8). Depending on the difference between the ambient temperature and the phase-transition temperature of the LC, the TLTC film can be thermally and reversibly changed from a transparent state to a strong light-scattering state. The thermally light-transmittance controlled film shows promise in a wide range of applications, such as smart window films and temperature sensors.

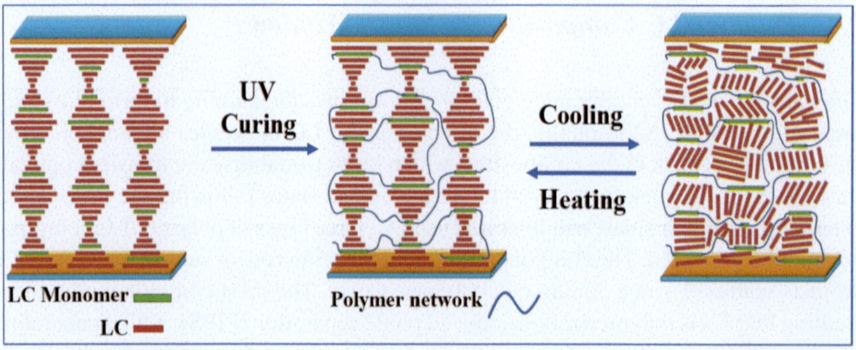

Fig. 7 The schematic illustration of the mechanism for the PSLC film. (Adapted from Ref. [50]. Copyright 2014, IOP)

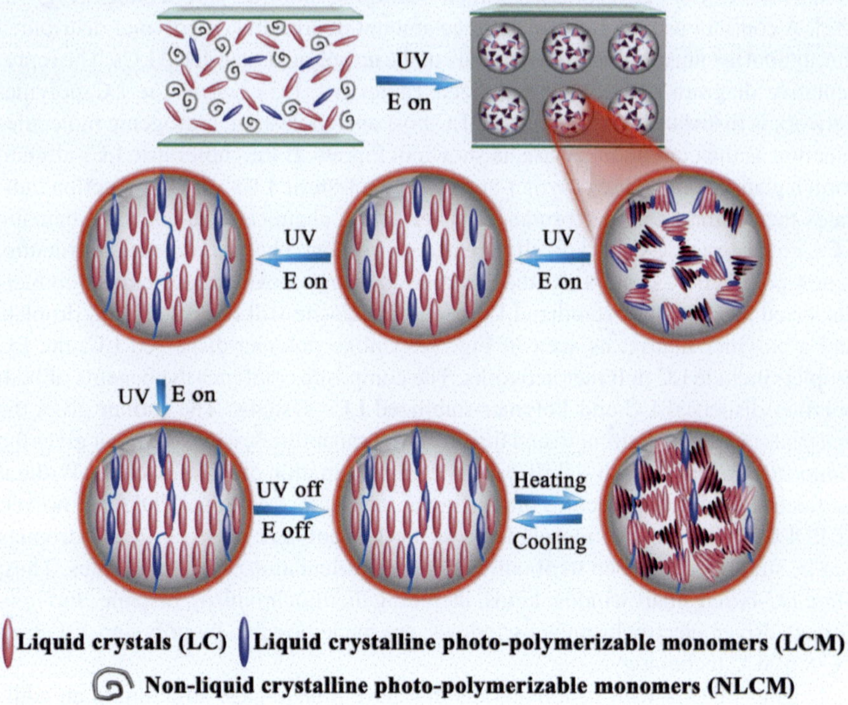

Fig. 8 Schematic diagram of preparation of polymer-dispersed and polymer-stabilized liquid crystals (PD&SLC) film. (Adapted from Ref. [51]. Copyright 2017, American Chemical Society)

1.4 Polymer/LC Composites for Smart Windows

The polymer and LC combination gives the composites their ability to form films and respond to stimuli. Additionally, the orientations of LC molecules are anchored by the polymer network in the composites, which helps to stabilise the transient optical states or induce a quick recovery of the initial optical states following the removal of external stimuli. For smart windows that use LC, three types of polymer/LC composites have been created. The components of PDLCs are micron- or submicron-sized LC droplets scattered over a continuous polymer matrix. The most common method for creating PDLCs is polymerization-induced phase separation (PIPS), which generates LC droplets by separating the mesogenic molecules from the matrix after the non-LC monomers are crosslinked to create the polymer matrix. After the LC droplets' and the polymer matrix's refractive indices mismatch or match, an external electric field switches the PDLCs between their scattering and transparent states (Fig. 9a) [52]. A continuous LC phase and a trace amount of crosslinked polymer distributed throughout the anisotropic fluid typically make up polymer-stabilized LCs. The representative diagram of polymer-stabilized cholesteric LCs, where the LC polymer network is distributed throughout the LC host and cholesteric mesogenic molecules function as the continuous phase, is shown in Fig. 9b. If the cholesteric LCs align to form a planar texture, the polymer-stabilized cholesteric LCs' Bragg reflection indicates their remarkable performance. When chiral chemicals are added to nematic LCs, cholesteric LCs are typically created, exhibiting a helical molecular structure. A new polymer/LCs composite that are derived from a coexisting system of polymer-dispersed and polymer-stabilized LCs. The composite still consists of LC droplets and a polymer matrix, as seen in Fig. 9c. Unlike polymer-dispersed LC, the LC droplets include LC polymer networks. The composite combines the benefits of both polymer-dispersed LC and Polymer-stabilized LCs systems. The former gives the materials mechanical strength and film-forming capabilities, while the latter gives the composite the capacity to stabilise the specific orientation of LC molecules. Without an electric field, LC molecules often align at random in traditional PDLC. However, in PDLC & PSLCs, the morphology of the LC polymer network in LC droplets can be suitably controlled to obtain the ordered orientation of LC molecules. Thus, more LC-based smart window kinds, including thermochromism, bistable, and low-voltage driven electrochromism windows, are made possible by polymer-dispersed LC & PSLC technology.

Doping PDLCs with nanomaterials that have a nano-cage structure is an additional strategy for low-voltage-driven smart windows. The polyhedral oligomeric silsesquioxane microstructure was effectively inserted into the polymer matrix by He et al. [53] to create low-voltage-driven smart windows with a high contrast ratio (>100). Zhu and colleagues [54] reported a new smart window through the doping of PSLC films with antimony doped tin oxide (ATO) nanoparticles (NPs), the schematic and digital images are shown in Fig. 10. Among the films that have been reported thus far, these films have the broadest waveband modulation function, spanning the visible and infrared spectrum (380–5500 nm). A reversible transition from a highly

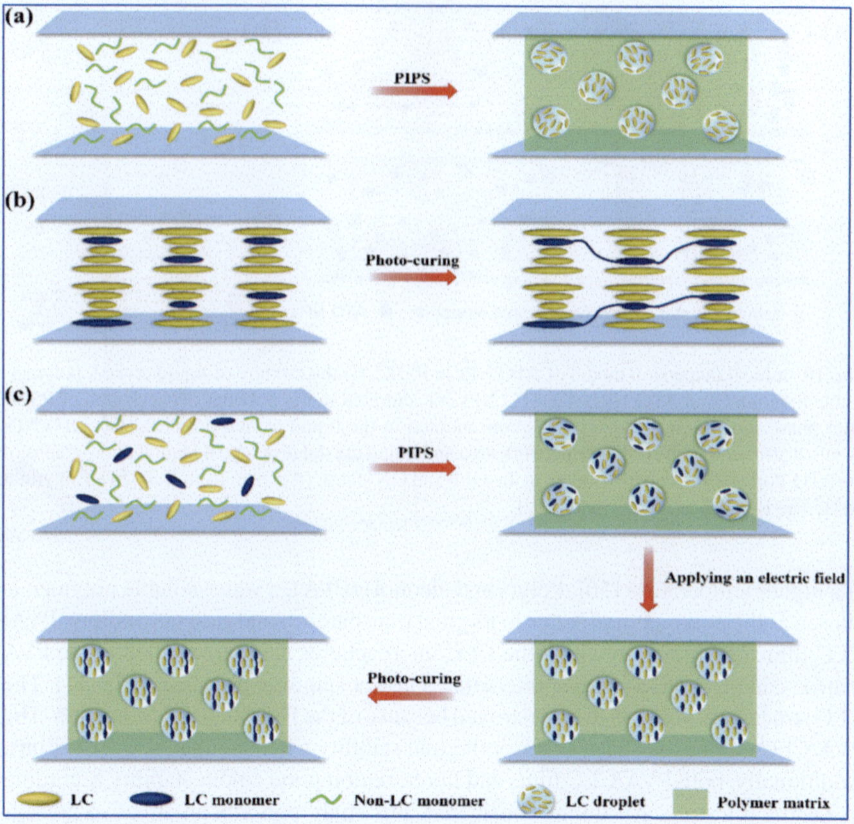

Fig. 9 Schematic illustration of the preparation and the microstructure of **a** PDLC, **b** PSLC, and **c** PDLC&PSLC. PIPS, polymerization induced phase separation. (Adapted from Ref. [52]. Copyright 2022, Wiley)

transparent (78.5%) state to a strong light scattering (10%) state in the visible region is possible for the as-prepared smart films. The localised surface plasmon resonance of ATO NPs also allows for the effective shielding of up to 80.7% of the infrared invisible light. Liu et al. [55] reported using graphene hybrids that were grown directly on quartz glass as building blocks for devices that used CLC. The unique design of graphene allows it to function as both a conductive and alignment layer, making the process of fabricating devices easier. The various domain orientations of graphene are thought to be responsible for the multiple helical axis orientations of CLCs and the creation of multi-domain structures, which results in a broad viewing angle for devices based on CLCs. The polarised optical microscope (POM) characterisation of polycrystalline monolayer graphene on quartz glass shows that multi-domain structures can be developed for CLCs. By combining a left-handed chiral dopant called 4′-(1,3-dimethyl-3-chloro)propoxy-4-cyanobiphenyl (NYCL) with a commercial nematic LC DLC-111, Chen created a CLC that formed a blue phase

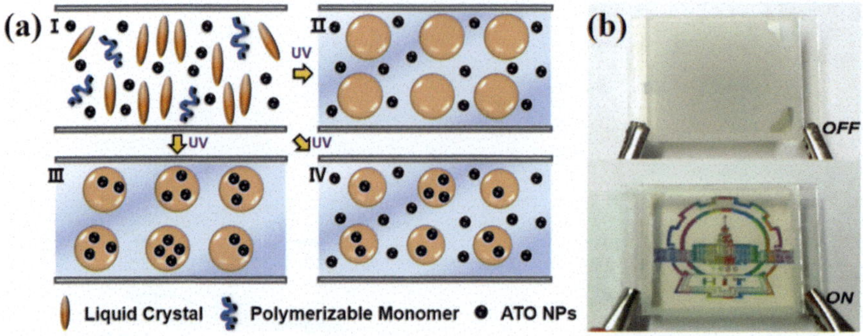

Fig. 10 **a** Ideal dispersion model of ATO NPs in PSLC, (I) the mixture of liquid crystal, polymerizable monomer and ATO NPs, (II) ATO NPs concentrated in the polymer network area of PSLC after phase separation, (III) ATO NPs concentrated in the liquid crystal droplets, (IV) ATO NPs evenly distributed in the polymer network area and liquid crystal droplets. **b** Digital photographs of the ATO-doped PSLC smart window in the OFF and ON states. (Adapted from Ref. [54]. Copyright 2021, The Royal Society of Chemistry)

at a higher temperature [56]. Poly(vinyl alcohol) (PVA), a water-soluble polymer, is chosen as the polymer matrix due to its great transparency and ease of handling. PVA/CLC films and fibres, in which the CLC microspheres are incorporated in the PVA matrix, can be created using spin-coating and wet spinning processes (Fig. 11). The CLC structure is not seen to be destroyed because of the PVA matrix's protection. The PVA/CLC composites have superior thermal stability when compared to CLC films. Additionally, both PVA/CLC films and fibres exhibit good colour stability according to electrical testing. All things considered, this study presents an effective nonsolvent and phase separation process for creating PVA/CLC films and fibres, creating opportunities for use in medical devices, smart fabrics, and display technologies.

2 Dual Responsive LC Smart Windows

The photochromic and thermochromic smart windows are more convenient and environmentally benign since they can automatically change their transmission status based on temperature or light intensity. This section describes a smart window with dual control mechanisms: thermally passive regulation and electrically active regulation, since both active and passive functions are advantageous. Multi-responsive LC smart windows that incorporate multiple driven techniques for potential future applications have attracted fascinating attention. A dual-responsive LC smart window is one that responds to two separate stimuli, like light and temperature or electric field and temperature, to achieve switching states (opaque and transparent states).

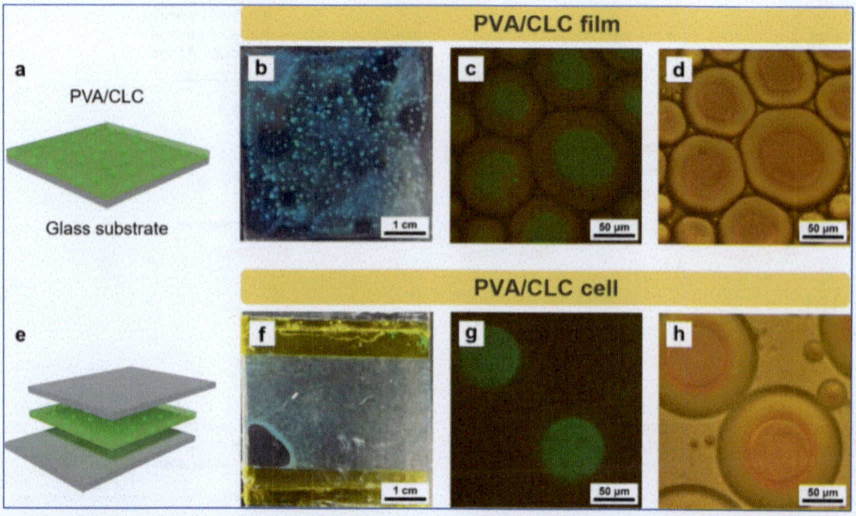

Fig. 11 Schematic illustrations of a PVA/CLC film (**a**) and a PVA/CLC cell (**e**). Photographs of a PVA/CLC film (**b**) and a PVA/CLC cell (**f**). Optical images of a PVA/CLC film and a PVA/CLC cell using reflection (**c, g**) and transmission modes (**d–h**). (Adapted from Ref. [54]. Copyright 2021, The Royal Society of Chemistry)

2.1 Photo-Thermal Responsive LC Smart Windows

The simplest method for creating thermal-/photo-responsive LC smart windows is to incorporate photo-responsive chiral dopants into LCs with the smectic A-Nematic phase transition. Photo-responsive materials can be grafted as a functional group in polymers for alignment layers in addition to being utilised as chiral dopants [57]. The Thermal responsive mode mainly depends on the phase change of LCs. In Photo responsive mode, the PSLC presents a transparent state due to the homeotropic alignment layer without irradiation. The azobenzene group undergoes photo-isomerization when the UV light (365 nm) is turned on, and it also causes a photothermal action. The LCs go through a phase transition as a result of the photothermal effect taking over, and the smart window changes from a transparent to an opaque state. The chiral smectic A phase of LCs in the manufactured cell can be transformed into the chiral nematic phase with heating, and vice versa with cooling, according to Yoon et al. [58]. Additionally, trans–cis photo-isomerization of azobenzene molecules in the presence of UV irradiation can alter the phase of LCs from the chiral smectic A to the chiral nematic phase (Fig. 12a and b). The push–pull azobenzene in the LC mixture has a fast thermal relaxation feature, which causes the LCs to quickly relax from the chiral nematic phase to the chiral smectic A phase by *cis–trans* back-isomerization of the azobenzene molecules once UV irradiation is removed.

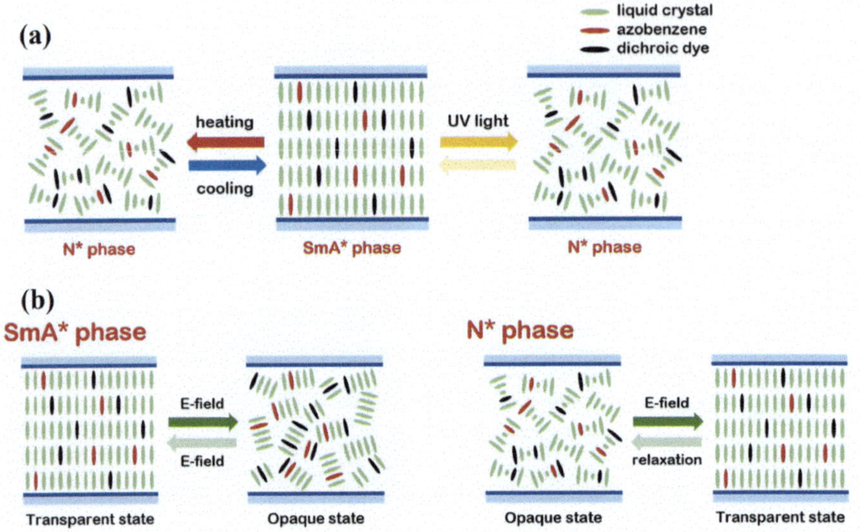

Fig. 12 **a** A schematic representation of self-shading light shutter, **b** electrical switching properties of the fabricated light shutter. (Adapted from Ref. [58], Copyright 2018, Wiley)

2.2 *Electrical-Thermal Responsive LC Smart Windows*

Extensive research efforts have focused on thermo- and (or) electro-responsive polymers among the stimuli-responsive polymers studied because they can be used as thermochromic (TC) and electrochromic (EC) materials, which leads to practical applications such as antiglare mirrors, smart windows, information displays, sensors, etc. [59]. The possibility of electrical-thermal responsive LC smart windows exists if a system can accomplish both thermal and electrical responsiveness. In order to provide temperature-dependent dynamic scattering for passive control, the operating concept relies on the variable strength of voltage-induced electrohydrodynamic flow. Furthermore, an applied AC voltage can be used to reversibly modulate and switch the transmission between the opaque and transparent states for active control. As a demonstration, the optically transparent state's distinctive Bragg reflection can be positioned in the near-infrared spectrum to partially block undesired thermal radiation. The design and operation of the thermo-responsive CLC window for adaptive smart switching are described by Sung et al. [60] While the reflecting portion of solar thermal radiation is at null voltage, the CLC glass stays transparent for incident visible light when the local temperature is comfortable or the incident solar radiation is low. As long as the voltage is switched off, this function is actually maintained regardless of the weather (Fig. 13a). After the CLC cell and an empty cell with the same effective area of 1.8 cm^2 were passed through an intense condition of sunshine, a black Bristol cardboard was used to test the viability. Following 210 min of exposure to sunlight on a sunny day, it was verified that the temperature behind the IR-reflective

CLC cell in the GP state was 40.8 °C, approximately 5 °C lower than the temperature (at 45.6 °C) at a nearby location behind the empty cell. On the other hand, Fig. 13b illustrates how the CLC smart window functions during uncomfortable high temperatures or on a hot, sunny day with intense solar irradiance. After the CLC absorbs enough thermal radiation to induce the DS state under the application of an external AC voltage, blocking the entry of sunlight (both visible and NIR) into the room, solar heating causes the electrohydrodynamic effect. In addition to the adjustable and switchable shading, the smart glass's dynamic scattering (DS) state allows for the protection of individual privacy. More importantly, the switching state correlates to different voltages at different temperatures, which is advantageous to people living in all latitudes. Single-response windows and dual-response windows often work well together, and the production techniques rely more on the advancements in single-response smart window technology. Therefore, creating more single-response preparation techniques is necessary.

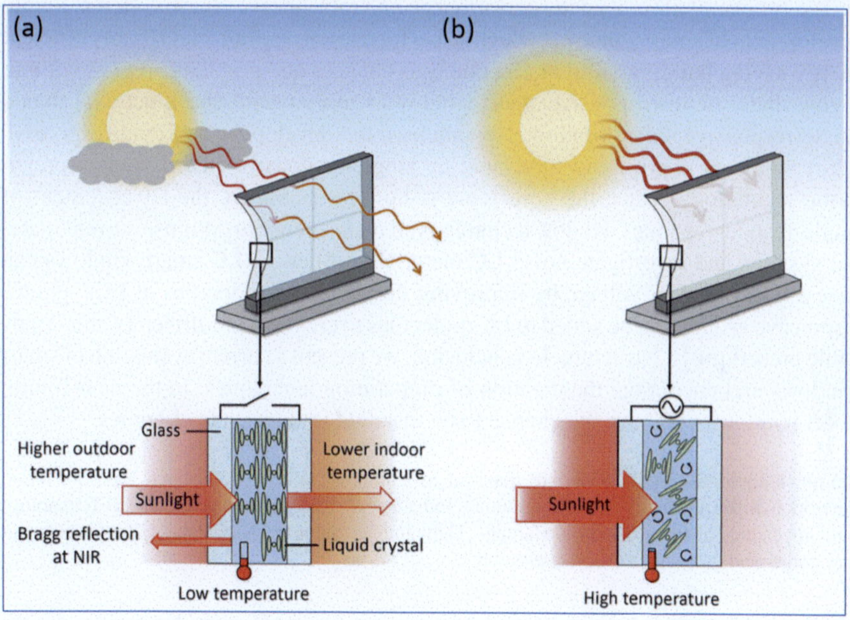

Fig. 13 **a** At a moderate temperature, the electrically unperturbed CLC exhibits the GP configuration, giving the transparency in the visible while partially reflecting the incident NIR radiation via Bragg reflection. **b** At a high surrounding temperature outdoors, solar heating from terrestrial radiation causes electro-hydrodynamic (EHD) flow in the CLC under the application of AC voltage. The CLC molecules in an unstable or irregular arrangement to procure the DS or opaque state will strongly scatter the incident sunlight. (Adapted from Ref. [60], Copyright 2021, Optica Publishing Group)

3 Conclusions

CLCs have received a lot of attention in the development of smart windows due to their remarkable ability to selectively reflect specific light wavelengths through Bragg reflection. This characteristic, along with their dynamic tunability in response to external stimuli like light, electric fields, and temperature, makes them ideal candidates for energy-efficient window solutions. For electrically, thermally, and photo-responsive windows, the chapter primarily provides an overview of the latest developments in LC smart window switching between transparent and opaque states. For each category, we outline the preparation methods and operational mechanisms with an emphasis on single- and dual-responsive approaches. The focus of recent advancements has been on enhancing the thermal and optical properties of CLC-based smart windows, which has increased their ability to regulate heat transfer, regulate solar radiation, and provide privacy without compromising visibility. The incorporation of polymer-stabilized CLCs and other developments in material design have significantly improved the performance and robustness of these systems. Future development of smart windows, which strive for carbon neutrality and highly efficient energy-saving buildings, still faces numerous chances and problems. It goes without saying that a multi-responsive smart window is more useful and functional than a single-responsive smart window. Nonetheless, the development of dual-responsive smart windows is primarily motivated by single-responsive smart window preparation technology. Thus, the core issue is how to encourage the advancement of single-responsive smart window technology. It makes sense to optimise current material systems and investigate novel LC materials further. ERLC smart windows can have a lower driving voltage by improving the dielectric properties of LCs; photo-responsive groups can be added to LC molecules to create photo-driven LC materials while preserving LC structure. In conclusion, we present a current evaluation of smart windows and encourage the creation of preparation techniques. In the near future, smart windows will have the chance to be installed in thousands of homes.

Acknowledgements All authors are thankful to Chitkara University Research and Innovation Network, Chitkara University, Patiala, Panjab, India. Authors are also thankful to KLE Technological University, Dr. M. S. Sheshgiri Campus, Udhyambag, Belagavi, Karnataka, India for constant encouragement and support for the research.

Funding Authors declare that they have not received any funding from any source or institutions or organizations.

Conflict of Interest All authors declare that they do not have any conflict of interest.

References

1. Timuçin, H., de Pieter, W.: The interaction between humans and buildings for energy efficiency: a critical review. Energy Res. Soc. Sci. **71**, 101828 (2021). https://doi.org/10.1016/j.erss.2020.101828
2. Xinle, L., Cai, S., Sun, B., Yang, C., Zhang, J., Liu, Y.: Chemically robust covalent organic frameworks: progress and perspective. Matter **3**(5), 1507–1540 (2020). https://doi.org/10.1038/s41467-018-05462-4
3. Nandi, S.S., Adimule, V., Kadapure, S.A., Kerur, S.S.: Rare earth-based nanocomposite materials for prominent performance supercapacitor: a review. Appl. Mech. Mater. **908**, 3–18 (2022). https://doi.org/10.4028/p-rff302
4. Zhou, M., Shang, L., Li, B., Huang, L., Dong, S.: Highly ordered mesoporous carbons as electrode material for the construction of electrochemical dehydrogenase-and oxidase-based biosensors. Biosen. Bioelectron. **24**, 442–447 (2008). https://doi.org/10.1016/j.bios.2008.04.025
5. Abuzeid, H.R., El-Mahdy, A.F., Kuo, S.W.: Covalent organic frameworks: design principles, synthetic strategies, and diverse applications. Giant **6**, 100054 (2021). https://doi.org/10.1016/J.GIANT.2021.100054
6. Shen, W., Li, G.: Recent progress in liquid crystal-based smart windows: materials, structures, and design. Laser Photonics Rev. **17**, 2200207 (2022). https://doi.org/10.1002/lpor.202200207
7. Xuanjie, W., Shankar, N.: Thermochromic materials for smart windows: a state-of-art review. Front. Energy Res. **9**, 800382 (2021). https://doi.org/10.3389/fenrg.2021.800382
8. Yang, Q., Luo, M., Liu, K., Cao, H., Yan, H.: Covalent organic frameworks for photocatalytic applications. Appl. Catal. B: Environ. **276**, 119174 (2021). https://doi.org/10.3389/fenrg.2021.800382
9. de Laurens, T.H., Julien, M.N.V., Dirk, J.B., Cees, W.M.B., Albertus, P.H.J.S.: Humidity-responsive liquid crystalline polymer actuators with an asymmetry in the molecular trigger that bend, fold, and curl. J. Am. Chem. Soc. **136**(30), 10585–10588 (2014). https://doi.org/10.1021/ja505475x
10. Zhang, R.C., Zhang, Z.B., Han, J.C., Yang, L., Li, J.J., Song, Z.C., Wang, T.Y., Zhu, J.Q.: Advanced liquid crystal-based switchable optical devices for light protection applications: principles and strategies. Light-Sci. Appl. **12**, 30 (2023). https://doi.org/10.1038/s41377-022-01032-y
11. Belmonte, A., Bus, T., Broer, D.J., Schenning, A.: Patterned full-color reflective coatings based on photonic cholesteric liquid-crystalline particles. ACS Appl. Mater. Interfaces **11**, 14376–14382 (2019). https://doi.org/10.1021/acsami.9b02680
12. Zheng, X.X., Jia, Y.N., Chen, A.H.: Azobenzene-containing liquid crystalline composites for robust ultraviolet detectors based on conversion of illuminance-mechanical stress-electric signals. Nat. Commun. **12**, 8 (2021). https://doi.org/10.1038/s41467-021-25178-2
13. Xie, X., Huang, P., Xu, M., Ding, Y., Qiu, L., Lu, H.: Tri-state switching of a high-order parameter, double-layered guest-host liquid-crystal shutter, doped with the mesogenic molecule 4HPB. Liq. Cryst. 1–6 (2021). https://doi.org/10.1080/02678292.2021.1885759
14. Davydova, T.E., Turchenko, E.A., Usachev, A.M.: Social and technological aspects of building materials production in the concept of a smart city. IOP Conf. Ser.: Mater. Sci. Eng. **775**, 012110 (2020). https://doi.org/10.1088/1757-899X/775/1/012110
15. Vinayak, A., Parashuram, L., Santosh, N., Rangappa, K.: Crystal structures, morphology, optical, supercapacitor properties of reduced graphene oxide doped cesium oxide@cobalt oxide bimetallic oxide nanostructures for visible light photocatalysis of Alizarin Red S dye. J. Energy Storage **104**, 114429 (2024). https://doi.org/10.1016/j.est.2024.114429
16. Ube, T., Yoshida, M., Kurihara, S., Ikeda, T.: Sunlight-driven smart windows with a wide temperature range of optical switching based on chiral nematic liquid crystals. ACS Appl. Mater. Interfaces **16**(22), 28638–28644 (2024). https://doi.org/10.1021/acsami.4c04370
17. Hari, K.B., Quan, L.: Liquid crystals: versatile self-organized smart soft materials. Chem. Rev. **122**, 4887–4926 (2022). https://doi.org/10.1021/acs.chemrev.1c00761

18. Vinayak, A., Rangappa, K., Santosh, N., Parashuram, L.: Fullerene C60 functionalized Sm_2O_3@Eu_2O_3 bimetallic oxides core shell nanocomposites for zeta potential and energy storage applications. J. Electroanal. Chem. **973**, 118663 (2024). https://doi.org/10.1016/j.jelechem.2024.118663
19. Vinayak, A., Kalpana, S., Vandna, S., Pankaj, K., Rangappa, K., Rajeev, J., Santosh, N.: Influence of core fluorination on the phase properties of fan-like azobenzene based supramolecules, their cis-trans photoisomerization and photoluminescence dynamics. Mater. Chem. Phys. **329**, 130140 (2024). https://doi.org/10.1016/j.matchemphys.2024.130140
20. Hari, K.B., Quan, L.: Light-driven liquid crystalline materials: from photo-induced phase transitions and property modulations to applications. Chem. Rev. **116**, 15089–15166 (2016). https://doi.org/10.1021/acs.chemrev.6b00415
21. Zheng-Wei, L., Victor, Y.Z., Wei, L.: Adjustable active–passive dual-frequency liquid crystal smart window based on thermally responsive chiral molecules. J. Mol. Liq. **415**, 126383 (2024). https://doi.org/10.1016/j.molliq.2024.126383
22. Saboor, S., Kirankumar, G., Venkata, R., Shantiswaroop, M., Kishor, S.K.: Thermal and cost assessment of various polymer-dispersed liquid crystal film smart windows for energy efficient buildings. Constr. Build. Mater. **263**, 120155 (2020). https://doi.org/10.1016/j.conbuildmat.2020.120155
23. Vinayak, A., Revaiah, R.G., Santosh, S.N., Adarsha, H.J.: Synthesis, characterization of Cr doped TeO_2 nanostructures and its application as EGFET pH sensor. Electroanalysis **33**, 579 (2021). https://doi.org/10.1002/elan.202060329
24. Adimule, V., Suryavanshi, A., Yallur, B.C., Santosh, S.N.: A facile synthesis of poly(3-octyl thiophene):Ni0.4$Sr_{0.6}TiO_3$ hybrid nanocomposites for solar cell applications. Macromol. Symp. **392**, 2000001 (2020). https://doi.org/10.1002/masy.202000001
25. Oh, S.W., Nam, S.M., Kim, S.H., Yoon, T.H., Kim, W.S.: Self-regulation of infrared using a liquid crystal mixture doped with push–pull azobenzene for energy-saving smart windows. ACS Appl. Mater. Interfaces **13**, 5028–5033 (2021). https://doi.org/10.1021/acsami.0c21033
26. Meng, W., Gao, Y., Hu, X., Tan, L., Li, L., Zhou, G., Yang, H., Wang, J., Jiang, L.: Photothermal dual passively driven liquid crystal smart window. ACS Appl. Mater. Interfaces **14**, 28301–28309 (2022). https://doi.org/10.1021/acsami.2c06428
27. Vinayak, A., Santosh, N., Kalpana, S., Rangappa, K.: Microstructure, zeta potential and enhanced photocatalytic dye degradation performances of mesoporous flower like Nb_2O_5@CeO_2 bimetallic oxide nanomaterials decorated with reduced graphene oxide. Inorg. Chem. Commun. **166**, 112592 (2024). https://doi.org/10.1016/j.inoche.2023.112592
28. Chen, C., Brigeman, A., Ho, T., Khoo, I.: Normally transparent smart window based on electrically induced instability in dielectrically negative cholesteric liquid crystal. Opt. Mater. Express **8**, 691–697 (2018). https://doi.org/10.1364/OME.8.000691
29. Guo, S.M., Liang, X., Zhang, C.H., Chen, M., Shen, C., Zhang, L.Y., Yuan, X., He, B.F., Yang, H.: Preparation of a thermally light-transmittance-controllable film from a coexistent system of polymer-dispersed and polymer-stabilized liquid crystals. ACS Appl. Mater. Interfaces **9**, 2942–2947 (2017). https://doi.org/10.1021/acsami.6b14967
30. Xianyu, M., Siyang, L., Shuo, C., Xuanzhe, S., Dekang, G., Jinbao, G.: Recent advances in smart windows based on photo-responsive liquid crystals featuring phase transition. Chem. Plus. Chem. **89**, 202300700 (2024). https://doi.org/10.1002/cplu.202300700
31. Seung-Won, O., Jong-Min, B., Sang-Hyeok, K., Tae-Hoon, Y.: Optical and electrical switching of cholesteric liquid crystals containing azo dye. RSC Adv. **7**, 19497 (2017). https://doi.org/10.1039/C7RA02294D
32. Javed, R.T., Hung-Yuan, L., Shin-Tson, W.: Photo- and electrical-responsive liquid crystal smart dimmer for augmented reality displays. Optical Express **27**(13), 18169 (2019). https://doi.org/10.1364/OE.27.018169
33. Adimule, V., Keri, R., Rajendrachari, S., Kendrekar, P., Kulkarni, C.V.: Highly photoluminescence and wide band gap insulating metal hybrid nanoparticles array of samarium-doped SrO: CoO: synthesis, characterizations and sensor characteristics. J. Mater. Sci.: Mater. Electron. **34**(5), 442 (2023). https://doi.org/10.1007/s10854-023-09899-2

34. Adimule, V., Yallur, B.C., Keri, R., Bathula, C., Batakurki, S.: Microstructure, photoluminescence and electrical properties of SmxGd(1–x): SrO hybrid nanomaterials synthesized via facile coprecipitation method. Electron. Mater. Lett. **19**(3), 278–297 (2023). https://doi.org/10.1007/s13391-022-00394-0
35. Zhang, H., Liu, J., Zhao, X., Gao, J., Ma, C., Zhao, Y., Yao, R., Miao, Z., Shen, W.: Electrically induced coloration of polymer-stabilized cholesteric liquid crystal films with broadband reflection capability for smart windows. Dye. Pigment. **203**, 110316 (2022). https://doi.org/10.1016/j.dyepig.2022.110316
36. Adimule, V., Sharma, V., Kumar, P., Keri, R., Sharma, K.: Perylene-3, 4, 9, 10-tetracarboxylic dianhydride dye intercalated liquid crystal molecules: synthesis, spectroscopic characterizations, and photoluminescence dynamics. J. Mol. Liq. 125990 (2024). https://doi.org/10.1016/j.molliq.2023.125990
37. Sharma, V., Kumar, P., Sharma, A., Raina, K.K.: Droplet configuration control with orange azo dichroic dye in polymer dispersed liquid crystal for advanced electro-optic characteristics. J. Mol. Liq. **233**, 122–130 (2017). https://doi.org/10.1016/j.molliq.2017.02.093
38. Yuanyuan, Z., Han, L., Mingliang, J., Guofu, Z.: Electrohydrodynamic instabilities for smart window applications. Liq. Cryst. **47**(4), 1–7 (2019). https://doi.org/10.1080/02678292.2019.1667313
39. Zhou, Y., Bukusoglu, E., Martínez-González, J.A., Rahimi, M., Roberts, T.F., Zhang, R., Wang, X., Abbott, N.L., De Pablo, J.J.: Structural transitions in cholesteric liquid crystal droplets. ACS Nano **10**, 6484–6490 (2016). https://doi.org/10.1021/acsnano.6b02174
40. Hu, X., de Haan, L.T., Khandelwal, H., Schenning, A.P., Nian, L., Zhou, G.: Cell thickness dependence of electrically tunable infrared reflectors based on polymer stabilized cholesteric liquid crystals. Sci. China Mater. **61**, 745–751 (2018). https://doi.org/10.1007/s40843-017-9262-3
41. Wang, Q., Zhang, X., Zhou, D., Chen, Y., Qu, C., Lu, C., Liu, Y.: Cholesteric liquid crystal doped with spiropyran and zinc ions for multi-state anti-counterfeiting. J. Mater. Chem. C **12**(20), 7225–7232 (2024). https://doi.org/10.1039/D4TC00522A
42. Chia-Cheng, L., Chun-Wei, S., Mei-Yung, C.: Mitigation of image blurring for performance enhancement in transparent displays based on polymer-dispersed liquid crystal. Displays **56**, 30–37 (2019). https://doi.org/10.1016/j.displa.2019.01.005
43. Xianyu, H., Faris, S., Crawford, G.P.: In-plane switching of cholesteric liquid crystals for visible and near-infrared applications. Appl. Opt. **43**, 5006–5015 (2004). https://doi.org/10.1364/AO.43.005006
44. Nemati, H., Liu, S., Zola, R.S., Tondiglia, V.P., Lee, K.M., White, T., Bunning, T., Yang, D.K.: Mechanism of electrically induced photonic band gap broadening in polymer stabilized cholesteric liquid crystals with negative dielectric anisotropies. Soft Matter **11**, 1208–1213 (2015). https://doi.org/10.1039/C4SM02174H
45. Adimule, V., Bathula, C., Sharma, K., Keri, R.: Study of structural, optical, dielectric properties poly-3-butyl thiophene reinforced with Co_3O_4 nanoparticles. Nano-Struct. Nano-Objects **38**, 101150 (2024). https://doi.org/10.1016/j.nanoso.2023.101150
46. Boychuk, A., Shibaev, V., Cigl, M., Hamplová, V., Novotná, V., Bobrovsky, A.: Large thermally irreversible photoinduced shift of selective light reflection in hydrazone-containing cholesteric polymer systems. Chemphyschem **24**, e202300011 (2023). https://doi.org/10.1002/cphc.202300011
47. Palto, S.P., Rybakov, D.O., Umanskii, B.A., Shtykov, N.M.: Spiral pitch control in cholesteric liquid crystal layers with hybrid boundary conditions. Crystals **13**, 10 (2023). https://doi.org/10.3390/cryst13101482
48. Shumeng, G., Xiao, L., Huimin, Z., Wenbo, S., Chunxin, L., Xiao, W., Cuihong, Z., Lanying, Z., Huai, Y.: An electrically light transmittance-controllable film with a low-driving voltage from a coexistent system of polymer-dispersed and polymer-stabilised cholesteric liquid crystals. Liq. Cryst. (2018). https://doi.org/10.1080/02678292.2018.1501820
49. Zou, X., Ji, H., Zhao, Y., Lu, M., Tao, J., Tang, P., Liu, B., Yu, X., Mao, Y.: Research progress of photo-/electro-driven thermochromic smart windows. Nanomaterials **11**(12), 3335 (2021). https://doi.org/10.3390/nano11123335

50. Sun, J., Wang, H., Wang, L., Cao, H., Xie, H., Luo, X., Xiao, J., Ding, H., Yang, Z., Yang, H.: Preparation and thermo-optical characteristics of a smart polymer-stabilized liquid crystal thin film based on smectic A–chiral nematic phase transition. Smart Mater. Struct. **23**, 125038 (2014). https://doi.org/10.1088/0964-1726/23/12/125038
51. Shu-Meng, G., Xiao, L., Cui-Hong, Z., Mei, C., Chen, S., Lan-Ying, Z., Xiao, Y., Bao-Feng, H., Huai, Y.: Preparation of a thermally light-transmittance-controllable film from a coexistent system of polymer-dispersed and polymer-stabilized liquid crystals. ACS Appl. Mater. Interfaces **9**(3), 2942–2947 (2017). https://doi.org/10.1021/acsami.6b13366
52. Wenbo, S., Guoqiang, L.: Recent progress in liquid crystal-based smart windows: materials, structures, and design. Laser Photonics Rev. **17**, 2200207 (2022). https://doi.org/10.1002/lpor.202200207
53. Zemin, H., Ping, Y., Huimin, Z., Yuzhen, Z., Yanfang, Z., Zhun, G., Cheng, M., Haiquan, Z., Zongcheng, M., Wenbo, S.: Silicon nanostructure-doped polymer/nematic liquid crystal composites for low voltage-driven smart windows. Nanotechnology **33**, 085205 (2021). https://doi.org/10.1088/1361-6528/ac3e29
54. Zhibo, Z., Ruicong, Z., Liangge, X., Jiajun, L., Lei, Y., Yanan, Y., Andrey, B., Jiaqi, Z.: Visible and infrared optical modulation of PSLC smart films doped with ATO nanoparticles. Dalton Trans. **50**, 10033–10040 (2021). https://doi.org/10.1039/D1DT00976F
55. Huihui, W., Bingzhi, L., Ling, W., Xudong, C., Zhaolong, C., Yue, Q., Guang, C., Huanhuan, X., Yanfeng, Z., Zhongfan, L.: Graphene glass inducing multidomain orientations in cholesteric liquid crystal devices toward wide viewing angles. ACS Nano **12**(7), 6443–6451 (2018). https://doi.org/10.1021/acsnano.8b02905
56. Tzu-Hsun, K., Hsun-Hao, H., Jui-Juin, C., Lin-Ruei, L., Hui-Yu, C., Jiun-Tai, C.: Fabrication of polymer/cholesteric liquid crystal films and fibers using the nonsolvent and phase separation method. Langmuir **40**(27), 14166–14172 (2024). https://doi.org/10.1021/acs.langmuir.4c01759
57. Zhi-Chao, J., Qing, L., Yao-Yu, X., Yue, Z.: Liquid crystal elastomers for actuation: a perspective on structure-property-function relation. Prog. Polym. Sci. **153**, 101829 (2024). https://doi.org/10.1016/j.progpolymsci.2024.101829
58. Seung-Won, O., Sang-Hyeok, K., Jong-Min, B., Tae-Hoon, Y.: Optical and thermal switching of liquid crystals for self-shading windows. Adv. Sustainable Syst. (2018). https://doi.org/10.1002/adsu.201700164
59. Chang, G., Ai-Bo, J., Yu-Mo, Z., Sean Xiao-An, Z.: Emerging electrochromic materials and devices for future displays. Chem. Rev. **122**(18), 14679–14721 (2022). https://doi.org/10.1021/acs.chemrev.2c00247
60. Guan-Fu, S., Po-Chang, W., Victor Ya, Z., Wei, L.: Electrically active and thermally passive liquid-crystal device toward smart glass. Photon. Res. **9**, 2288–2295 (2021). https://doi.org/10.1364/PRJ.427104

Impact on the Dynamic Responses of Nanoparticles Embedded Cholesteric Liquid Crystal Composites

Rishi Kumar, Ramneek Kaur, and K. K. Raina

Abstract Scientists are increasingly investigating chiral liquid crystal phases as flexible frameworks for helical assembly structures with the dispersion of various nanomaterials. This chapter will cover current advances in liquid crystal, providing many examples and critical methodologies for coupling nanomaterials with liquid crystal chirality, opening up new options for novel material design in the display industry. This chapter will provide an overview of how the introduction of nanomaterials alters the characteristics of cholesteric liquid crystals (CLCs). It will explore how nanomaterial dispersion affects the dynamic responses of CLCs, with a focus on the link between structural features, dispersion methodologies, and the resulting dynamic behavior. It will also describe the effects on the electro-optical performance of liquid crystal devices made by the dispersion of a variety of organic and inorganic nanomaterials with different sizes, shapes, and compositions in liquid crystals. Furthermore, this chapter will review and delve into thermotropic liquid crystal phases. It will demonstrate their potential use in nanotechnology, photonics, and material science, ultimately leading to the development of new functional materials.

Keywords Cholesteric Liquid Crystals (CLCs) · Switching properties · Morphology · Electro-optical · Carbon Nanotubes (CNTs) · Liquid Crystal Composites

---

R. Kumar (✉)
Department of Physics, Guru Nanak College Budhlada, Mansa, Punjab, India
e-mail: rishikumar.phd@gmail.com

R. Kaur
P. G. Department of Physics, Mata Gujri College, Fatehgarh Sahib, India

K. K. Raina
M.S. Ramaiah University of Applied Sciences, Bangalore, India

P. Kumar et al. (eds.), *Cholesteric Liquid Crystals*, Engineering Materials,
https://doi.org/10.1007/978-981-96-3821-5_7

1 Introduction

Cholesteric liquid crystals (CLCs) are a novel class of liquid crystals (LCs) which are characterized by helical molecular arrangements. Their electro-optical properties make them suitable for various applications in display technologies to sensors [1–15]. Recent advances in nanotechnology [16–33] have opened new avenues for high-contrast display devices made from nanoparticles (NPs) dispersed CLC composites. This chapter explores the study of the effect of nanomaterial dispersion on the dynamic responses of CLCs. This also highlights the structural correlation between the NPs and CLC templates.

1.1 Background of Cholesteric Liquid Crystals (CLCs)

CLCs are the chiral phases of LCs known for their distinctive helical structures, in which the arrangement of molecules creates the helix (Fig. 1) that are layered in a helical twisted ordered system.

The helical pitch is defined as the distance over which the molecules complete a full twist. This determines the wavelength of reflected light, making them suitable in tunable color filters. CLCs are also preferred in a variety of applications, such as temperature-sensitive displays, smart windows, and sensors [22–25] due to the following key properties:

- The main property of CLCs is their temperature sensitivity. With the temperature change (Fig. 2), the pitch of the helix gets affected and either expands or contracts, thereby altering the wavelength of the reflected light.
- This key property makes them suitable for thermochromic LC devices. Furthermore, this optical property of CLCs can be tailored to reflect the incident light in different spectrum regions, such as visible to ultraviolet (UV), making them versatile in reflective display technology.
- Another property of CLCs is that they can maintain their optical textures after removing an external electric field. This makes them suitable for energy-efficient static displays in which the image does not require constant refreshing [22–25].

1.2 Why Are Nanomaterials Preferred for the Dispersion into Liquid Crystals?

- Nanomaterials are often preferred for dispersion [26–30] in CLCs because of their unique ability to improve the electro-optic performance (Fig. 3) of LC display devices. When dispersed nanomaterials such as carbon nanotubes, graphene, and quantum dots are uniformly dispersed within the LC matrix, it improves the

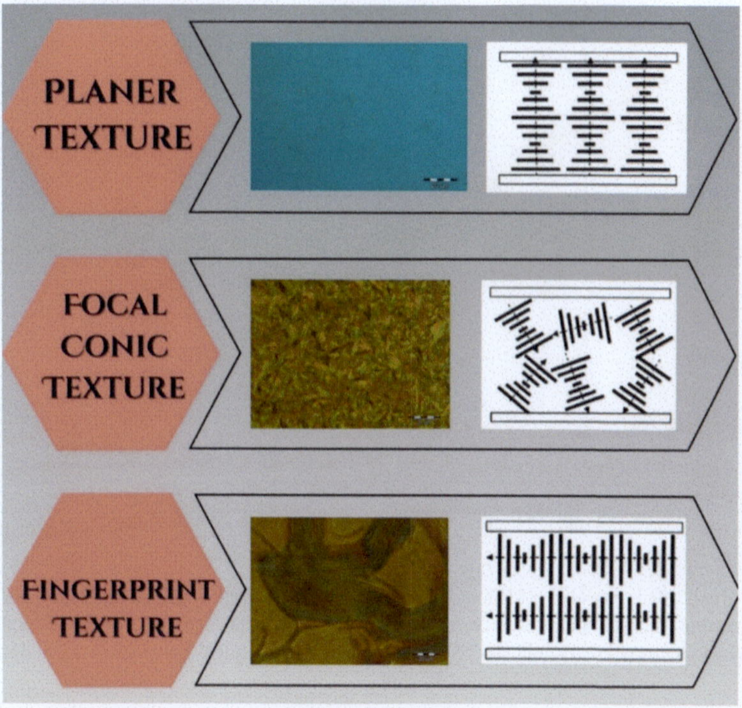

Fig. 1 The image displays three types of textures with corresponding microscopic images and diagrams. The first row shows "Planer Texture" with a blue microscopic image and a diagram of parallel lines. The second row features "Focal Conic Texture" with a fan-like microscopic texture and a diagram of intersecting lines. The third row presents "Fingerprint Texture" with an optical fingerprints curve lines in texture

electro-optical contrast of LC display devices by enhancing their ability to control the transmission of light.

- The high surface-area-to-volume ratio of NPs ensures stable and consistent interaction between LCs and nanomaterials, which plays a crucial role in creating high-quality displays.
- Nanomaterials such as graphene and carbon nanotubes (CNTs) can improve the electrical conductivity of LC composite materials used in the display device. This enhanced conductivity is especially important in high-performance displays where fast switching response and high contrast are required. By improving the electrical conductivity without increasing the power consumption, these carbon-based nanomaterials can make LCDs more energy efficient and faster response switching displays.
- Metallic NPs (Ag, Au ZnO, etc.) can also play a crucial role in improving the optical switching in LC composites. For example, CdSe quantum dots can be used to find the optimal tuning in the high-contrast displays. Therefore these

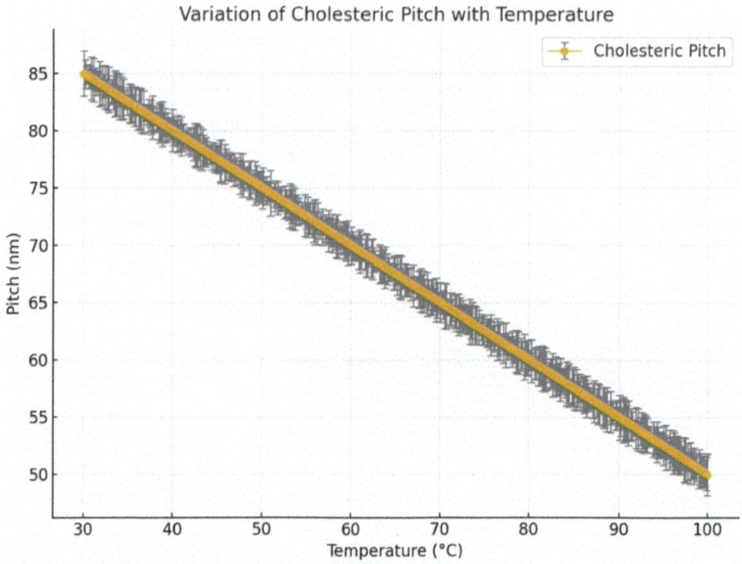

Fig. 2 Graph titled "Variation of Cholesteric Pitch with Temperature" showing a downward trend. The x-axis represents temperature in degrees Celsius, and the y-axis represents pitch in nanometers. A yellow line with error bars indicates that as temperature increases, the cholesteric pitch decreases

nanocomposites can be engineered to emit specific wavelengths of light so that they can be used in displays with broader photoluminescence (PL) contrast.

- Additionally, nanomaterials help reduce unwanted light scattering within the chiral LC matrix, which improves image clarity and reduces glare. Hence these composites can be used in antiglare coatings in displays.
- Furthermore, SWCNTs or MWCNTs can improve the mechanical properties of LC devices. The inclusion of these nanomaterials not only increases the flexibility of LC displays but also makes them more durable. This opens up their possibilities and uses for the development of flexible foldable displays for next-generation devices. The enhanced durability of these nano-composites ensures that these flexible displays can withstand mechanical stress without compromising the switching performance of displays.
- Finally, with the ability to tailor the properties of LCs with nanomaterial dispersion, the researchers have expanded their use in a wide range of switchable displays or smart privacy windows that switch from opaque to transparent state with very few external stimuli voltage.

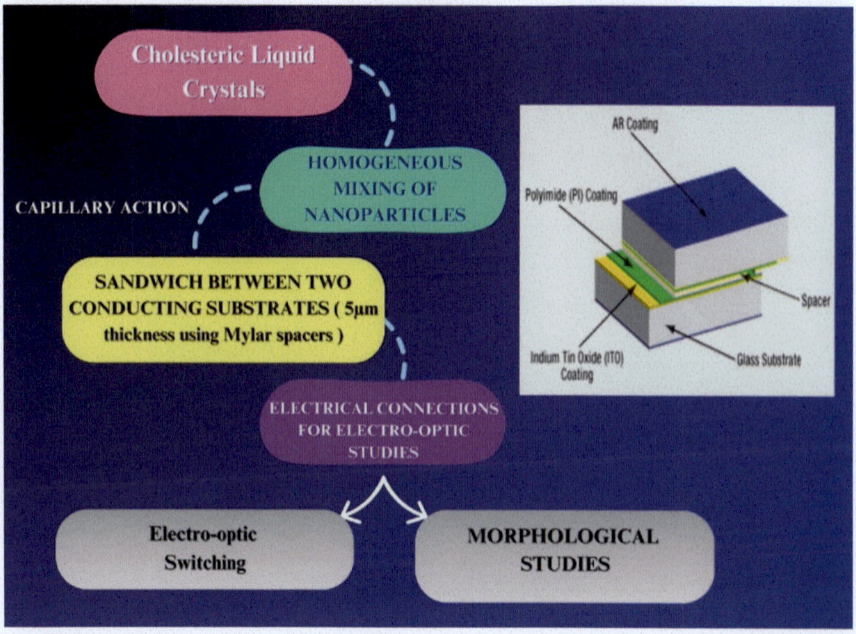

Fig. 3 Diagram illustrating the process of using cholesteric liquid crystals for electro-optic and morphological studies. The process involves homogeneous mixing of nanoparticles, which are sandwiched between two conducting substrates with 5 μm thickness using Mylar spacers. This setup is used for electrical connections in electro-optic studies

2 Mechanisms of Interaction

2.1 *Van Der Wall's Interaction Between Nanomaterials and CLCs*

Figure 4 shows that the Van der Waals interactions play an important role in the interactions in nanomaterials dispersed CLC composites [28–33] that can significantly affect the alignment and orientation of the LC molecules. These forces are particularly significant in nanoscale composite systems, where the relative surface-area-to-volume ratio is very high. Van der Waals forces can facilitate the alignment of the CLC molecular orientations near the surfaces of nanomaterials. This molecular alignment of CLC in the presence of nanomaterial helps in tuning the optical properties of display devices.

Also, the dispersion of nanomaterials in CLC composites can modify the local environment of CLC molecules through Van der Waals interactions, leading to alterations in the helicity of the LC. This can also contribute to the change in optical anisotropy and birefringence of CLC Composites.

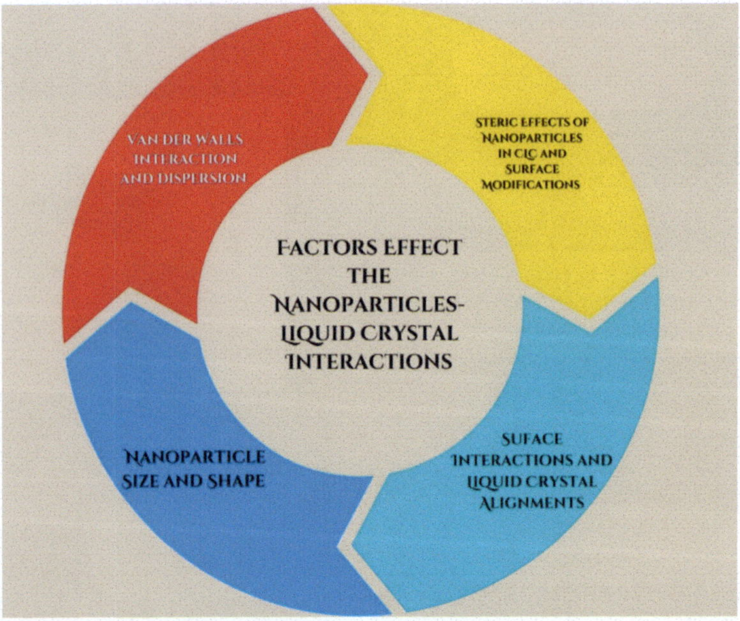

Fig. 4 A circular diagram illustrating factors affecting nanoparticles-liquid crystal interactions. The circle is divided into four colored segments: red for Van der Waals interaction and dispersion, yellow for steric effects of nanoparticles in liquid crystals and surface modifications, blue for nanoparticle size and shape, and light blue for surface interactions and liquid crystal alignments

- The Proper functionalization of nanomaterials can enhance their stability [29, 30] in CLC composites by altering their van der Waals interactions. This uniform distribution of NPs in the composite sample can improve the optical performance of devices. Hence this can lead to a significant improvement in tunable birefringence and optical switching capabilities of display devices.
- Van der Waals interactions between CLC molecules and NPs can influence the thermal stability of the composite by providing a network for the NPs, preventing phase separation in the desired mesophases.
- The mechanical properties of CLC composites can be enhanced through the incorporation of NPs [30], where van der Waals forces contribute to enhancing the mechanical properties such as stress transfer between the nanomaterials and the LC matrix.

2.2 *Steric Effects of Nanoparticles in CLC Composites*

Steric hindrance in the composites refers to the presence of a physical obstruction in the molecular orientation of matrix components due to the size and shape of the particles. In the context of CLCs, the incorporation of NPs introduces steric effects that can significantly influence the properties and behaviors of the LC matrix. NPs can also disrupt the helical structure of CLCs owing to steric effects, leading to hinder the molecular orientation of CLC molecules that can affect the changes in the pitch of the cholesteric phase. This directly impact on the wavelength of the reflected light and hence alter the optical characteristics of NPs doped CLC composites.

2.3 *Influence of Nanoparticle Size and Shape on the Interaction*

The size and shape of nanomaterials affects the degree of interaction with CLCs. Spherical NPs may induce isotropic effects, whereas elongated or rod-like nanomaterials can change the anisotropic behaviors, influencing the overall orientation and response time of chiral nematic LCs. The size of the NPs can also alter the helical pitch of the cholesteric phase. Smaller NPs may induce less disruption in the molecular arrangement of CLC molecules with in the composite that allow for finer control of the pitch, whereas larger NPs can cause more substantial alterations in the helical structure that directly reflect the optical properties [28–33].

Hence, the careful selection of NP size and shape is crucial for fine-tuning the optical properties of CLC composites. This precise manipulation allows the development of novel composite materials with tailored functionalities, enhancing their performance in various applications such as displays, sensors, and photonic devices [8–10].

3 Methods of Dispersion of Nanomaterials in CLCs

3.1 *Ultrasonication*

Ultra sonication is a most powerful and preferable technique widely used for fine dispersion of NPs into CLCs. The efficient mixing and dispersion were ensured by employing high-frequency sound waves to generate high-energy waves in a CLC medium. At low pressure, cavitation bubbles are formed and quickly burst during the ensuing high-pressure phase, producing shock waves that have the ability to break the aggregation of NPs. Consequently, they are homogeneously and effectively dispersed throughout the CLC matrix. Researchers may vigorously control the size and distribution of NPs in the CLC matrix and modify the optical characteristics, including light scattering, absorption, and color responses, by tuning the parameters, such as power, duration, and temperature.

3.2 Stirring and Shear Mixing

The fundamental techniques used for the uniform dispersion of NPs into CLCs are stirring and mixing. Stirring possesses the mechanical forces to the LC and NP mixture using a stirring rod or mechanical mixer. The controlled stirring speed and time also support the homogeneous dispersion of NPs to the CLC composites, and finally, change in the optical and mechanical properties of the resulting composite material based on the dispersion.

4 Impact on Dynamic Responses

Figures 5 and 6 describe the impact of NP dispersion on the dynamic responses such as viscosity, switching response time of LC molecules, optical anisotropy, thermal stability, and phase transition properties of the CLC composites [10].

4.1 Viscosity Changes

The dispersion of nanomaterials in CLCs often results in changes in viscosity, which can affect the response time of the LC and hence it plays a very crucial role in determining their device performance characteristics and overall functionality in various applications [11–13].

Sometime, the incorporation of higher concentrations of NPs in CLC composites is problematic for their use in LC display technologies. Because, the dispersion of NPs increases the viscosity of CLC composites which may lead to slower switching times between different optical switching states of CLCs. This critical response can particularly become problematic for faster switching displays for gaming applications that need fast transitions. The increased viscosity may cause noticeable lag or blur motion. Hence these CLC composites are potentially limiting the suitability of the device for fast-paced visual device applications. Thus, a controlled value of viscosity can offer different advantages of NPs dispersed CLCs in display technology. For instance, in tunable optical filters, a more gradual transition between states is often required and demandable in defense applications. These tunable optical filters made from CLC composites are used in telecommunications and spectroscopy due to their slower switching response. The increased viscosity of these CLC composites allows them for finer control over the spectral characteristics of the tunable filters. So the engineers and researchers in the field of display technology must carefully balance these interrelated factors to optimize LC devices performances for the specific applications.

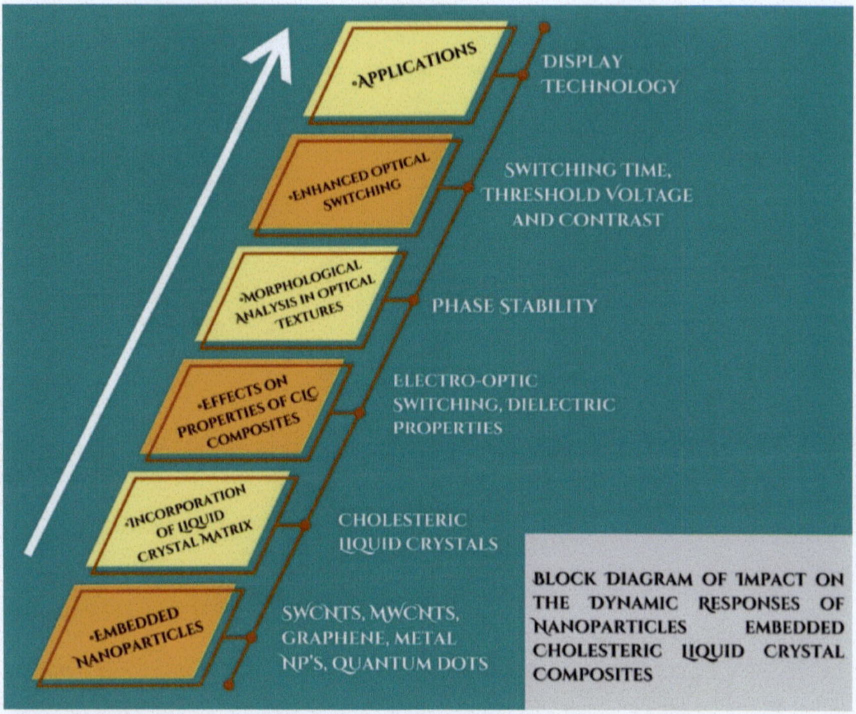

Fig. 5 Diagram illustrating the impact of nanoparticles embedded in cholesteric liquid crystal composites. The diagram is structured as an ascending arrow with six blocks. From bottom to top: "Embedded Nanoparticles" includes SWCNTs, MWCNTs, graphene, metal nanoparticles, and quantum dots; "Incorporation of Liquid Crystal Matrix" relates to cholesteric liquid crystals; "Effects on Properties of CLC Composites" covers electro-optic switching and dielectric properties; "Morphological Analysis in Optical Textures" addresses phase stability; "Enhanced Electro-Optic Switching" involves switching time, threshold voltage, and contrast; "Applications" focuses on display technology

4.2 Optical Switching Behavior

The optical switching behaviour of NPs [31–33] doped LCs can be significantly influenced with the incorporations of various nanomaterials. Nanomaterials, such as carbon nanotubes (SWCNTs, MWCNTs), graphene, metal NPs (Ag, Au, ZnO), and quantum dots (CdS, CdSe), can alter the physical and chemical properties of the LC matrix. When these NPs are embedded into a CLC composite system, their helical pitch gets modified, this affects the refractive index, and dielectric anisotropy of these composites. The interaction between nanomaterials and chiral-doped LCs occurs at the molecular level, leading to complex phenomena that affect the overall optical switching behaviour [29–31]. As a result, the electro-optical properties, including the response time, threshold voltage, and contrast ratio, can be enhanced or tailored for the specific display applications in the industry.

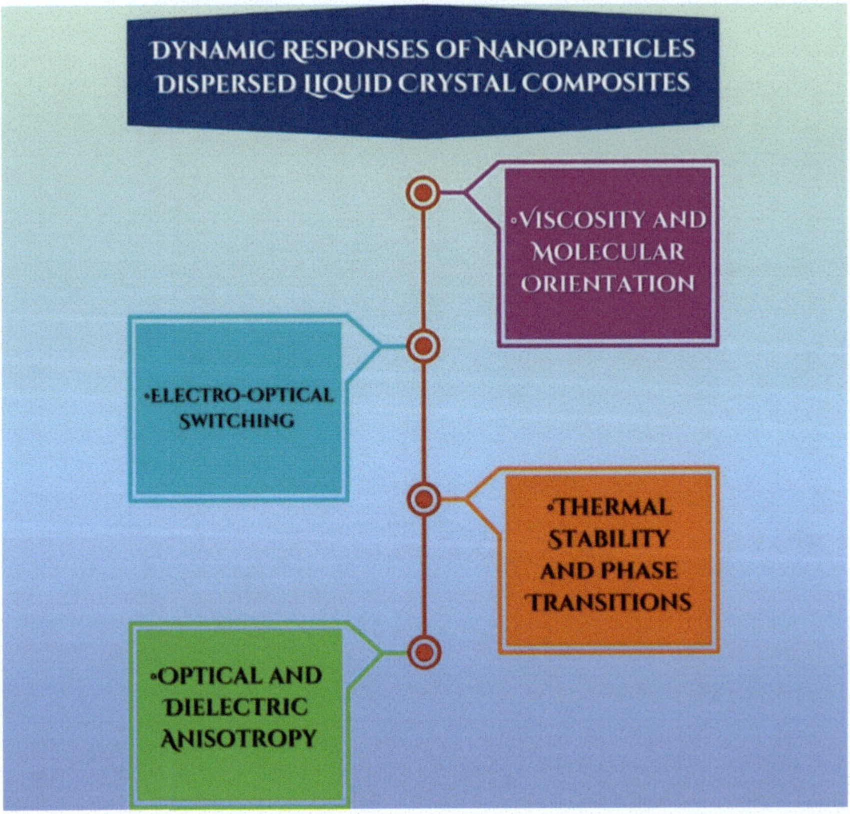

Fig. 6 Diagram titled "Dynamic Responses of Nanoparticles Dispersed Liquid Crystal Composites" with four key factors: "Viscosity and Molecular Orientation", "Thermal Stability and Phase Transitions", "Electro-Optical Switching", and "Optical and Dielectric Anisotropy"

For instance, the addition of plasmonic (silver) NPs into the CLC matrix can induce localized surface plasmon resonance, which can enhance light absorption and scattering within CLCs in a helical arrangement. Similarly, carbon-based nanomaterials such as CNTs or graphene can improve the electrical conductivity of the CLC mixture, potentially reducing the threshold or driving voltage required for molecular switching. By carefully selecting and optimizing the type, concentration, and distribution of nanomaterials within CLC composites, researchers can develop advanced novel functional materials with improved optical switching characteristics for application in displays, photonics, and optoelectronic devices [13].

4.3 Optical Anisotropy

The incorporation of NPs into CLCs can significantly alter their optical properties because of changing their refractive index with the unique interactions between the NPs and LC molecules. As NPs are dispersed within the CLC matrix, they can disrupt the helical structure of the LCs, leading to changes in the way of propagation of light through the material and hence therefore alter the refractive index of the composite material. The extent of this disruption depends on various factors, including the size, shape, and concentration of the NPs within the CLC matrix, as well as their chemical composition with the LC host.

The altered refractive index resulting from NP inclusions has profound implications for light modulation in CLC-based devices [18, 19]. By carefully selecting the type and concentration of NPs, researchers can fine-tune the optical anisotropy of CLCs to achieve the desired specific wavelength. This ability to manipulate light propagation through NP-doped CLCs opens new possibilities for the sensors and display technologies in the industry.

4.4 Thermal Stability and Phase Transitions

Nanomaterials can also influence the thermal stability and phase-transition temperatures of CLC composites [21]. The extent of these shifts in transition temperature depends on various factors, including the size, shape, and surface properties of the NPs, as well as their concentration within the CLC matrix.

These temperature range shifts have profound implications for applying CLC-NP composites in the operating range of thermal sensors, potentially expanding or limiting their usability under different environmental conditions. Here the precise control of the phase transition temperatures through NP doping in CLCs could enable them for the development of more sensitive or tailored responsive composites. Furthermore, understanding and manipulating the shifting of phase transition temperatures can provide valuable insights into the fundamental physics of CLC-NP interactions. This can contribute to the broader field of soft matter science and nanotechnology.

5 Recent Advances in the Field of Nanoparticles Dispersed CLC Composites

Figure 7 provides a detailed analysis of the properties that affect the CLC with the dispersion of various NPs. Pathinti et al. [22] examined the optical and dielectric properties of CLC containing green-synthesized TiO_2 NPs. The CLC mixture was synthesized by embedding a 27 wt.% R-811 chiral dopant with an E7 nematic LC. At room temperature, polarizing optical microscopy (POM) observations revealed red region reflection, with enhanced alignment as the dispersion concentration increased. Both pure and TiO_2 NP-dispersed CLC samples underwent various analyses, including UV–visible absorbance, FE-SEM, EDAX, XRD, TEM, photoluminescence (PL), and FTIR. In comparison to pure CLC, the NP-dispersed CLC samples demonstrated heightened PL intensity without altering the emission peak position and reduced UV absorbance as NPs concentration increased. Moreover, the TiO_2 NP-dispersed CLC exhibited a smaller optical band gap than the pure CLC. FTIR analysis highlighted the effect of NP dispersion on the CLC molecular dynamics. Dielectric spectroscopy showed a reduction in the relaxation frequency of the TiO_2-dispersed CLC system, suggesting enhanced optical and dielectric properties.

In a 2023 study, Pathinti and Ramadevi Suguru [23] developed an electrically adjustable smart window utilizing CLC infused with green-synthesized zinc oxide (ZnO) NPs. The device exhibits three states: semi-transparent (planar), opaque (focal conic), and fully transparent (homeotropic), controlled by varying the voltage. It also manages the ambient temperature through sunlight regulation. The addition of ZnO NPs to the chiral nematic LC mixture affects light transmission, reducing it in the focal conic state and enhancing it in the homeotropic state as the ZnO NP concentration increases. Specifically, transmittance dropped from approximately 87 to 4% in the focal conic state (2.4 V/μm) and rose from 4% to about 90% in the homeotropic state (7 V/μm), with a response time under 1 ms. The window demonstrated robust stability, maintaining a consistent spectral transmittance after 20 switching cycles. This technology shows promise for privacy and smart window applications, potentially reducing artificial lighting requirements. The researchers tested indoor temperature control using both pure CLC and ZnO-doped CLC devices with a solar simulator. This technology is expected to find applications in vehicles and buildings for temperature regulation and energy conservation.

Akman and Bukusoglu et al. 2023 [24] investigated the self-organization of colloids at fluid interfaces that enable elastic interactions. While earlier research concentrated on micron-sized or molecular entities at aqueous LC interfaces, this study explored the self-assembly of NPs of intermediate size. The research focused on surface-modified silica NPs (ranging from 50 to 500 nm) adsorbed at LC-water interfaces, and their positions were analyzed by electron microscopy after polymerization. The findings revealed that electric double-layer forces and elastic forces from the LC strain are the primary drivers of the NP assembly. These forces can be manipulated to guide self-assembly based on the sub-interface symmetry of the confined CLCs. At high ionic concentrations, NPs were strongly concentrated at defects, while intermediate ionic strengths resulted in partial enrichment into cholesteric fingerprint patterns with binary interaction strength calculations. This study underscores the importance of ion partitioning at LC-aqueous interfaces in assembly formation, suggesting potential applications in sensors, microelectronics, and photonics.

In 2023, Luo et al. [25] investigated the crucial role of responsive energy transfer (ET) control in the widespread application of lanthanide-doped nanomaterials. This study introduces the mechanisms of ET and the photonic effect of chiral nematic

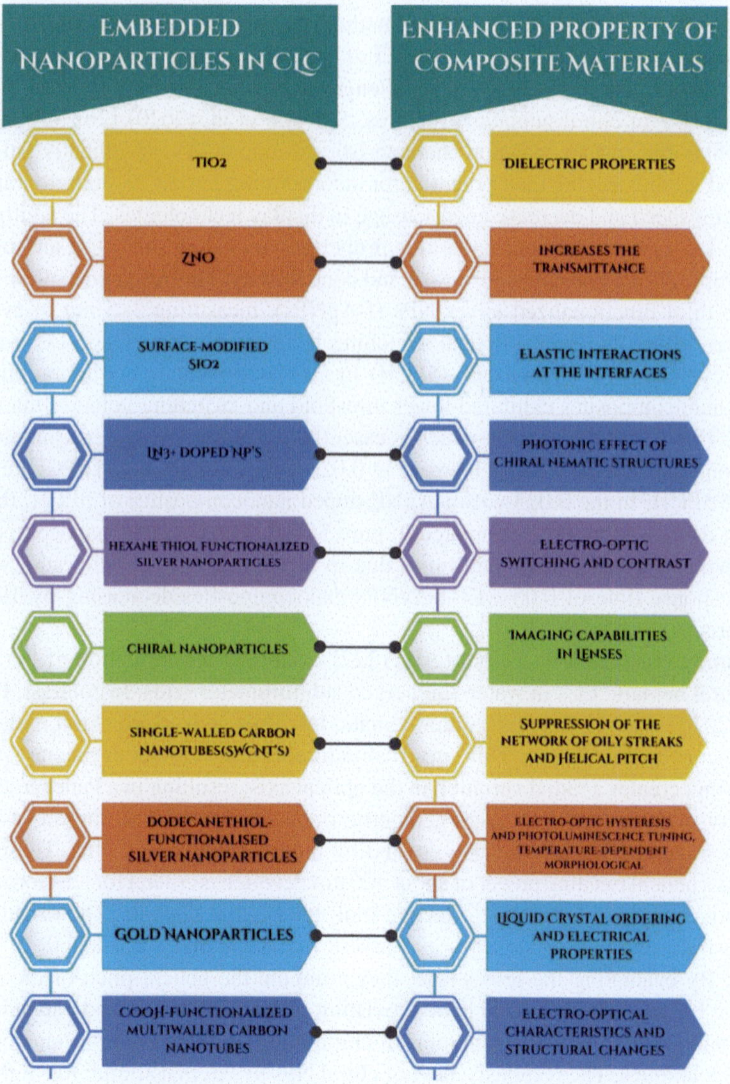

Fig. 7 The image is a flowchart comparing embedded nanoparticles in cholesteric liquid crystals (CLC) with their enhanced properties in composite materials. On the left, it lists NPs: TiO_2, ZnO, surface-modified SiO_2, Ln3 + doped NPs, hexane thiol functionalized silver nanoparticles, chiral nanoparticles, single-walled carbon nanotubes (SWCNTs), dodecanethiol-functionalized silver nanoparticles, gold nanoparticles, and COOH-functionalized multiwalled carbon nanotubes. On the right, it describes their enhanced properties: dielectric properties, increased transmittance, elastic interactions at interfaces, photonic effects of chiral nematic structures, electro-optic switching and contrast, imaging capabilities in lenses, suppression of oily streaks and helical pitch, electro-optic synthesis and photoluminescence tuning, LC ordering and electrical properties, and electro-optical characteristics and structural changes. Each NP is connected to its corresponding property with color-coded arrows

structures on ET. It also highlights the methods for regulating chiral nematic structures and the resulting alterations in the ET of lanthanide-doped nanomaterials. Additionally, this research discusses the challenges and promising opportunities associated with ET in chiral nematic structures. Shivaraj et al. (2020) [26] investigated the dipolar and surface plasmon characteristics of metal NPs, specifically gold and silver NPs, emphasizing their potential for incorporation into LC systems to improve switching speed and decrease energy usage in display technologies. The addition of NPs to LC systems influences various properties, including dielectric and optical anisotropy, elastic constants, viscosity, and conductivity. This research explores how hexane thiol-functionalized silver NPs (f-AgNPs), measuring 3–5 nm in average diameter, affect the electro-optical switching behavior of the nematic LC 4-*trans*-pentyl-cyclohexyl cyanobenzene (5PCH) in a twisted nematic configuration. The 5PCH nanocomposites exhibited lower threshold and switching voltages than pure 5PCH. The V-T graph's slope, which is essential for multiplexing performance and high contrast, showed a 230% increase in 0.025 wt% f-AgNP-doped 5PCH relative to pure 5PCH. In the 0.025 wt% f-AgNP-doped nanocomposites of 5PCH, the ion current decreased by 60% compared to pure 5PCH. The f-AgNPs nanocomposites demonstrated quicker switching times due to a reduced visco-elastic ratio, with a total response time of 0.05 wt% f-AgNPs nanocomposites decreasing by 19% in comparison to pure 5PCH.

A study conducted by Perera (2021) [27] examined the formation of flat films by achiral nematic LCs in water-submerged submillimeter grids. In contrast, Popov et al. (2017) showed that underwater nematic LC films with added chiral molecules form biconvex lens shapes. This curved structure, combined with degenerate planar anchoring, creates a radial variation in the optical axis, resulting in a Pancharatnam–Berry-type phase lens that alters geometric optical imaging. The current research explores nematic LC microlenses created by incorporating chiral NPs. These NPs possess a helical twisting power of about 400 μm^{-1}, which is crucial for lens formation and surpasses that of the most effective molecular chiral dopants. The researchers demonstrate imaging capabilities and assess the shape and focal length of these lenses. By evaluating the lens shape, they compute the helical pitch of the chiral nematic LC and determine the helical twisting power of chiral ligand-capped NPs. This method requires only nanograms of chiral NPs, significantly less than conventional techniques. The responsiveness of chiral NPs to external stimuli like light and magnetic fields indicates potential uses in developing adjustable optical properties for microlens arrays.

In 2018, Samoilov et al. [28] examined optical transmission and selective reflection data from single-walled carbon nanotube (CNT) suspensions in the photoactive nematic material ZhK-440, which included the mesogenic chiral dopant M5. At low CNT concentrations ($C \approx 0.01$–0.05%), CNTs tended to accumulate at oily streaks (cholesteric topological defects) and hinder the formation of oily streak networks. They found that the optical density D had a non-linear relationship with CNT concentration, reaching a minimum at $C \approx 0.05$–0.08%. This behavior was linked to a structural transition from loosely arranged, highly anisotropic aggregates (aligned with the anchoring direction on rubbed polyvinyl alcohol) to more densely packed aggregates. The location of this minimum and the selective reflection maximum (helical pitch) were responsive to partially reversible UV-induced *trans–cis-trans* isomerization effects. The addition of CNTs had minimal impact on the UV-controlled variation in helical pitch. Middha et al. (2016) [30] studied the influence of chirality on molecular organization in a nematic LC matrix doped with dodecanethiol-capped silver NPs (Ag NPs). The temperature-dependent morphological behavior was examined using polarized optical microscopy. The thermo-optic and electro-optic switching behavior of the liquid crystalline helix was verified using an active chiral dopant. The induction of chirality in the Ag NP-doped nematic matrix under planar boundary conditions led to enhanced dielectric behavior over the frequency range of 50 Hz to 1 MHz. A reduction in the Fredericks transition threshold and a nearly tenfold increase in the dielectric constant with temperature were observed. Additionally, a significant increase in photoluminescence and irreversible memory was noted in the chiral medium. These notable optical and electro-optical properties suggest that silver NP-doped chiral LCs have significant potential for use in optical sensors and electro-optic LC devices with memory.

In 2014, Infusino et al. [31] examined the impact of dispersing gold NPs (Au NPs) in a chiral nematic LC. Their study revealed changes in the LC host's order, as demonstrated by polarized optical microscopy and X-ray diffraction measurements. Furthermore, a comparative study using dielectric and voltametric spectroscopies on pure LC and Au NP-doped LC indicated that the presence of Au NPs not only altered the LC's order but also affected its electrical properties. Notably, ion conductivity was significantly decreased, and electrophoresis phenomena were observed when the applied field exceeded a certain threshold value.

Middha et al. (2016) [32, 33] investigated the effects of incorporating COOH-functionalized multi-walled carbon nanotubes into chiral nematic LCs within a planar-aligned cell. They examined the electro-optical switching characteristics (Fig. 8) and structural changes of the doped samples under varying dopant concentrations and surface anchoring forces. The resulting material demonstrated enhanced conductivity, refractive index, and dielectric properties. Anisotropic alterations due to CNT doping were verified through polarized fluorescence spectroscopy. The study revealed significant changes in the Fredericks transition's onset voltage and the development of electro-optic memory, which was stabilized by the nanotubes. These findings suggest potential practical applications for the material.

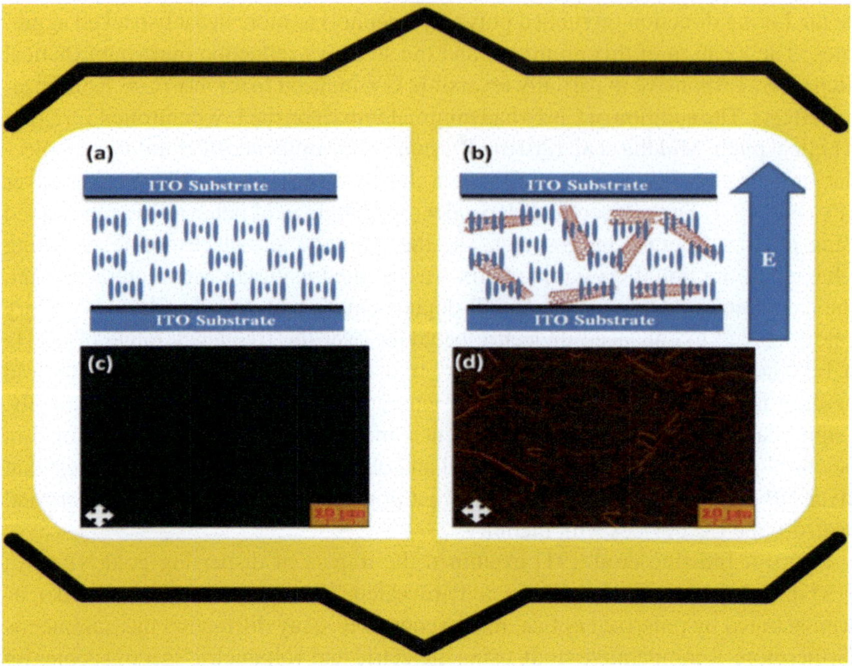

Fig. 8 The image is divided into two main sections, each with a diagram and a corresponding microscopic optical texture. The left section **a** shows a schematic of switching of aligned structures in 'Switch ON State' between two ITO substrates in the absence of CNT network, with a black microscopic image (**c**) below it. The right section **b** depicts 'Switch ON State' between two ITO substrates in the presence of CNT's network between ITO substrates with an electric field (E) indicated by an arrow, and a microscopic image **d** showing a network of lines. Both microscopic images have a scale bar labeled "10 μm" in the bottom right corner

6 Applications of Nanoparticles Embedded CLC Composites

NPs-doped CLC have been explored in various research fields due to their unique optical, electro-optical, and thermal properties. The combination of the nanoscale features of NPs and the LC properties of cholesteric phases creates novel applications (as shown in Fig. 9) in fields such as displays, sensors, photonics, and energy conversion. The significant applications of CLCs are discussed below.

6.1 Tunable Optical Devices

Researchers have modified the optical properties of CLCs by incorporating NPs as the NPs interact with the LC matrix, affect the pitch and ordering of the CLC structure to

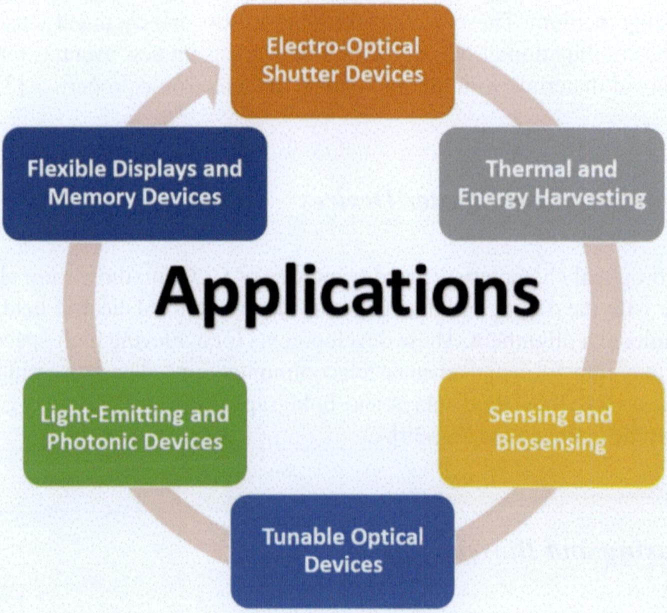

Fig. 9 Diagram illustrating various applications of nanoparticles embedded cholesteric liquid crystal composites in display technology

control light reflection and transmission. Further, the tunability of optical filters with reflection peaks can be obtained using CLC materials, tunable pitch of CLCs enables enhanced color reproduction and contrast for their advanced applications in displays, smart windows, optical communication, and sensing devices. The enduring research work in this field is advancing it to uncover new opportunities for these materials, potentially revolutionizing numerous aspects of optics and photonics [34].

6.2 Light-Emitting and Photonic Devices

The advancement of CLC materials for their applications in light emitting and photonic devices is going on by doping quantum dots, gold, or silver NPs into CLCs. These NPs interact with the LC matrix in complex ways, leading to improved emission properties and the ability to manipulate the photonic band structure of the CLCs. QDs doped CLC mixtures obtained adjustable colors and enhanced efficiency for their application in light-emitting devices, photonic crystals with unique band gaps, enabling the creation of high-efficiency optical switches and sensors. The improved surface plasmon resonance effects were observed with the doping of metallic NPs, such as gold and silver, which were found precisely sensitive to detect the variations in surrounding medium. These are useful for biosensors and chemical sensors applications. The combination of NPs and CLCs thus opens up new avenues for creating multifunctional materials with tailored optical and electronic properties [35, 36].

6.3 Electro-Optical Shutter Devices

The electro-optical characteristics and responses of CLC mixtures were also found to improve with the doping of NPs due to the change in local electric field distribution and molecular alignment. These developments for achieving high-speed, energy efficiency may transform and advance telecommunications, data processing, display technologies, including 3D displays and holographic devices and next-generation flexible display applications [37–39].

6.4 Sensing and Biosensing

Various researchers tailored the sensing properties of CLC by carefully selecting the type, size, and concentration of NPs in CLCs to detect specific chemical or biological agents. For example, metal NPs can improve the plasmonic behaviour of CLCs for colorimetric finding of certain analytes, while semiconductor NPs can show the photoluminescent properties for the detection pf fluorescence-based sensing. Additionally, NPs are orderly arranged in 3D structure in CLCs due to the self-assembled arrangement of CLCs. This arrangement supports in achieving better optical and electronic properties advantageous for sensing detections. Therefore, the next-generation sensors in various fields, including environmental monitoring, medical diagnostics, and food safety can be introduced using NPs doped CLC systems [40, 41].

6.5 Flexible Displays

Flexible displays are an exciting area of application for CLC composites due to their ability to create lightweight, durable, and energy-efficient screens. CLC composites can be used in reflective displays, where the LCs' helical structure interacts with light. By manipulating the alignment of the LCs using external fields, the display can change its color or brightness without requiring a backlight. NPs embedded in the CLC matrix can enhance the response time and stability of the material, further improving the performance of flexible displays. CLC composites can be employed in e-paper technologies for flexible, low-power, and high-contrast displays that are easy to read in sunlight. Flexible displays can be incorporated into wearables like smartwatches, fitness trackers, and clothing, offering dynamic, high-contrast, and low-power displays. Flexible CLC composite displays could lead to new innovations in foldable or rollable phones, televisions, and other portable devices [42].

6.6 Memory Devices

CLC composites have the potential to be used in non-volatile memory devices due to their ability to maintain stable states under various conditions. CLCs exhibit bistable or multistable states, which means they can hold distinct configurations (e.g., reflecting light at different wavelengths or having different orientations). This characteristic can be harnessed for memory applications, where the CLC state can represent binary data. NPs, such as metallic or magnetic particles, can be embedded into the CLC structure to further stabilize the material's states and provide additional control over switching times and memory retention. CLC composites can be used in rewritable data storage devices, where information can be stored in the form of specific color or light-reflective states of the CLCs. The bistable nature of CLCs makes them suitable for non-volatile memory devices that retain data without requiring power, similar to technologies like flash memory or phase-change memory. CLC-based memory devices can operate at optical wavelengths, making them suitable for high-speed data processing and storage in photonic computing systems [43].

7 Future Direction for the Research on Nanoparticles Embedded CLCs for Display Technology

Future research on NPs embedded in CLC for display applications has the potential to revolutionize optical and mechanical properties, leading to significant advancements in next-generation technologies [44]. To fully harness this potential, several key areas of investigation must be pursued. One of the primary challenges lies in ensuring the stability and uniformity of NP dispersion within the CLC matrix. The effective distribution of NPs is crucial for improving the performance and longevity of devices, as poorly dispersed particles can lead to inconsistent electro-optical behaviour, reduced device efficiency, and diminished durability. Surface functionalization techniques, such as the attachment of polymers or surfactants to NPs, can help enhance their interaction with the LC host, minimizing aggregation and ensuring a more uniform dispersion throughout the composite material. NPs synthesis is another critical area of research. While much of the existing work has focused on metallic NPs like gold and silver, as well as carbon-based materials such as graphene and carbon nanotubes, future studies should explore a broader range of NPs with diverse chemical compositions and sizes. The use of semiconducting quantum dots, magnetic NPs and hybrid

NPs—which possess both magnetic and optoelectronic properties—could significantly enhance the electro-optical characteristics of CLC composites. Now a days, there is a great demand for simulation and computational modeling-based research works to optimize the design of NP-CLC systems and can accelerate the development of new materials with specific desired properties for applications in displays, sensors, and other devices.

The future of NPs-embedded chiral LCs for display applications is filled with exciting possibilities. By addressing challenges related to NP dispersion, synthesis, and optical tuning, as well as exploring new avenues such as multifunctional composites, enhanced mechanical properties, and integration with emerging technologies, researchers can unlock the full potential of these materials. Additionally, focusing on sustainability and scalable manufacturing will be crucial for ensuring that these advanced materials can be produced efficiently and in an environmentally responsible manner. With the help of computational modelling and simulation tools, researchers can expedite the development of these complex systems, bringing us closer to realizing the next generation of optical devices, flexible electronics, and wearable technologies.

8 Conclusion

The review of the chapter indicates the incorporation of NPs into CLCs has significant potential for enhancing and tailoring the material properties. The review explores the impact of NP dispersions into CLCs and analyzes the effects on dynamic responses. The key findings in this chapter underscore the profound impact of nanomaterial dispersion on CLC properties including viscosity, optical switching behavior, optical anisotropy, and thermal stability. The size, shape, and concentration of NPs have been identified as critical factors influencing CLC composite behavior. While substantial progress has been discussed in the understanding and utilizing these nanomaterials dispersed CLC composites but still, the challenges persist in optimizing various material parameters for specific display applications. Future research directions for the researchers may focus on developing more precise control over NP dispersion and exploring novel nanomaterial-CLC combinations. As researchers continue to explore and optimize these NP interactions, the potential for innovative solutions in advanced technologies expands, paving the way for more efficient and versatile applications in display technology in the market.

References

1. Shivakumar, U., Mirzaei, J., Feng, X., Sharma, A., Moreira, P., Hegmann, T.: Nanoparticles: complex and multifaceted additives for liquid crystals. Liq. Cryst. **38**(11–12), 1495–1514 (2011). https://doi.org/10.1080/02678292.2011.605477
2. Gahrotra, R., Sharma, V., Dogra, A.R., Malik, P., Kumar, P.: Performance augmentation of bistable cholesteric liquid crystal light shutter-effect of dichroic dye on morphological and electro-optical characteristics. Opt. Mater. **127**, 112243 (2022). https://doi.org/10.1016/j.optmat.2022.112243
3. Adimule, V., Keri, R., Rajendrachari, S., Kendrekar, P., Kulkarni, C.V.: Highly photoluminescence and wide band gap insulating metal hybrid nanoparticles array of samarium-doped SrO: CoO: Synthesis, characterizations and sensor characteristics. J. Mater. Sci.: Mater. Electron. **34**, 442 (2023). https://doi.org/10.1007/s10854-023-09899-2
4. Liu, X., Qin, L., Zhan, Y., Chen, M., Yu, Y.: Phototuning of Structural Colors in Cholesteric Liquid Crystals, vol. 78, pp. 478–489, Acta Chimica Sinica. Science Press (2020). https://doi.org/10.6023/A20040103
5. Gahrotra, R., Sharma, V., Dogra, A.R., Malik, P., Kumar, P.: Temperature dependent morphological and electro-optical characteristics of dye doped cholesteric liquid crystal. ECS Trans. **107**(1), 5481 (2022). https://doi.org/10.1149/10701.5481ecst
6. Gahrotra, R., Sharma, V., Malik, P., Kumar, P.: Textural and electro-optical study of cholesteric liquid crystal for light shutter. In: AIP Conference Proceedings, vol. 2220, no. 1. AIP Publishing (2020). https://doi.org/10.1063/5.0001809
7. Gahrotra, R., Sharma, V., Malik, P., Kumar, P.: Microscopic studies of chiral nematic liquid crystal doped with dichroic dye. In: AIP Conference Proceedings, vol. 2357, no. 1. AIP Publishing (2022). https://doi.org/10.1063/5.0080701
8. Gao, Y., Luo, Y., Lu, J.: High-reflective templated cholesteric liquid crystal filters. Molecules **26**(22) (2021). https://doi.org/10.3390/molecules26226889
9. Middha, M., Kumar, R., Raina, K.K.: Effects of chirality on optical and electro-optic behavior of nematic liquid crystals doped with functionalized silver nanoparticles. J. Mol. Liq. **219** (2016). https://doi.org/10.1016/j.molliq.2016.03.087
10. Manohar, R., Shukla, J.P.: Refractive indices, order parameter and principal polarizability of cholesteric liquid crystals and their homogeneous mixtures. J. Phys. Chem. Solids **65**(10), 1643–1650 (2004). https://doi.org/10.1016/j.jpcs.2004.03.012
11. Yoshida, H., Tanaka, Y., Kawamoto, K., Kubo, H., Tsuda, T., Fujii, A., et al.: Nanoparticle-stabilized cholesteric blue phases. Appl. Phys. Express **2**(12) (2009). https://doi.org/10.1143/APEX.2.121501
12. Manohar, R., Gupta, M., Shukla, J.P.: Phase transition studies of some cholesteric liquid crystals and their mixtures using dielectric, optical transmittance and density measurement techniques. J. Phys. Chem. Solids **61**, 1465–1473 (2000). https://doi.org/10.1016/S0022-3697(00)00014-7
13. Kumar, R., Raina, K.K.: Electrically modulated fluorescence in optically active polymer stabilised cholesteric liquid crystal shutter. Liq. Cryst. **41**(2) (2014). https://doi.org/10.1080/02678292.2013.851287
14. Lorenz, A., Zimmermann, N., Kumar, S., Evans, D.R., Cook, G., Kitzerow, H.S.: Doping the nematic liquid crystal 5CB with milled $BaTiO_3$ nanoparticles. Phys. Rev. E Stat. Nonlinear, Soft Matter Phys. **86**(5) (2012). https://doi.org/10.1103/PhysRevE.86.051704
15. Schmidtke, J., Stille, W.: Fluorescence of a dye-doped cholesteric liquid crystal film in the region of the stop band: theory and experiment. Eur. Phys. J. B. **31**(2), 179–194 (2003). https://doi.org/10.1140/epjb/e2003-00022-x
16. Adimule, V., Sharma, V., Kumar, P., Keri, R., Sharma, K.: Perylene-3, 4, 9, 10-tetracarboxylic dianhydride dye intercalated liquid crystal molecules: synthesis, spectroscopic characterizations and photoluminescence dynamics. J. Mol. Liq. 125990 (2024). https://doi.org/10.1016/j.molliq.2024.125990

17. Adimule, V., Yallur, B.C., Gowda, A.H.: Advanced sensors based on carbon nanomaterials. Carbon Nanomater.-Based Sens. 259–268 (2022). https://doi.org/10.1016/B978-0-323-91174-0.00004-4
18. Qi, H., Hegmann, T.: Liquid crystal-gold nanoparticle composites. Liquid Cryst. Today. **20**(4), 102–114 (2011). https://doi.org/10.1080/1358314X.2011.610133
19. Manhas, N., Adimule, V., Lakshminarayana, P., Keri, R., Kumar, L.S., Bathula, C.: Utilization of CO^{2+} doped Eu_2O_3 nanostructures and their effect on the dielectric and photoisomerization of liquid crystals. Phys. Fluids **35**, 10 (2023). https://doi.org/10.1063/5.0171578
20. Li, W., Zhu, M., Ding, X., Li, B., Huang, W., Cao, H., et al.: Studies on electro-optical properties of polymer matrix/LC/SiO_2 nanoparticles composites. J. Appl. Polym. Sci. **111**(3), 1449–1453 (2009). https://doi.org/10.1002/app.28874
21. Adimule, V., Bathula, C., Sharma, K., Keri, R.: Study of structural, optical, dielectric properties poly-3-butyl thiophene reinforced with CO_3O_4 nanoparticles. Nano-Struct. Nano-Objects **38**, 101150 (2024). https://doi.org/10.1016/j.nanoso.2024.101150
22. Pathinti, R.S., Gollapelli, B., Jakka, S.K., Vallamkondu, J.: Green synthesized TiO_2 nanoparticles dispersed cholesteric liquid crystal systems for enhanced optical and dielectric properties. J. Mol. Liq. **15**(336), 116877 (2021). https://doi.org/10.1016/j.molliq.2021.116877
23. Pathinti, R.S., Tatipamula, A.K., Vallamkondu, J.: ZnO nanoparticles dispersed cholesteric liquid crystal based smart window for energy saving application. J. Alloy. Compd. **963**, 171198 (2023). https://doi.org/10.1016/j.jallcom.2023.171198
24. Akman, A., Bukusoglu, E.: Understanding directed assembly of concentrated nanoparticles at energetically heterogeneous interfaces of cholesteric liquid crystal droplets. J. Colloid Interface Sci. **1**(649), 772–784 (2023). https://doi.org/10.1016/j.jcis.2023.06.143
25. Luo, Y., Liu, Q., He, P., Li, L., Zhang, Z., Li, X., Bao, G., Wong, K.L., Tanner, P.A., Jiang, L.: Responsive regulation of energy transfer in lanthanide-doped nanomaterials dispersed in chiral nematic structure. Adv. Sci. **10**(27), 2303235 (2023). https://doi.org/10.1002/advs.202303235
26. Sj, S., Gupta, R.K., Kumar, S., Manjuladevi, V.: Enhanced electro-optical response of nematic liquid crystal doped with functionalised silver nanoparticles in twisted nematic configuration. Liq. Cryst. **47**(11), 1678–1690 (2020). https://doi.org/10.1080/02678292.2020.1755901
27. Perera, K., Nemati, A., Mann, E.K., Hegmann, T., Jakli, A.: Converging microlens array using nematic liquid crystals doped with chiral nanoparticles. ACS Appl. Mater. Interfaces. **13**(3), 4574–4582 (2021). https://doi.org/10.1021/acsami.0c21044
28. Samoilov, A.N., Minenko, S.S., Lisetski, L.N., Soskin, M.S., Torgova, S.I., Lebovka, N.I.: Anomalous optical properties of photoactive cholesteric liquid crystal doped with single-walled carbon nanotubes. Liq. Cryst. **45**(2), 250–261 (2018). https://doi.org/10.1080/02678292.2017.1314560
29. Middha, M., Kumar, R., Raina, K.K.: Photoluminescence tuning and electro-optical memory in chiral nematic liquid crystals doped with silver nanoparticles. Liq. Cryst. **43**(7), 1002–1008 (2016). https://doi.org/10.1080/02678292.2016.1155772
30. Middha, M., Kumar, R., Raina, K.K.: Effects of chirality on optical and electro-optic behavior of nematic liquid crystals doped with functionalized silver nanoparticles. J. Mol. Liq. **1**(219), 631–636 (2016). https://doi.org/10.1016/j.molliq.2016.03.087
31. Infusino, M., De Luca, A., Ciuchi, F., Ionescu, A., Scaramuzza, N., Strangi, G.: Optical and electrical characterization of a gold nanoparticle dispersion in a chiral liquid crystal matrix. J. Mater. Sci. **49**, 1805–1811 (2014). https://doi.org/10.1007/s10853-013-7868-6
32. Middha, M., Kumar, R., Raina, K.K.: Improved electro-optical response of induced chiral nematic liquid crystal doped with multi-walled carbon nanotubes. Ferroelectrics **495**(1), 75–86 (2016). https://doi.org/10.1080/00150193.2016.1136765
33. Middha, M., Kumar, R., Raina, K.K.: Memory effects in chiral nematic liquid crystals doped with functionalised single walled carbon nanotubes. Liq. Cryst. **42**(7), 1028– (2015). https://doi.org/10.1080/02678292.2015.1015180
34. Ryabchun, A., Bobrovsky, A.: Cholesteric liquid crystal materials for tunable diffractive optics. Adv. Opt. Mater. **6**(15), 1800335 (2018). https://doi.org/10.1002/adom.201800335

35. Tamaoki, N.: Cholesteric liquid crystals for color information technology. Adv. Mater. **13**(15), 1135–1147 (2001). https://doi.org/10.1002/1521-4095(200108)13:15%3c1135::AID-ADMA1135%3e3.0.CO;2-S
36. Alberto, B., Tom, B., Dirk, J.B., Albert, P.H.J.: Schenning, patterned full-color reflective coatings based on photonic cholesteric liquid-crystalline particles. ACS Appl. Mater. Interfaces **11**(15) (2019).https://doi.org/10.1021/acsami.9b02680
37. Chauhan, G., Malik, P., Deep, A.: Morphological, dielectric, electro-optic and photoluminescence properties of titanium oxide nanoparticles enriched polymer stabilized cholesteric liquid crystal composites. J. Mol. Liq. **376**, 121406 (2023). https://doi.org/10.1016/j.molliq.2023.121406
38. Gulsum, K., Oguz, K.: Electro-optical, dielectric and morphological properties of a cholesteric liquid crystal light shutter: the influence of azo dye and quantum dot nanoparticles. J. Electron. Mater. **51**(12), 6864–6876 (2022). https://doi.org/10.1007/s11664-022-09913-z
39. Liu, H., Guo, Z.H., Xu, V, Jia, L., Pan, C., Lin, Z., Pu, X.: Triboelectric-optical responsive cholesteric liquid crystals for self-powered smart window, E-paper display and optical switch. Sci. Bullet. **66**(19), 1986–1993 (2021). https://doi.org/10.1016/j.scib.2021.05.016
40. Chang, B.Y., Chen, F.L., Hsu, C.C., Chen, W.H., Shang, L.D., Hsiao, Y.C.: Cholesteric nanostructure biosensors on flexible paper fibres for disease detection through label-free techniques and image analysis. Liq. Cryst. **51**(2), 215–222 (2023). https://doi.org/10.1080/02678292.2023.2290216
41. Lee, H.G., Munir, S., Park, S.Y.: Cholesteric liquid crystal droplets for biosensors. ACS Appl. Mater. Interfaces **8**(39) (2016). https://doi.org/10.1021/acsami.6b09624
42. Kim, D.Y., Lee, K.M., White, T.J., et al.: Cholesteric liquid crystal paints: in situ photopolymerization of helicoidally stacked multilayer nanostructures for flexible broadband mirrors. NPG Asia Mater. **10**, 1061–1068 (2018). https://doi.org/10.1038/s41427-018-0096-4
43. Tan, C., Liu, Z., Huang, W., Zhang, H.: Non-volatile resistive memory devices based on solution-processed ultrathin two-dimensional nanomaterials. Chem. Soc. Rev. **44**, 2615–2628 (2015). https://doi.org/10.1039/C4CS00399C
44. Kumar, R., Singh, R., Singh, S., Raina, K.K.: Ceramic nanoparticles doped liquid crystals: a review of material properties for display applications. Adv. Ceram. Versatile Interdiscip. Appl. (2022). https://doi.org/10.1016/B978-0-323-89952-9.00018-X

Tunable Photoluminescence of Nanoparticles Doped Cholesteric Liquid Crystals Composites for Photonic Applications

P. Priscilla, Arvind K. Gathania, Jai Prakash, Supreet, Ajay Kumar, Riccardo Castagna, Sandeep Kumar, and Gautam Singh

Abstract Cholesteric liquid crystals (CLCs) show selective reflection, circular dichroism and optical rotation properties due to their intrinsic self-assembled helical superstructures. Due to such unique optical properties, CLCs remain in limelight for their utility in displays, photonic devices and sensors. Interestingly, they are also observed to show photoluminescence (PL). However, the tunability of PL in terms of spectral peak and intensity is not ample and in turn limits their applications in PL based devices. In the recent past, the doping of nanoparticles (NPs) into CLCs and polymer stabilised CLCs (PSCLCs) has been greatly investigated in order to

---

P. Priscilla · G. Singh (✉)
Department of Applied Physics, Amity Institute of Applied Sciences, AmityUniversity Uttar Pradesh, Noida, India
e-mail: gsingh6@amity.edu

A. K. Gathania
Department of Physics and Photonic Science, National Institute of Technology, Hamirpur, Himachal Pradesh, India

J. Prakash
Department of Physics, Aligarh Muslim University, Aligarh, India

Supreet
Department of Physics, Amity School of Applied Sciences, Amity University Haryana, Gurgaon, India

A. Kumar
Department of Physics, Dharma Samaj College (Raja Mahendra Pratap Singh State University, Aligarh), Aligarh, Uttar Pradesh, India

R. Castagna
CNR, Institute of Heritage Science, Sesto Fiorentino, FI, Italy

URT-CNR, Universit'a Di Camerino (UNICAM), Camerino, MC, Italy

S. Kumar
Raman Research Institute, Sadashivanagar, Bengaluru, India

Department of Chemistry, Nitte Meenakshi Institute of Technology, Yelahanka, Bengaluru, India

P. Kumar et al. (eds.), *Cholesteric Liquid Crystals*, Engineering Materials,
https://doi.org/10.1007/978-981-96-3821-5_8

improve upon the PL of CLCs/PSCLCs. Therefore, in this chapter, we have provided a comprehensive overview of the impact of doping of various types of NPs on the PL of CLCs/PSCLCs. For instance, enhancement in the PL intensity, quantum yield and spectral peak and reduction in the full width and half maximum in NPs doped CLC/PSCLCs composites would be reviewed in a greater detail. We believe that the ability to fine-tune the PL of CLCs/PSCLCs using NPs would certainly pave the way forward for their potential use in the numerous next-generation high-performance luminescent displays, tunable photonic devices and sensing applications.

Keywords Cholesteric liquid crystals · Nanoparticles · Composites · Photoluminescence · Photonic devices

1 Introduction

In the late nineteenth century, the discovery of liquid crystal (LC) by an Austrian botanist Friedrich Reinitzer marked the beginning of a revolutionary period in the field of material science and technology. As Reinitzer was investigating the properties of cholesterols extracted from carrots, he inadvertently found that the cholesteryl benzoate (Fig. 1a) was showing two different melting points [1–3]. Since the LC phase was first observed in this material, it became significant historically and was eventually classified as cholesteric liquid crystal (CLC) also known as chiral nematic liquid crystal (N*) [4]. Such a fascinating and unique class of LC material exhibits a helical structure due to the presence of molecular chirality [5, 6]. For the sake of understanding, one can imagine the CLC molecules to be arranged in sheets, wherein within each sheet, molecular orientation resembles that of nematic phase. Moreover, molecules in each sheet rotate by a constant angle giving it a helicoidal organization wherein the director (**n**) remains perpendicular to the helical axis shown in Fig. 1b.

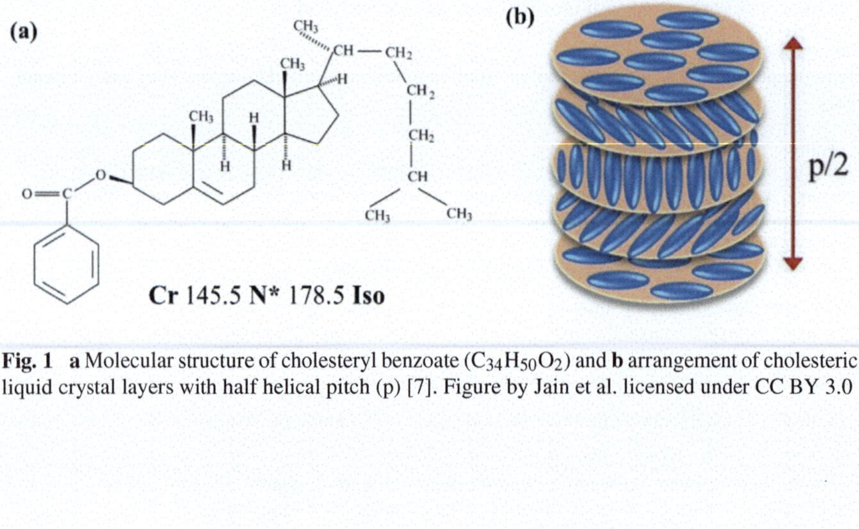

Fig. 1 **a** Molecular structure of cholesteryl benzoate ($C_{34}H_{50}O_2$) and **b** arrangement of cholesteric liquid crystal layers with half helical pitch (p) [7]. Figure by Jain et al. licensed under CC BY 3.0

The helical configuration is characterized by helical pitch (p), and which is defined as the distance traversed by the director (**n**) to attain a complete 360° rotation. Since these sheets are twisted either left or right-handed so they reflect a circularly polarized light with the same handedness as that of the helix. For example, a right-handed helix will reflect right circularly polarized light [8]. The handedness of the cholesteric helix is determined by a widely accepted convention which is given as: the positive pitch p corresponds to right-handed helical superstructure and negative pitch gives left-handed helix [9]. The helical pitch is a critical parameter of the CLC material as it determines the wavelength of light being selected for reflection usually in the order of the wavelength of visible light [10–12]. Helical pitch can be adjusted by tuning the concentration of chiral dopant. The relation between pitch of CLC sample and the light's wavelength (λ) is given as [13, 14]

$$p = \frac{1}{(HTP) \times c} \tag{1}$$

$$\lambda = np \tag{2}$$

where, c is the concentration of chiral dopant, HTP is the helical twist power and n stands for the refractive index. When filled in between untreated glass substrates, cholesteric phase exhibits oily streak texture when viewed under cross polarized microscope. Originally, Lehmann described this texture as "*ölige Streifen*" and eventually referred as "*stries huileuses*" by Friedel. These oily streaks appear as a network of defect lines that is dispersed within the helical regions. They have a complicated structure having a dependency on elasticity and surface anchoring. When the size of the pitch becomes comparable to the cell gap, there occurs a distortion in the helical superstructure of CLCs due to their elastic interaction with the substrates. This causes a slight variation in the intensity of the sample when rotated between cross polarizers. In extreme cases, such elastic deformation produces a uniform planar nematic texture despite being composed of chiral molecules. For such instances, cholesteric pitch exceeds twice the cell gap. Another texture exhibited by CLCs is the fan-shaped (or fan-like) texture very similar to the smectic A texture. A point to be noted here that materials with strong twist exhibit fan-like texture and those with slightly smaller twists show focal conic texture [9]. CLCs exhibit varied properties like optical rotation, selective reflection, reflect circularly polarized light at a specific wavelength, and responsiveness towards temperature change and external field making them inevitable materials for device applications [15–19]. Application of CLCs ranges on a broad spectrum like reflective displays [20], tunable lasers [21], biosensors [11], rewritable color recordings [22], smart windows [23, 24], etc. (Fig. 2).

An advanced form of CLCs known as polymer stabilized cholesteric liquid crystals (PSCLCs) are also utilized in recent times which are produced by the introduction of a small percentage of chiral LC monomers and a photo-initiator into the traditional CLC material and eventually being cured by the UV-light [28]. Addition of polymers into CLC mixture locks the helical arrangement thereby preserving the

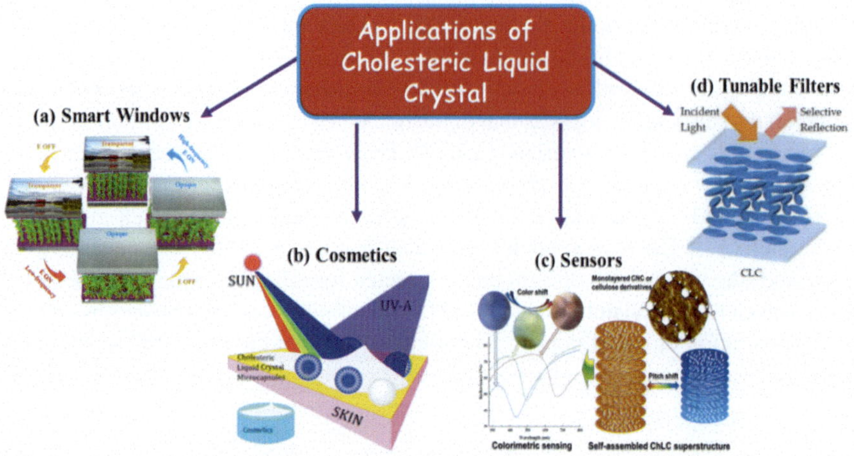

Fig. 2 Applications of cholesteric liquid crystal material **a** Smart windows [24], **b** cosmetics [25], **c** sensors [26] and **d** tunable filters [27]

selective reflection properties even during stress or any change in external conditions. If the LC monomers are polymerized in planar state, the formed polymer networks align parallel to the substrate thereby stabilizing the planar state [29]. Application of external electric field produces a switch of PSCLC molecules from planar to either focal conic or homeotropic state. The removal of external field reorients the molecules to its initial planar state preventing them from forming a metastable focal conic state that can be ascribed to the elastic restoring force produced by polymer networks. In another case, if the LC monomers are stabilized in homeotropic state, the polymers would be aligned perpendicular to the substrate. Here, the alignment switches from homeotropic to focal conic with increase in temperature. The focal conic state can be reverted back to its homeotropic state by the application of electric field [30]. Application of PSCLCs closely align with those of CLCs but extend further to include additional functionalities, such as their use in anti-counterfeiting documents, infrared shielding for the fabrication of energy saving displays and other innovative domains [31]. To expand the presence of CLCs/PSCLCs in advanced, small size, and economical technologies, it became necessary to modify their inherent properties which led researchers to focus on the newly discovered materials called nanoparticles (NPs).

In recent years, NPs have become an exciting frontier in research and development due to their small size (1–100 nm), varied shapes (Fig. 3), high surface area to volume ratio, quantum effects, enhanced optical, magnetic, thermal properties, etc. [32–35]. There are various types of NPs such as polymeric, metal oxides, metallic, semiconducting (i.e., quantum dots, rods), and carbon [36]. Their utilization in various fields such as electronics [37], optoelectronics [38], energy [39], material science [40], medicine and healthcare [41, 42] etc. make them indispensable materials of this century.

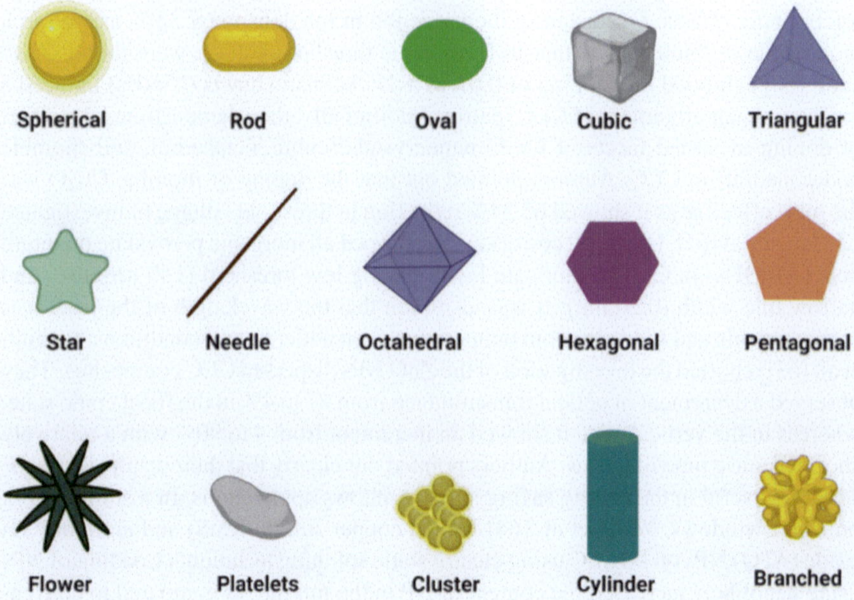

Fig. 3 Various shapes of nanoparticles [43]. Originally published by and used with permission from Dove Medical Press Ltd.

In liquid crystal nanoscience, topical area of research in the field of LCs, the properties of LCs are tuned by the dopant NPs and vice-versa. However, in this chapter, we specifically emphasize on the role of NPs as dopant to modulate the intrinsic properties of LCs. For instance, the doping of NPs into LCs has demonstrated the significant influence on various properties, such as dielectric permittivity, switching response time, threshold voltage, molecular alignment, conductivity, phase transition temperature, etc., across the diverse LC phases like nematic, smectic and their chiral analogues [44–56]. Such tuning in the physical properties of LCs has eventually sparked a renaissance in this topical research field. Under this framework, the impact of doping of NPs on the properties of CLC/PSCLC materials has also been investigated and highlighted the unique effects and their potential applications. Studies have demonstrated a lowered threshold voltage, increased dielectric constant, enhanced photoluminescence (PL) effect, induced vertical alignment by the doping of NPs in CLCs/PSCLCs [57–64]. Firstly, Kurochkin et al. [57] reported that the doping of small quantity of ferroelectric NPs (FNPs) in CLC exhibited a 45% decrease in the driving voltage and a twofold increment in the dielectric anisotropy ($\Delta\varepsilon$) of the CLC matrix. This was attributed to the strong dipole moment of FNP eventually producing a very strong electric field around it. Jeng et al. [58] reported that the dispersion of polyhedral oligomeric silsesquioxane (POSS) NPs induced a vertical alignment in the CLC material due to the adsorption of NPs on the surface of the ITO substrates. In a different work, Wu and group [59] demonstrated that the incorporation of 0.5 wt% graphene nanoplatelets in CLC matrix suppresses the ionic behavior due to their ion trapping effects. For instance, the reduction in ion density by 32%, as well as a remarkable decrement in planar to focal conic threshold voltage were found. Zhao et al. [60] exhibited that doping of 0.01 wt% Nickel nano bowls (NiNBs) induced a uniform planar alignment of CLC matrix. Additionally, they demonstrated the effect of doping of varied facets of Cu_2O nanocrystals (cubic, octahedral, and rhombic dodecahedral) in CLC. Authors pointed out that the doping of rhombic Cu_2O was the most effective as it showed 65.31% reduction in threshold voltage of investigated CLC material [61]. Chen and coworkers [62] doped all inorganic perovskite quantum dots (AIPQDs) in CLC to fabricate lasers having low threshold (150 nJ/pulse) and narrow line width (0.20 nm). It was also seen that the wavelength of the laser was highly tunable and stable in room temperature even under humid conditions. Pathinti et al. [63] reported the investigation of the ZnO NPs dispersed CLC composites. They observed a decrement in optical transmittance from 87 to 4% in the focal conic state, whereas in the vertical state it showed an increment from 4 to 90% with a relatively short response time of <1 ms. Authors pointed out clearly that their results would be certainly useful in the energy saving smart windows applications. In a similar work on smart windows, Wang et al. [64] loaded copper sulfide (CuS) and antimony tin oxide (ATO) NPs on PSCLC using electro static spinning technique. Loading of NPs using nanofibers increased the content of NP in the mixture as compared to conventional direct doping approaches. They successfully developed a multi-field responsive PSCLC smart window. This smart window was able to provide a precise regulation of visible and near infrared light for effective temperature management by modulating the applied electric field. The bandwidth reflectance got enhanced from 256 to 507 nm through nanofiber loading and by refining the polymerization conditions. Authors also explained that such inclusion of nanofibers embedded with NPs significantly improved the system's infrared shielding capacity and electrical performance. Furthermore, both thermal and electrical stability of the smart window has been improved by the nanofibers, also their precise distribution ensures the development of a highly flexible and dependable infrared anti-counterfeiting smart window.

It appears from the literature review that much has been done in terms of modulation of optical, dielectric, electrical and electro-optical properties of CLCs/PSCLCs by doping of various types of NPs. However, currently there is a greater search for photoluminescent (PL) based LC devices. So, in our opinion, it would be worth providing a review on the tunable PL of NPs doped CLCs/PSCLCs composites useful for photonic applications. Therefore, in this chapter, we would provide a comprehensive overview on the impact of doping of various types of NPs on the optical properties (especially PL) of CLC/PSCLC materials (Table 1). Moreover, the future perspectives of this work would also be discussed.

Table 1 Effect of various NPs on the photoluminescence (PL) of CLC materials

Nanoparticles	Size	CLC	Result	References
Cadmium selenide (CdSe) nanorods	Length: 40 nm Dia: 6 nm	NLC's: ZCE-5099LA, ZCE-5352LA, and MJ991597 Chiral dopants: benzoic acid, 4-(*trans*-4-pentylcyclohexyl)-, R1011	Enhanced PL intensity by 3times	[65]
Octadecylamine (ODA)-functionalised Single walled carbon nanotubes (SWCNTs)	Length: 0.5–2 μm, Dia: 2–10 nm	NLC: ZLI-4151, Chiral Dopant: CB15	Enhanced fluorescence emission at 0.3 wt%	[66]
Multi-walled carbon nanotubes (MWCNT)	Length: 5–9 μm, Dia: 110–170 nm	NLC: ZLI-4151, Chiral Dopant: CB15	Decrease in PL emission peak	[67]
Dodecanethiol functionalized silver nanoparticles (AgNPs)	5–15 nm	NLC: ZLI-4151, Chiral Dopant: CB15	Increased intensity of PL peak and shift in peak position	[68]
Dodecane thiol (DDT)-functionalised silver nanoparticles	5–15 nm	NLC: ZLI-4151, Chiral Dopant: CB15	Enhanced intensity of PL peak till 0.015 wt% AgNPs	[69]
Graphene quantum dots (GQDs)	2.6 nm	NLC: HTG135200 Chiral Dopant: S5011	PL peak intensity enhancement and red shift in amplified spontaneous emission	[70]
Gold nanoparticles (AuNPs)	NA	NLC: E7 Chiral Dopant: S811	Enhanced PL intensity peak at band edge and defect spectrum dip	[71]
Octadecylamine-capped CdSe/ZnS quantum dot	6 nm	Lab made CLC	Increased PL intensity for high concentrations (0.5 wt%)	[72]
Titanium Oxide (TiO_2) NPs	25.01 nm	NLC: E7 Chiral Dopant: R-811	Enhanced PL intensity peak for increasing concentration	[73]

(continued)

Table 1 (continued)

Nanoparticles	Size	CLC	Result	References
CuInZnSe-QDs	2 nm	NLC: SLC1717 Chiral Dopant: CB15	2–threefold rise in peak intensity for higher concentration (0.2 and 0.4 wt%)	[74]
CdSe QDs	6–7 nm	NLC: BL036, Chiral Dopant: CB15, Polymer used: NOA65	Shift in spectral speak by 32 nm wavelength	[75]
CdSe QDs	3.8 nm	LC: 8CB Chiral Dopant: CB15 Polymer used: PMMA	A 75% rise in PL peak intensity at 0.25 wt% of QDs in PSCLC	[76]
TiO_2 NPs	~50–70 nm	NLC: BL036 Chiral Dopant: CB15 Polymer used: NOA65	Enhanced PL intensity	[77]

2 Preparation of Nanoparticles (NPs)—Cholesteric Liquid Crystal (NPs-CLC) and NPs-Polymer Stabilized CLC (NPs-PSCLC) Composites

CLC is generally prepared by the addition of a chiral compound to a nematic liquid crystal (NLC) material. For this a suitable nematic host generally nCB, or any other NLC material with well-defined mesogenic property is selected. Thereafter, a chiral dopant is carefully chosen based on its compatibility with the host NLC material and also on the desired optical properties needed for the application. The addition of chiral dopant imparts the required helical pitch. For the preparation of PSCLC, CLC mixture is combined with polymerizable monomer and a photoinitiator. The photo initiator facilitates polymerization when being exposed to UV light. The intensity and duration of UV exposure, as well as the temperature during polymerization can be optimized to achieve specific optical properties. NPs either synthesized in lab or purchased from companies are used as dopant in the CLC/PSCLC host. These NPs are usually dissolved in organic solvents like chloroform, dichloromethane, ethanol etc. prior to their dispersion in the host CLC/PSCLC system. Some NPs gets readily dissolved in the solvent and do not require sonication; however, some NPs need sonication for getting a uniform dispersion in the solvent. A certain amount of CLC/PSCLC is weighed in a vial (or petri dish) initially and therein a certain concentration (as required for the work) of NPs solution is added into the CLC/PSCLC. The prepared mixture is generally sonicated for 1–2 h to obtain uniform dispersion of NPs in the host CLC/PSCLC material, whereafter they are kept in vacuum oven for the evaporation of the solvent. The temperature of the oven is maintained as per the boiling point of the solvent used in the system. After the evaporation of the solvent, the prepared mixture is then filled into LC sample cells at the isotropic temperature through capillary action.

3 Review of Photoluminescence (PL) NPs-CLC and NPs-PSCLC Composites

Here, we will first review the studies focused on NPs doped CLC composites, highlighting their key findings and the applications. Afterwards, we will review the research involving NPs doped PSCLC composites emphasizing the advancements in the properties of this new class of composite materials.

One of the earliest works on the effect of NPs on the photoluminescent property of CLC was reported by Huang et al. [65]. In this work, authors embedded fluorescent Cadmium Selenide (CdSe) nanorods (L:40 nm, D: 6 nm) in three different CLCs of varying helical pitches and observed an inverse relation between PL enhancement and helical pitch. An enhancement in the PL intensity was seen in all three samples however, CLCs with shorter pitches exhibited the highest increment of PL signal (>3 times). This was attributed to the nanomaterial attracting and light scattering properties of the oily streak defects within CLCs. The authors predicted that their work could be beneficial in modulating the PL signals in many optical devices. Middha and co-workers [66] investigated the effect of doping octadecylamine (ODA) functionalized single-walled carbon nanotubes (SWCNTs) in CLC matrix. The doping of functionalized SWCNTs (Length: 0.5–2 μm, Dia: 2–10 nm) produced an electro-optical memory that stayed intact for a period of 2 months for the composite with highest concentration of 0.3 wt%. The emission and excitation fluorescence spectra were also explored for all samples under investigation (Fig. 4).

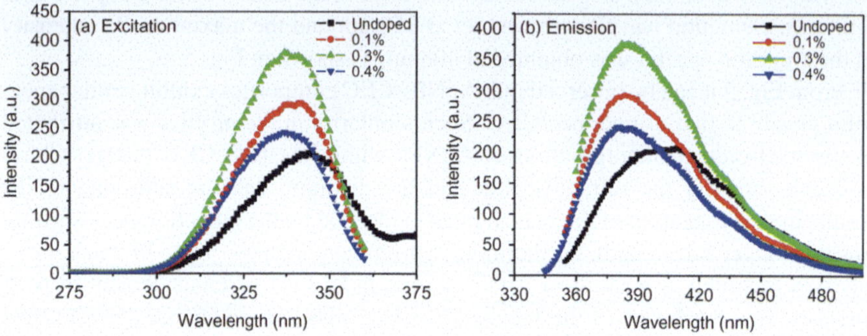

Fig. 4 **a** Excitation and **b** emission fluorescence spectra for various concentrations of ODA capped SWCNTs doped CLC samples. Reproduced with permission from Ref. [66]

It offers profound insight regarding fluorescence emission considering the relative orientation in the supramolecular structure of CLC molecules with varying SWCNT concentrations. Furthermore, the authors also computed the fluorescence anisotropy which showed an increasing trend with rising concentration except for the highest concentration of 0.4 wt%, wherein the value decreased significantly. The highly anisotropic nature of CNTs also enhanced the anisotropy of CLC material which was confirmed through the fluorescence anisotropy. Additionally, the authors also found a decrease in threshold voltage due to its dependence on optical anisotropy and ac conductivity. They also observed an increased electro-optic memory for 0.3 wt% which was ascribed to the stability produced by CNT dispersion in CLC material. Afterwards, these authors conducted a similar work by using COOH functionalized multi-walled carbon nanotubes (MWCNTs, Length: 5–9 μm, Dia: 110–170 nm) as dopants in CLC material [67]. It was observed that increasing the doping concentration of MWCNTs produces a strong decrement in the intensity of the PL emission peak of CLC matrix. They also observed that the emission peaks showed minimal horizontal shift which accounts for the slight variation in the color of the reflected light for distinct samples. No shift in reflected wavelength made the authors believe that there occurs a decrement in the helical pitch of the sample. Incorporation of MWCNTs into the CLC matrix produces alterations in the order of the supramolecular arrangement. Authors also reported a ~13% increment in the value of refractive index which they believed could be a promising optical material for futuristic device applications.

In another work [68], this group also explored the effect of doping dodecanthiol-capped silver nanoparticles (Ag NPs, size ~5–15 nm) in CLC as well as in simple NLC matrix and a comparison was made between them. AgNPs doped in CLC exhibited an increase in dielectric constant at lower frequency by 100% and in NLC matrix a 200% increase was observed. This was correlated with the conductivity wherein, a four-fold rise in conductivity was seen in AgNPs-NLC composite however, almost negligible for AgNPs-CLC composite. Authors also conducted PL analysis of the doped and undoped samples by ensuring consistent experimental parameters such as slit width and excitation step across all samples. For the analysis of the doped chiral mixtures, excitation wavelength was set at 350 nm and the maximum PL intensity in the emission spectra was obtained at 408 nm as shown in Fig. 5.

From Fig. 5 it can be observed that AgNPs-CLC composites exhibit similar spectrum profile to that of undoped CLC. Such similarity in the profiles was attributed to the well-ordered incorporation of AgNPs within the host CLC matrix. Chiral molecules modify the refractive index of the medium, and the addition of NPs results in the elongation of the helical pitch evidenced by the reflection peak shifting towards longer wavelength. Furthermore, a prominent increase in the PL peak intensity was recorded as a result of AgNP doping in CLC. This enhancement in peak intensity was attributed to the localized surface plasmon resonance (SPR) effect shown by AgNPs in the chiral structure of CLC. The resonance occurs when the incident photon frequency matches the natural oscillation frequency of the surface electrons counteracting with positive nuclei's restoring force. Therefore, authors believed that the shift in PL peak spectrum could be attributed to the change in

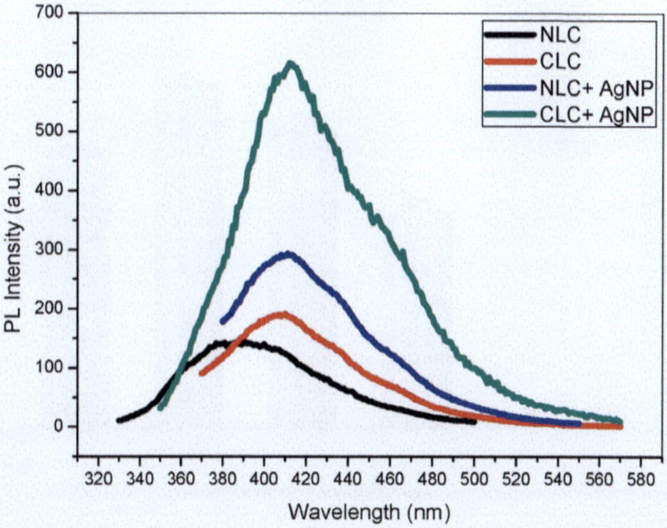

Fig. 5 PL emission spectra of undoped and doped NLC and CLC samples. Reproduced with permission from Ref. [68]

plasmon frequency induced by the modification of refractive index due to the addition of chiral molecules. Authors proposed that the results would be useful in the fabrication of optical sensors and other functional materials for memory devices. Based on the findings of previous study, authors in a similar work [69] reported the increment in PL intensity and electro-optical memory in CLC when doped with AgNPs. Figure 6a demonstrates the response of PL spectra with the application and withdrawal of external ac voltage. It can be seen that the peak intensity reaches it maximum for 0.015 wt% AgNPs concentration whereafter increasing the concentration produces a reduction in the peak intensity level. The authors attributed such enhancement in intensity to the localized SPR effect by AgNPs for concentrations ranging from 0.005 to 0.0015 wt%.

The light incident on the surface of AgNPs generates SPR due to the oscillation of electron cloud inside the NPs which syncs with the incoming oscillating electric field of an electromagnetic wave. This phenomenon causes a rise in the electromagnetic field generated on the surface of AgNPs leading to the formation of strong optical fields by surface plasmons amplifying the fluorescence peak intensity of the CLC composites. Authors also observed a shift in the PL peak towards higher wavelengths with increasing concentration ascribed to the elongation of helical pitch in CLC. As can be seen in Fig. 6a, the application of external bias modifies the peak intensity in all samples including undoped. They believed that the PL intensity remains constant until the application of a certain threshold voltage, whereafter it increases to a certain voltage and then again becomes constant. The threshold voltage differed for every sample that was investigated. And once voltage was removed from the sample, the PL intensity did not return to its initial state. This memory effect is shown in Fig. 6b for

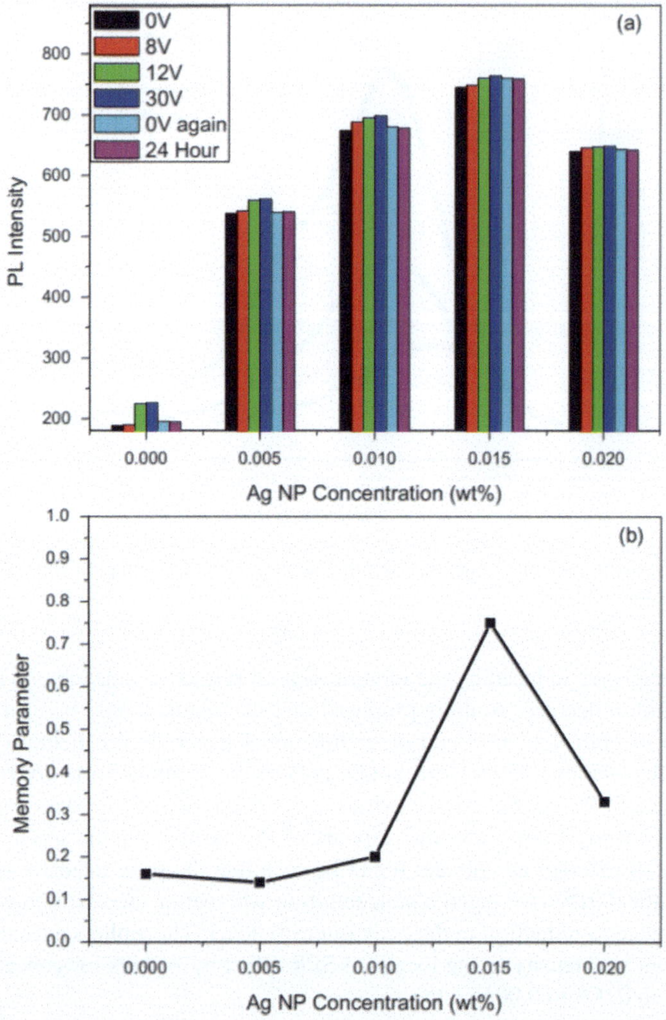

Fig. 6 **a** Photoluminescence response of silver nanoparticles (AgNPs) doped CLC with the application and withdrawal of ac electric field. **b** Variation of the memory parameter as a function of dopant concentration in AgNPs-doped CLC samples. Reproduced with permission from Ref. [69]

varying AgNP concentrations which was in agreement with the concept of electron cloud oscillation. Authors claimed that the induction of exceptionally high electric field in doped samples causes an effective delay in the relaxation of LC molecules back to their scattering state. Such optical hysteresis produced by the intrinsic field is increased for 0.005–0.0015 wt% concentrations. Additionally, the authors proposed that the AgNPs were able to organize themselves uniformly into the defects within CLC structures in such a way that the mixture gets stabilized in their presence. As a result, it causes a distortion in the helical structure of CLC causing a delay in the reorientation of molecules thereby preventing its return to the original configuration. The decrement in memory parameter for 0.020 wt% composites is ascribed to the agglomeration of AgNPs. The exceptional results made the authors believe that this system could be used in the fabrication of tuneable optical functional materials that exhibit electro-optical memory.

Cao et al. [70] reported the observation of tuneable amplified spontaneous emission (ASE) for a mixture of CLC doped with graphene quantum dots (GQDs, size ~2.6 nm). The ASE for the composite was observed that the edge of the bandgap. The overlapping of CLC's long wavelength edge of photonic bandgap with emission peak produced significant enhancement in the PL intensity of the composite. A red shift in the ASE peak position with increasing temperature for GQDs-CLC composite was also observed which corresponds to the dependence of CLC's helical pitch with temperature. The PL peak intensity of composite reduced radically for temperatures >80 °C due to the deformation in the photonic bandgap. Furthermore, the authors extended their study on PL intensity with varying pump energy and analyzed its effect on full width half maxima. Liu et al. [71] studied the effect of doping gold nanoparticles (AuNPs) and laser dye (Pyrromethene 597, PM_{597}) on the CLC assembled with defect-mode. Authors prepared the defect mode by stripping one glass substrate and exposing the alignment layer (PVA) to the air. Two cells were prepared in such manner and then these were stacked together face to face and a finite gap is maintained by using Mylar spacers having thickness of 3 μm. Mixture of LC, laser dye and AuNPs were filled in cell gap hence forming the defect layer. The absorption peak of laser dye was found to be around 532 nm, and the fluorescence emission peak was found at approx. 575 nm. As the CLC functions as one-dimensional photonic crystal authors believed that incorporation of laser dye would produce strong fluorescence. For PL studies, both the undoped and doped CLC matrix were excited by a laser light of 532 nm. In the undoped CLC state, the peak was obtained near the band edge of 617–620 nm having an intensity of 1641 au. The parameter full width at half maxima (FWHM) were found to be around 8 nm. For 575 nm, the peak intensity for undoped sample was recorded to be 1146 au. Comparing these with AuNPs doped samples, authors found that for 532 nm laser light, the peak intensity was recorded to be around 2265 au with same FWHM value. And for 575 nm laser light, AuNPs doped CLC composite, fluorescence peak intensity was 1791 au. The doped sample also exhibited a slight red-shift in the peak by 4 nm. Such enhancement in peak intensity has been observed at both the defect dip and in the band edges and attributed to the localized surface plasmon resonance effect formed in the AuNP.

Hegde and group [72] investigated the impact of dispersing octadecylamine-capped CdSe/ZnS QDs (size ~6 nm) on the optical property of CLC matrix. Figure 7a demonstrates the effect of CdSe/ZnS QDs on CLC matrix with increase in doping concentration (Mix 1 → 0.1 wt%, Mix 2 → 0.25 wt% and Mix 3 → 0.5 wt%). For lower concentration, much effect was not seen in the PL intensity however, at higher concentrations the peak intensity of PL increased dramatically that can be attributed to the surface property of QDs.

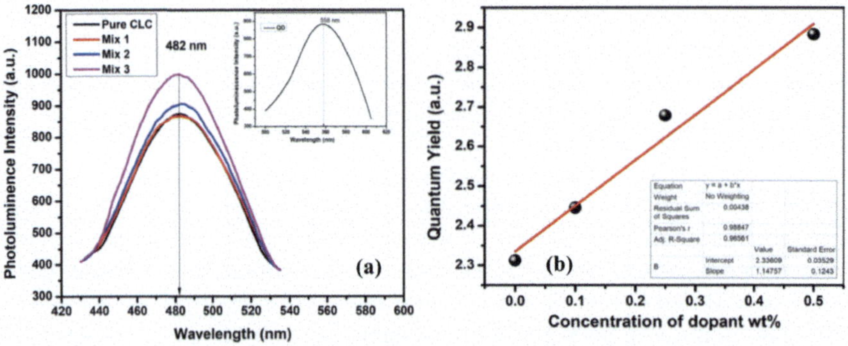

Fig. 7 **a** Photoluminescence intensity (in arbitrary unit) of pure CLC and CLC–QDs composites and **b** Variation of quantum yield as a function of dopant concentration. Reproduced with permission from Ref. [72]

The authors proposed that the incident light source generates a huge electromagnetic field on the surface of QDs resulting in the enhancement of the PL intensity. Additionally, the phonons of QDs (quantized vibrations of the atoms within a QD) and photons of incident beam (i.e. the excitation source) interact constructively causing an enhanced PL emission. The transition of electrons from highest occupied molecular orbital (HOMO) to the lowest occupied molecular orbital (LUMO) could happen for both QD and CLC material if the dispersion of QD causes an excitation of CLC molecules thereby adding on to the overall rise in the PL intensity occurring in the composites. No significant shift was reported in the PL intensity peak in comparison to Pure CLC peak seen at 482 nm. The authors have also calculated quantum yield with dopant concentration as shown in Fig. 7b and reported an increase in quantum yield with increasing dopant concentration. The FWHM shows a reduction with increasing concentration with respect to Pure CLC. For instance, for Pure CLC the value of FWHM was ~75 nm which decreased to ~74, ~73 and ~69 for 0.1, 0.25 and 0.5 wt% concentrations, respectively. Authors predicted that in this system, emission peak is seen due to the surface defect level transitions because for FWHM < 25 nm, peak corresponds to near band emissions however, if FWHM > 25 nm it represents surface defect level transitions. As the broadening of the peak decreases there occurs an increased PL intensity that minimizes the contribution from surface state emissions. The decrease in FWHM for QD-CLC composite suggests that QDs facilitate radiative emissions originating from energy bands instead of the defect level within CLC material. They also conducted UV absorbance study and demonstrated a reduction in the intensity of absorption as well as reduction in band gap by the doping of QDs in CLC mixture. It was proposed that such composite material could be useful in display devices that require high luminescence. Expanding on these insights, Pathinti et al. [73] where they reported the improvement in optical and dielectric properties of CLC dispersed with greenly synthesized TiO_2 NPs (size ~25.01 nm). A low cost, less-toxic and simple approach has been selected by the authors for the preparation of TiO_2 NPs wherein green synthesis was achieved using syzygium cumini (jamun) leaf extract as a stabilizing agent.

UV–Vis confirmed the absorption peak for all samples (T1: 0.1 wt%, T2: 0.25 wt% and T3: 0.5 wt%) at 329 nm hence for PL studies excitation wavelength was fixed at this peak absorption value. From the PL spectra (Fig. 8), it can be seen that the emission peak for Pure CLC and composites are seen at 392 nm. Beyond a specific concentration (>0.1wt%), dispersion of more TiO_2 NPs produced a significant enhancement in the PL intensity (Fig. 8 inset). The presence of electromagnetic field on the surface of NP greatly affects the transfer of electron between CLC and NP thereby increasing the PL intensity. As the composite gets exposed to electromagnetic radiations, they generate a cluster of intense electromagnetic fields around NP surface. Additionally, the authors also attributed to the increase in intensity to the constructive interference between incoming photons and the low energy phonons of NPs. As known generally that in a NP-CLC system electronic transition happens from HOMO to LUMO. Furthermore, the low band gap of this green synthesized TiO_2 NPs may also contribute in the rise in PL intensity. No change in the PL emission wavelength was observed for the NP-CLC composite. Since there occurred a decrease in UV-absorbance of CLC with increasing concentration of NP, PL emission increased. Authors predicted that this would cause a significant increase in the PL quantum yield thereby suggesting the strong impact of TiO_2 NPs on the PL strength of CLC. Authors hoped that such composites could be employed in display applications requiring high luminescence. Following this, Chemingui et al. [74] also explored the effect of doping CuInZnSe-QDs (size ~2 nm) in CLC matrix and reported the enhancement in PL intensity peak with increasing concentration of QDs. An asymmetric distribution in emission peak was observed which pointed towards the existence of multiple peaks in both Pure CLC and composites. Correspondingly, the authors fitted the PL spectra using Gaussian function and found the formation of two different emission peaks, an intense violet peak at ~410 nm wavelength and a weak blue peak at ~467 nm. Confirming the absorbance of electromagnetic energy at a particular wavelength and thereby emitting it at higher wavelengths. In all samples, there occurred no change in the peak wavelength, as well as there was no broadening or narrowing of the peaks except for a substantial increment in the PL intensity. For 0.2 wt% concentration, there occurred two-fold increment in peak intensity and in 0.4 wt% concentration a threefold increment was observed. Moreover, the authors also discussed the reduction in FWHM value with increasing concentration supporting the PL intensity peak enhancement. It was proposed that such effective tuning of optical property could open a new avenue in the fabrication of advanced anisotropic materials.

Exploring the arena of NPs doped PSCLC composites, Rishi et al. [75] synthesized photosensitive CdSe QDs (6–7 nm) to explore its effect on the luminescence of PSCLC. In this report, they mentioned that for undoped sample, the spectral peak of fluorescence polarization was seen in deep blue region which got shifted to visible yellow region with the addition of CdSe QDs. Increasing the concentration of CdSe QDs from 0.02 to 0.06 wt% produced a 32 nm spectral shift. For 0.02 wt% concentration, the value of circularly polarization has shown an increase by almost four times in contrast to undoped sample. Furthermore, the authors also recorded electrically

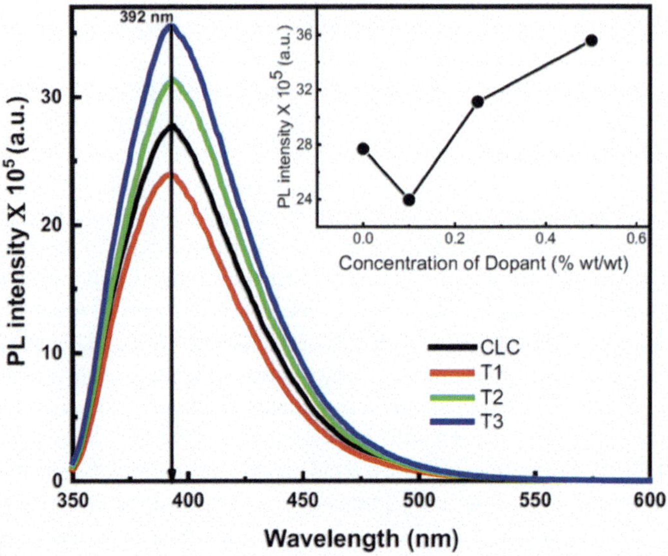

Fig. 8 Photoluminescence spectra of pure CLC, T1, T2 and T3 composites at excitation wavelength 329 nm in ITO films. Inset shows the variation in PL intensity with TiO_2 NPs concentration. Reproduced with permission from Ref. [73]

tuned PL spectra (Fig. 9). It can be seen that increasing the applied electric field causes a gradual reduction in the relative fluorescence intensity and reaches a state of saturation at higher electric fields. Doping causes a reduction in the driving field to 2.6 V/μm for 0.02 wt% concentration from 4.17 V/μm in the undoped sample indicating the swift orientation of LC molecules at homeotropic alignment when doped with 0.02 wt% CdSe QDs. Now increasing the concentration to 0.06 wt% exhibited an enhancement in driving field and attributed to the aggregation of QDs at higher concentration. Such aggregation caused a reduction in the fluorescence intensity corroborated well with the reduced optical contrast seen in optical textures. Authors proposed them as promising candidates for advancing liquid crystal displays with high contrast and reduced power consumption.

Chauhan et al. [76] depicted the utilization of PSCLC material as low power light shutter when doped with CdSe QDs (size ~3.8 nm). Reduction in dielectric permittivity by ~36%, lowered threshold and saturation voltages, enhanced dielectric anisotropy by ~68.32% for 0.1 wt% concentration and transformation of PSCLC material from planar to homeotropic state at low applied voltages were observed. The effect of CdSe QDs on the PL spectra of PSCLC was also explored and shown in Fig. 10.

The authors demonstrated that the samples were excited at 300 nm and gaussian fit was performed on the PL plots. PL spectra was obtained at 399 nm for all samples and no red or blue spectral shift happened with the doping of CdSe QDs in PSCLC system (Fig. 10). A 0.1 wt% composite didn't show much impact on the PL intensity,

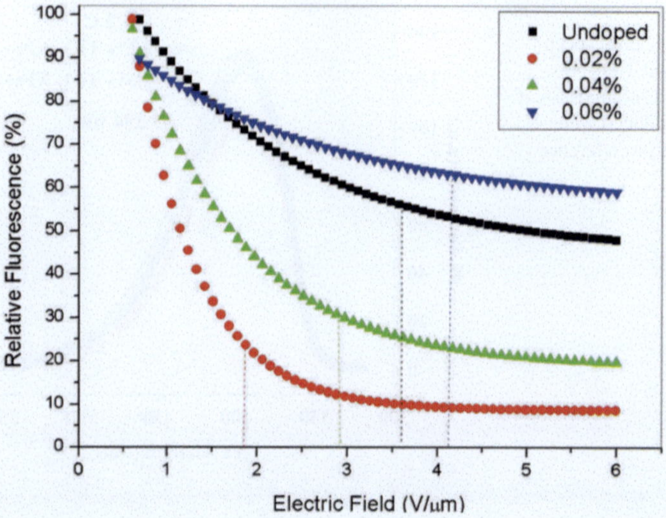

Fig. 9 Electrically controlled fluorescence scan of CdSe QDs dispersed PSCLC cells. Reproduced with permission from Ref. [75]

Fig. 10 PL spectra of PSCLC, 0.1 and 0.25 wt% CdSe QDs doped PSCLC. Reproduced with permission from Ref. [76]

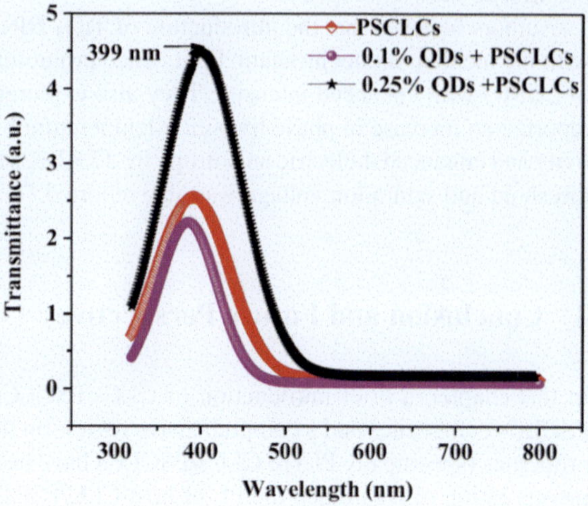

however in 0.25 wt% composite, a significant rise of 75% in peak intensity was reported. Such observed enhancement could be due to the addition of extra radiative levels in PSCLC material by the doping of CdSe QDs. Authors believed that CdSe QDs-PSCLC composites would be useful in the preparation of low power operated optical devices that require high luminescence for their effective functioning. In another work, authors reported the influence of titanium dioxide (TiO_2, ~50–70 nm) NPs on the dielectric, electro-optic and PL properties of PSCLC [77]. In PL study,

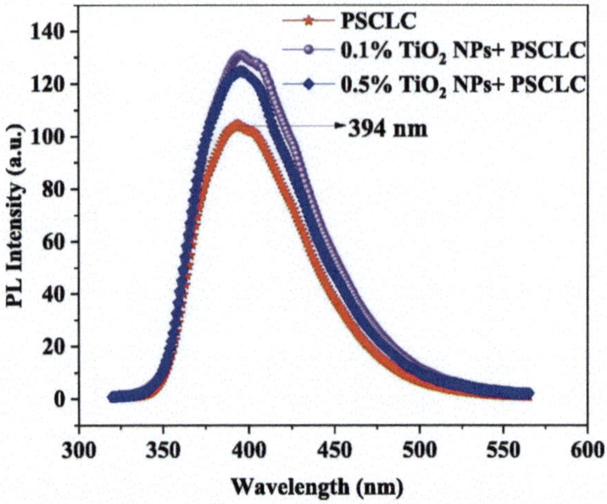

Fig. 11 Photoluminescence spectra of PSCLC and TiO_2 NPs doped PSCLC composites at 30 °C. Reproduced with permission from reference [77]

authors recorded the emission spectra at an excitation wavelength of 300 nm with a slit width of 5 nm. For both concentrations of 0.1 and 0 0.5 wt%, PL intensity got enhanced and they observed the emission peak at 394 nm with no shift in peak position as shown in Fig. 11.

Authors believed that the introduction of TiO_2 NPs facilitates the LC interaction with the incident monochromatic light hence promoting the transition from HOMO to LUMO with enhanced intensity. They also explored on dielectric properties and reported an increase in phase transition temperature by ~3 °C, reduction in permittivity and enhanced dielectric anisotropy by 27.83%. Moreover, the reduction in both threshold and saturation voltage was also reported.

4 Conclusion and Future Perspectives

In this chapter, a brief introduction of CLCs/PSCLCs, preparation of NPs-CLCs/PSCLCs composites and a comprehensive review on the effect of NPs on the optical properties (specifically PL) of CLCs/PSCLCs have been provided. Interestingly, the incorporation of NPs tuned the PL of host CLC/PSCLC materials in various ways. For instance, the enhancement in the PL intensity was found in multiple studies that establish the robustness of NPs in providing high luminescence in CLC/PSCLC matrix thereby making them suitable for display applications. The formation of strong electromagnetic field on the surface of metallic NPs when illuminated with incident light source eventually causes an enhancement in the PL intensity peak. Additionally, researchers have reported the spectral shift of PL peak with the dispersion of NPs and attributed it to the elongation of CLC's helical pitch. These results clearly provide an emphasis on the significant impact that doping of NPs bring in the properties of CLC/PSCLC materials. Tuning the helical pitch, PL intensity and FWHM of CLC/

PSCLC material are some of the targeted changes produced by the doping of NPs. It is worth pointing out here that these nanocomposites would be certainly beneficial in the fabrication of display and other optical devices that require high luminescence. Also, the reports on the impact of NPs on the PL of PSCLCs are less compared to CLCs and hence there is a good scope for further research in this area.

Besides yielding intriguing results on NPs-CLCs/PSCLCs composites for high performance luminescent LC devices, it is essential to carefully investigate their toxicity, stability and durability. The toxicity issue could be mitigated by using the eco-friendly NPs such as TiO_2, carbon dots, graphene quantum dots, MoS_2 QDs, etc. Additionally, further experimental and theoretical research on these CLCs/PSCLCs nanocomposites should be carried out in order to examine the stability and durability of the fluorescent devices based on them.

Acknowledgements Priscilla P thanks UGC for the fellowship awarded under the scheme Savitribai Jyotirao Phule Single Girl Child Fellowship (SJSGC 2022–2023). Jai Prakash and Gautam Singh are grateful to the Council of Science and Technology (CST), U. P., India for financial assistance under CST, U. P. funded research project (Project ID: 1960). Authors Supreet and Gautam Singh acknowledge the support provided under SERB SURE Grant No. SUR/2022/004020 of Govt. of India.

References

1. Di Lisi, G.A., De Luca, J.J.: An Introduction to Liquid Crystals, 1st edn. Morgan and Claypool Publishers, United Kingdom (2019). https://doi.org/10.1088/2053-2571/AB2A6F
2. Geelhaar, T., Griesar, K., Reckmann, B.: 125 years of liquid crystals—a scientific revolution in the home. Angew. Chem. Int. Ed. **52**, 8798–8809 (2013). https://doi.org/10.1002/ANIE.201301457
3. Kelker, H.: History of liquid crystals. Mol. Cryst. Liq. Cryst. **21**, 1–48 (1973). https://doi.org/10.1080/15421407308083312
4. Chilaya, G.S., Lisetski, L.N.: Cholesteric liquid crystals: physical properties and molecular-statistical theories. Mol. Cryst. Liq. Cryst. **140**, 243–286 (1986). https://doi.org/10.1080/00268948608080157
5. Mitov, M.: Cholesteric liquid crystals in living matter. Soft Matter **13**, 4176–4209 (2017). https://doi.org/10.1039/C7SM00384F
6. Blinov, L.M.: Structure and Properties of Liquid Crystals, 1st edn. Springer Dordrecht, Netherlands (2011). https://doi.org/10.1007/978-90-481-8829-1
7. Jain, A.K., Deshmukh, R.R.: An overview of polymer-dispersed liquid crystals composite films and their applications. In: Liquid Crystals and Display Technology. IntechOpen (2020). https://doi.org/10.5772/INTECHOPEN.91889
8. de Gennes, P.G., Prost, J.: The Physics of Liquid Crystals, 2nd edn. Clarendon Press, United Kingdom (1993)
9. Dierking, I.: Textures of Liquid Crystals. Wiley-VCH (2003)
10. Coates, D.: Development and applications of cholesteric liquid crystals. Liq. Cryst. **42**, 653–665 (2015). https://doi.org/10.1080/02678292.2015.1020454
11. Lee, H.G., Munir, S., Park, S.Y.: Cholesteric liquid crystal droplets for biosensors. ACS Appl. Mater. Interfaces **8**, 26407–26417 (2016). https://doi.org/10.1021/ACSAMI.6B09624

12. Kawamoto, H.: The history of liquid-crystal display and its industry. In: 3rd Region-8 IEEE History of Electro-Technology Conference: The Origins of Electrotechnologies, Conference Proceedings (2012). https://doi.org/10.1109/HISTELCON.2012.6487587
13. Tamaoki, N.: Cholesteric liquid crystals for color information technology. Adv. Mater. **13**, 1135–1147 (2001). https://doi.org/10.1002/1521-4095(200108)13:15%3c1135::AID-ADMA1135%3e3.0.CO;2-S
14. Lin, T.-H., Fuh, A.Y.-G., Ko, S.-W., Huang, S.-H.: Measurement of helical twisting power based on axially symmetrical photo-aligned dye-doped liquid crystal film. Opt. Express **17**, 15926–15931 (2009). https://doi.org/10.1364/OE.17.015926
15. Mitov, M.: Cholesteric liquid crystals with a broad light reflection band. Adv. Mater. **24**, 6260–6276 (2012). https://doi.org/10.1002/ADMA.201202913
16. Gahrotra, R., Sharma, V., Dogra, A.R., Malik, P., Kumar, P.: Performance augmentation of bistable cholesteric liquid crystal light shutter- effect of dichroic dye on morphological and electro-optical characteristics. Opt Mater. **127**, 112243 (2022). https://doi.org/10.1016/j.optmat.2022.112243
17. Gahrotra, R., Sharma, V., Malik, P., Kumar, P.: Textural and electro-optical study of cholesteric liquid crystal for light shutter. In: AIP Conference Proceedings, vol. 2220, no. 1. AIP Publishing (2020). https://doi.org/10.1063/5.0001809
18. Mulder, D.J., Schenning, A.P.H.J., Bastiaansen, C.W.M.: Chiral-nematic liquid crystals as one -dimensional photonic materials in optical sensors. J. Mater. Chem. C Mater. **2**, 6695–6705 (2014). https://doi.org/10.1039/C4TC00785A
19. Belyakov, V.A., Dmitrienko, V.E., Orlov, V.P.: Optics of cholesteric liquid crystals. Sov. Phy. Uspekhi **22**, 64–88 (1979). https://doi.org/10.1070/PU1979V022N02ABEH005417
20. Tson, S., Wu, D., Yang, K.: Reflective Liquid Crystal Displays, 1st edn. Wiley, United Kingdom (2001)
21. Manabe, T., Sonoyama, K., Takanishi, Y., Ishikawa, K., Takezoe, H.: Toward practical application of cholesteric liquid crystals to tunable lasers. J. Mater. Chem. **18**, 3040–3043 (2008). https://doi.org/10.1039/B802461H
22. Li, J., Bisoyi, H.K., Tian, J., Guo, J., Li, Q.: Optically rewritable transparent liquid crystal displays enabled by light-driven chiral fluorescent molecular switches. Adv. Mater. **31**, 1807751 (2019). https://doi.org/10.1002/ADMA.201807751
23. Wang, L., Bisoyi, H.K., Zheng, Z., Gutierrez-Cuevas, K.G., Singh, G., Kumar, S., Bunning, T.J., Li, Q.: Stimuli-directed self-organized chiral superstructures for adaptive windows enabled by mesogen-functionalized graphene. Mater. Today **20**, 230–237 (2017). https://doi.org/10.1016/j.mattod.2017.04.028
24. Li, X., Zhang, M., Zhang, C., Niu, R., Ma, H., Sun, Y.: A bistable ion-doped cholesteric liquid crystal smart window with a small amount of polymer. Opt. Mater. (Amst) **138**, 113659 (2023). https://doi.org/10.1016/J.OPTMAT.2023.113659
25. Oh, H., Kikuchi, H., Lee, J.H., Kim, S.J., Lee, J.B., Cho, M.S., Lee, M.Y., Okumura, Y., Hong, J.H., Hong, S.K.: Ultraviolet light screen using cholesteric liquid crystal capsules on the basis of selective reflection. RSC Adv. **11**, 25471–25476 (2021). https://doi.org/10.1039/D1RA03499E
26. Guo, M., Xiao, R., Prempeh, N., Liu, D., Fu, J., Cai, Y., Oumaima, A., Iimaa, T., Surenjav, U.: Cellulose-derived cholesteric liquid crystal: towards smart and sustainable colorimetric sensors. Ind. Crops. Prod. **192**, 116007 (2023). https://doi.org/10.1016/J.INDCROP.2022.116007
27. Sun, C., Lu, J.: Optical filters based on cholesteric, blue and sphere mesophases. Polymers **14**, 4898 (2022). https://doi.org/10.3390/POLYM14224898
28. Mitov, M., Nouvet, E., Dessaud, N.: Polymer-stabilized cholesteric liquid crystals as switchable photonic broad bandgaps. Euro. Phys. J. E: Soft Matter Biol. Phys. **15**(4), 413–419 (2004). https://doi.org/10.1140/epje/i2004-10058-4
29. Broer, D.J.: Polymer-Stabilized Cholesteric Textures: Materials and applications, 1st edn. CRC Press (1996). https://doi.org/10.1201/9781482272796-5
30. Ye, Y., Guo, L., Zhong, T.: A review of developments in polymer stabilized liquid crystals. Polymers **15**, 2962 (2023). https://doi.org/10.3390/POLYM15132962

31. Lu, Y., Yang, D., Qin, J., Zhong, Z., Lv, D., Gao, H., Du, X., Zhao, Y., He, Z., Cao, H., Yang, Z., He, W., Wang, D.: Polymer-stabilized cholesteric liquid crystals with multi-state reversibility for applications in advanced smart adjustable anti-counterfeiting and energy-saving displays. Chem. Eng. J. **499**, 156338 (2024). https://doi.org/10.1016/J.CEJ.2024.156338
32. Adimule, V., Yallur, B.C., Keri, R., Bathula, C., Batakurki, S.: Microstructure, photoluminescence and electrical properties of SmxGd (1–x): SrO hybrid nanomaterials synthesized via facile coprecipitation method. Electron. Mater. Lett. **19**, 278–297 (2023). https://doi.org/10.1007/s13391-022-00394-0
33. Maalathi, M.C., Yallur, B.C., Ambika, M.R., Adimule, V.: Influence of nano particles on optical properties of Cu-MOFs. Adv. Mat. Res. **1173**, 23–33 (2022). https://doi.org/10.4028/p-vn4hd4
34. Strambeanu, N., Demetrovici, L., Dragos, D., Lungu, M.: Nanoparticles: definition, classification and general physical properties. In: Nanoparticles' Promises and Risks: Characterization, Manipulation, and Potential Hazards to Humanity and the Environment, pp 3–8. Springer International Publishing (2015). https://doi.org/10.1007/978-3-319-11728-7_1
35. Xu, L., Liang, H.W., Yang, Y., Yu, S.H.: Stability and reactivity: positive and negative aspects for nanoparticle processing. Chem. Rev. **118**, 3209–3250 (2018). https://doi.org/10.1021/ACS.CHEMREV.7B00208
36. Ealias, A.M., Saravanakumar, M.P.: A review on the classification, characterisation, synthesis of nanoparticles and their application. In: IOP Conf Ser Mater Sci Eng, p. 032019. Institute of Physics Publishing (2017). https://doi.org/10.1088/1757-899X/263/3/032019
37. He, Z., Zhang, Z., Bi, S.: Nanoparticles for organic electronics applications. Mater. Res. Express **7**, 012004 (2020). https://doi.org/10.1088/2053-1591/AB636F
38. Trave, E., Bello, V., Enrichi, F., Mattei, G., Borsella, E., Carpanese, M., Falconieri, M., Abate, C., Herlin-Boime, N., Jursiokova, K., Costa, F., Costa, L., Gini, L.: Towards controllable optical properties of silicon-based nanoparticles for applications in opto-electronics. Opt. Mater. (Amst) **27**, 1014–1019 (2005). https://doi.org/10.1016/J.OPTMAT.2004.08.055
39. de Aberasturi, D.J., Serrano-Montes, A.B., Liz-Marzán, L.M.: Modern applications of plasmonic nanoparticles: from energy to health. Adv. Opt. Mater. **3**, 602–617 (2015). https://doi.org/10.1002/ADOM.201500053
40. Jamkhande, P.G., Ghule, N.W., Bamer, A.H., Kalaskar, M.G.: Metal nanoparticles synthesis: an overview on methods of preparation, advantages and disadvantages, and applications. J. Drug. Deliv. Sci. Technol **53**, 101174 (2019). https://doi.org/10.1016/J.JDDST.2019.101174
41. Zhang, L., Gu, F.X., Chan, J.M., Wang, A.Z., Langer, R.S., Farokhzad, O.C.: Nanoparticles in medicine: therapeutic applications and developments. Clin. Pharmacol. Ther. **83**, 761–769 (2008). https://doi.org/10.1038/SJ.CLPT.6100400
42. Lin, W.: Introduction: nanoparticles in medicine. Chem. Rev. **115**, 10407–10409 (2015). https://doi.org/10.1021/ACS.CHEMREV.5B00534
43. Hamida, R.S., Ali, M.A., Redhwan, A., Bin-Meferij, M.M.: Cyanobacteria-a promising platform in green nanotechnology: a review on nanoparticles fabrication and their prospective applications. Int. J. Nanomedicine **15**, 6033–6066 (2020). https://doi.org/10.2147/IJN.S256134
44. Choudhary, A., Singh, G., Biradar, A.M.: Advances in gold nanoparticle-liquid crystal composites. Nanoscale **6**, 7743–7756 (2014). https://doi.org/10.1039/C4NR01325E
45. Garbovskiy, Y., Glushchenko, A.: Ferroelectric nanoparticles in liquid crystals: recent progress and current challenges. Nanomaterials **7**, 361 (2017). https://doi.org/10.3390/nano7110361
46. Singh, G., Fisch, M., Kumar, S.: Emissivity and electrooptical properties of semiconducting quantum dots/rods and liquid crystal composites: a review. Rep. Prog. Phys. **79**, 056502 (2016). https://doi.org/10.1088/0034-4885/79/5/056502
47. Adimule, V., Sharma, V., Kumar, P., Keri, R., Sharma, K.: Perylene-3, 4, 9, 10-tetracarboxylic dianhydride dye intercalated liquid crystal molecules: synthesis, spectroscopic characterizations and photoluminescence dynamics. J. Mol. Liq. **413**, 125990 (2024). https://doi.org/10.1016/j.molliq.2024.125990
48. Supreet, Singh, G.: Recent advances on cadmium free quantum dots-liquid crystal nanocomposites. Appl. Mater. Today. **21**, 100840 (2020). https://doi.org/10.1016/j.apmt.2020.100840

49. Bisoyi, H.K., Kumar, S.: Liquid-crystal nanoscience: an emerging avenue of soft self-assembly. Chem. Soc. Rev. **40**, 306 (2010). https://doi.org/10.1039/B901793N
50. Kumar, A., Priyam, Meena, H., Prakash, J., Wang, L., Singh, G.: Recent advances on semiconducting nanomaterials–ferroelectric liquid crystals nanocomposites. J. Phys.: Condens. Matter **34**, 013004 (2022). https://doi.org/10.1088/1361-648X/ac2ace
51. Yadav, S.P., Yadav, K., Lahiri, J., Singh Parmar, A.: Ferroelectric liquid crystal nanocomposites: recent development and future perspective. Liq. Cryst. Rev. **6**, 143 (2018). https://doi.org/10.1080/21680396.2019.1589400
52. Kumar, A., Singh, D.P., Singh, G.: Recent progress and future perspectives on carbon-nanomaterial-dispersed liquid crystal composites. J. Phys. D Appl. Phys. **55**, 083002 (2022). https://doi.org/10.1088/1361-6463/AC2CED
53. Singh, S.: Impact of dispersion of nanoscale particles on the properties of nematic. Cryst. **9**, 475 (2019). https://doi.org/10.3390/cryst9090475
54. Kumar, A., Singh, G.: Recent advances and future perspectives of photoluminescent liquid crystals and their nanocomposites for emissive displays and other tunable photonic devices. J. Mol. Liq. **386**, 122607 (2023). https://doi.org/10.1016/j.molliq.2023.122607
55. Manhas, N., Adimule, V., Lakshminarayana, P., Keri, R., Kumar, L.S., Bathula, C.: Utilization of CO^{2+} doped Eu_2O_3 nanostructures and their effect on the dielectric and photoisomerization of liquid crystals. Phys. Fluids **35** (2023). https://doi.org/10.1063/5.0171578
56. Neha, Singh, G., Kumar, S., Malik, P., Supreet: Recent trends and insights into carbon dots dispersed liquid crystal composites. J. Mol. Liq. **384**, 122225 (2023). https://doi.org/10.1016/j.molliq.2023.122225
57. Kurochkin, O., Buchnev, O., Iljin, A., Park, S.K., Kwon, S.B., Grabar, O., Reznikov, Y.: A colloid of ferroelectric nanoparticles in a cholesteric liquid crystal. J. Opt. A: Pure Appl. Opt. **11**, 024003 (2009). https://doi.org/10.1088/1464-4258/11/2/024003
58. Hung, Y.-H., Hwang, S.-J., Chen, S.-C., Jeng, S.-C.: Cholesteric liquid crystal devices with nanoparticle aggregation. Opt. Express **18**(21), 22572–22577 (2010). https://doi.org/10.1364/OE.18.022572
59. Wu, P.-C., Lisetski, L.N., Lee, W.: Suppressed ionic effect and low-frequency texture transitions in a cholesteric liquid crystal doped with graphene nanoplatelets. Opt. Express **23**(9), 11195–11204 (2015). https://doi.org/10.1364/OE.23.011195
60. Zhao, D., Zhou, W., Cui, X., Tian, Y., Guo, L., Yang, H.: Alignment of liquid crystals doped with nickel nanoparticles containing different morphologies. Adv. Mater. **23**, 5779–5784 (2011). https://doi.org/10.1002/ADMA.201102611
61. Zhao, D., Xu, L., Shang, Y., Li, X., Guo, L.: Facet-dependent electro-optical properties of cholesteric liquid crystals doped with Cu_2O nanocrystals. Nano Res. **11**, 4836–4845 (2018). https://doi.org/10.1007/S12274-018-2072-2
62. Chen, L.J., Dai, J.H., De Lin, J., Mo, T.S., Lin, H.P., Yeh, H.C., Chuang, Y.C., Jiang, S.A., Lee, C.R.: Wavelength-tunable and highly stable perovskite-quantum-dot-doped lasers with liquid crystal lasing cavities. ACS Appl. Mater. Interfaces **10**, 33307–33315 (2018). https://doi.org/10.1021/ACSAMI.8B08474
63. Pathinti, R.S., Tatipamula, A.K., Vallamkondu, J.: ZnO nanoparticles dispersed cholesteric liquid crystal based smart window for energy saving application. J. Alloys Comp. **963**, 171198 (2023). https://doi.org/10.1016/J.JALLCOM.2023.171198
64. Wang, X., Lu, Y., Qiu, T., Yang, D., Zhao, Y., Wang, D., Zhang, Y., He, Z., Miao, Z., Luan, Y.: Polymer stabilized cholesteric liquid crystals enhanced by functional nanofibers loaded with CuS NPs and ATO NPs for efficient smart windows and advanced infrared camouflage/anti-counterfeiting technologies. Appl. Mater. Today **42**, 102540 (2025). https://doi.org/10.1016/J.APMT.2024.102540
65. Huang, C.C., Chou, T.R., Chen, J.W., Chao, C.Y.: Enhancement of photoluminescence intensity of CdSe nanorods doped in cholesteric liquid crystals. Brazil. J. Phys. **45**, 41–46 (2015). https://doi.org/10.1007/S13538-014-0284-9
66. Middha, M., Kumar, R., Raina, K.K.: Memory effects in chiral nematic liquid crystals doped with functionalised single-walled carbon nanotubes. Liq. Cryst. **42**, 1028–1035 (2015). https://doi.org/10.1080/02678292.2015.1015180

67. Middha, M., Kumar, R., Raina, K.K.: Improved electro-optical response of induced chiral nematic liquid crystal doped with multi-walled carbon nanotubes. Ferroelectrics **495**, 75–86 (2016). https://doi.org/10.1080/00150193.2016.1136765
68. Middha, M., Kumar, R., Raina, K.K.: Effects of chirality on optical and electro-optic behavior of nematic liquid crystals doped with functionalized silver nanoparticles. J. Mol. Liq. **219**, 631–636 (2016). https://doi.org/10.1016/J.MOLLIQ.2016.03.087
69. Middha, M., Kumar, R., Raina, K.K.: Photoluminescence tuning and electro-optical memory in chiral nematic liquid crystals doped with silver nanoparticles. Liq. Cryst. **43**, 1002–1008 (2016). https://doi.org/10.1080/02678292.2016.1155772
70. Cao, M., Yang, S., Zhang, Y., Song, X., Che, Y., Zhang, H., Yu, Y., Ding, G., Zhang, G., Yao, J.: Tunable amplified spontaneous emission in graphene quantum dots doped cholesteric liquid crystals. Nanotechnology **28**, 245202 (2017). https://doi.org/10.1088/1361-6528/AA67CA
71. Liu, Y.S., Lin, H.C., Xu, H.L.: The surface Plasmon resonance effect on the defect-mode cholesteric liquid crystals doped with gold nanoparticles. IEEE Photonics J. **10**(5), 4700407 (2018). https://doi.org/10.1109/JPHOT.2018.2821560
72. Pathak, G., Hegde, G., Prasad, V.: Octadecylamine-capped CdSe/ZnS quantum dot dispersed cholesteric liquid crystal for potential display application: Investigation on photoluminescence and UV absorbance. Liq. Cryst. **48**, 579–587 (2021). https://doi.org/10.1080/02678292.2020.1799085
73. Pathinti, R.S., Gollapelli, B., Jakka, S.K., Vallamkondu, J.: Green synthesized TiO2 nanoparticles dispersed cholesteric liquid crystal systems for enhanced optical and dielectric properties. J. Mol. Liq. **336**, 116877 (2021). https://doi.org/10.1016/J.MOLLIQ.2021.116877
74. Chemingui, M., Yang, D., Wang, Y., Ayeb, H., Cai, W., Tang, A., Drevensek-Olenik, I., Zhang, X., Xu, J.: The influence of quantum dots on the optical properties of a room temperature cholesteric liquid crystal. Liq. Cryst. **49**, 2095–2107 (2022). https://doi.org/10.1080/02678292.2022.2103850
75. Kumar, R., Raina, K.K.: Optical and electrical control of circularly polarised fluorescence in CdSe quantum dots dispersed polymer stabilised cholesteric liquid crystal shutter. Liq. Cryst. **43**, 994–1001 (2016). https://doi.org/10.1080/02678292.2016.1155771
76. Chauhan, G., Malik, P., Malik, P., Deep, A.: Improved performance of cadmium selenide quantum dots-doped polymer stabilized cholesteric liquid crystals for light shutter. Liq. Cryst. **50**, 2540–2551 (2023). https://doi.org/10.1080/02678292.2023.2264249
77. Chauhan, G., Malik, P., Deep, A.: Morphological, dielectric, electro-optic and photoluminescence properties of titanium oxide nanoparticles enriched polymer stabilized cholesteric liquid crystal composites. J. Mol. Liq. **376**, 121406 (2023). https://doi.org/10.1016/J.MOLLIQ.2023.121406

Electro-Optical and Dielectric Properties of Nanomaterial Dispersed Cholesteric Liquid Crystal Composites

Harsh Sharma, Garima Chauhan, Manpreet Kaur, Ronak Kumar Sharma, Khushboo, and Praveen Malik

Abstract Cholesteric Liquid Crystals (CLCs) exhibit unique features of supramolecular periodic helical structure and chirality in molecules. The chiral structure of CLCs enables them to reflect the incident light, resulting in brilliant colors selectively. Under the influence of external stimuli, the order of magnitude of the helical pitch gets modified. These modifications alter the physical properties of system that are used in various applications. This chapter presents a comprehensive study of the combined theoretical and applied aspects of dielectric and electro-optical properties of chiral nematic. The special attention is devoted to the optical properties of nanoparticles dispersed CLCs and other dopants in CLC system with a focus on their applications in optical shutters, photonics, and switchable mirrors. Furthermore, this chapter aims to gain collaborations between physicists and engineers to integrate novel chiral photo switches, shutters, and the development of optical materials, including display technology, biomedical imaging, and optical communication systems.

Keywords Cholesteric liquid crystals · Photonics · Optical shutters · Response time

H. Sharma · M. Kaur · R. K. Sharma · Khushboo · P. Malik (✉)
Dr. B R Ambedkar National Institute of Technology, Jalandhar, Punjab, India
e-mail: malikp@nitj.ac.in

H. Sharma
e-mail: Harshsharmamuradnagar@gmail.com

M. Kaur
e-mail: manpreetk.ph.21@nitj.ac.in

R. K. Sharma
e-mail: Sharmaronak3006@gmail.com

Khushboo
e-mail: khushurpuk@gmail.com

G. Chauhan
Institute of Foreign Trade and Management, Muradnagar, India
e-mail: Chauhangarima656@gmail.com

P. Kumar et al. (eds.), *Cholesteric Liquid Crystals*, Engineering Materials,
https://doi.org/10.1007/978-981-96-3821-5_9

1 Introduction

An intriguing class of liquid crystal (LC) material is cholesteric liquid crystals (CLCs), also called chiral nematic LCs. One identifying feature of these compounds is their helical molecular arrangement, in which the orientation of the molecule twists along a helical axis, resulting in periodic layers [1–5]. This distinctive structure differentiates them from nematic liquid crystals (NLCs). The helical structure of CLCs is essentially formed by the chirality transferred from the chiral dopant to the NLCs phase, by which the cholesteric phase of LC appears [6].

A key factor that governs the optical characteristics of materials is the helical pitch, defined as the length over which the molecules undergo a complete twist [6–9]. The reflected wavelength (λ) is directly proportional to the helical pitch (p) under irradiation of polarized light normally on CLCs in accordance with Bragg's law. The result of this phenomenon is the selective reflection of particular light wavelengths [9, 10]. The orderly arrangement of molecules is upset, though, when the CLCs transition from their planar state (P) to the focal-conic (FC) state. The incoming polarized light is scattered by the randomly orientated molecules in this FC state, thereby removing the selective reflection in the P state [11, 12].

The remarkable potential of CLCs to revolutionize numerous fields has brought them considerable interest as a new generation of materials. These encompass light shutters, transparent displays, smart windows with color alteration that do not require curtains, see-through displays, augmented reality displays, etc. [13–17]. Their adaptability to new display technologies results from their exceptional optical characteristics and tunable behavior.

The optical behavior of cholesteric materials is primarily influenced by absorption, transmission, scattering, diffraction, and reflection; all determined by the molecular arrangement in their P state, FC state, and homeotropic (H) states. In the P state, CLC molecules maintain a helical structure within the LC cell. In contrast, in the FC state, the helix becomes disrupted, with molecules aligning at an angle to the cell substrate. Over the past five decades, significant experimental research has been conducted to study the existence and behavior of these states, particularly in how they affect the optical properties of CLCs [18, 19]. As a result, extensive research has been done on electrically and thermally induced changes to the optical properties of CLCs [20–27].

A variety of approaches have been devised to achieve energy-efficient CLC devices, encompassing polymer dispersed liquid crystals (PDLCs) [28–34] and polymer-stabilized liquid crystals (PSLCs) [35–39] under field application and specialized surface treatments. The CLC materials were additionally used for polymer-stabilized cholesteric texture (PSCT) light shutters, functioning in normal and reverse modes. The integration of nanomaterials within LCs has garnered considerable interest due to their capacity to enhance device performance. Various types of nanoparticles, including semiconducting, ferroelectric and metallic varieties, are currently under investigation for their potential to alter the optical, electro-optic, and dielectric characteristics of CLCs. This methodology seeks to improve the operational abilities and effectiveness of CLC-based technologies by utilizing the inherent distinctive characteristics in these nanomaterials [40–43].

In the present chapter, we discussed the working principle of CLCs and the effect of various doping materials such as nanomaterials, quantum dots (QDs), polymers, and dyes on the electro-optical, photoluminescence, absorbance, and dielectric properties of CLC.

2 Working Principle

CLCs undergo transitions between the P state, FC state, and H state in their switching mechanism, as shown in Fig. 1. The transition from the P state to the FC state is usually induced by applying a square wave pulse of alternating current (AC) across the cell.

Upon the application of a high voltage exceeding a specific threshold, the CLC transitions to the H state, wherein the orientation of LC molecules is perpendicular to the surface. Upon rapid deactivation of the voltage, the CLC will return to the P state.

Nevertheless, if the voltage is diminished progressively, the LCs transition into the FC state rather than the P state. In this process, the transition from the P state to the H state is accomplished through an indirect path that involves the intermediate FC state. The nature of the molecular realignment can result in a relatively slow transition time between these states, particularly from H state to P state [44]. The ability of these devices to switch between transparent (P state) and light-scattering (FC state) modes is advantageous because it enables to regulate the light that is transmitted and reflected.

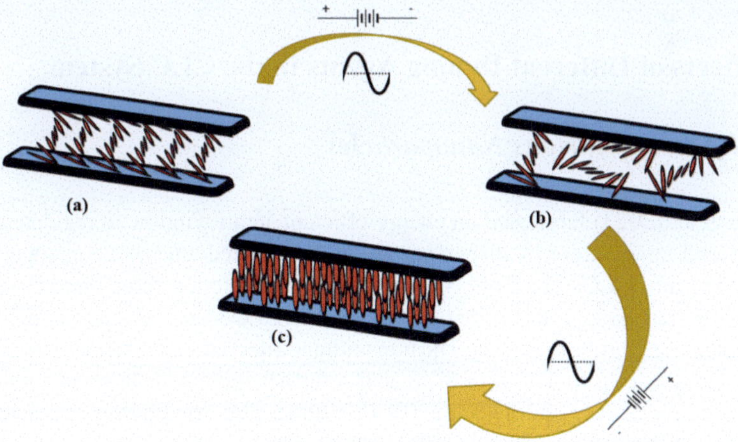

Fig. 1 Different cholesteric states with varying electric field (**a**) P state, (**b**) FC state and (**c**) H state

3 Applications of CLC Systems

CLCs have a wide range of applications across various fields due to their unique properties. In light shutters, CLCs offer significant advantages for light control, transitioning between transparent and opaque states in response to external stimuli like electric fields or temperature, making them ideal for adaptive lighting systems and display technologies that require rapid adjustments. Their energy efficiency is particularly beneficial for applications such as smart windows and privacy screens, allowing users to control exposure easily. In sensing applications, CLCs exhibit temperature-dependent color changes, functioning as effective temperature sensors, and can detect gases, pH variations, pressure, humidity, and other substances through colorimetric and optical responses. They are also utilized in bio-sensing to identify biological agents. In smart windows, CLCs provide dynamic light management, transitioning between transparent and reflective states for enhanced comfort, glare reduction, and energy savings while integrating with smart home systems for automated adjustments. In biomedicine, CLCs serve as sensitive biosensors for early disease detection and can facilitate controlled drug release in response to stimuli. Their self-organizing structure makes them suitable for tissue engineering, promoting cell development, and they can enhance wound dressings and biomedical imaging. CLCs are also used in reflective displays, such as electronic paper and smart labels, taking advantage of their ability to reflect specific wavelengths for vivid colors without backlighting and find applications in holographic displays and 3D visualization systems. Additionally, CLCs are employed in optical devices like filters and switches due to their selective reflection, and also can be integrated into smart textiles that change color or pattern in response to environmental stimuli, with potential applications in fashion and wearables.

4 Effects of Different Doping Agents in the CLC System

4.1 Effects of Various Nanoparticles

Pathinti et al. (2023) fabricated an energy-efficient smart window to regulate transparency and temperature by dispersing green-synthesized zinc oxide nanoparticles (ZnO NPs) into CLC. The ZnO NPs-doped CLCs exhibited three distinct optical states with increasing voltage: P state (translucent), FC state (opaque), and H state (transparent), as shown in Fig. 2a. The transmittance decreased in the FC state (2.4 V/μm) and reached up to 4% from 87%, and a further increment in transmittance is observed in the H state, from 4% to 90% (7 V/μm). This condition is characterized by a quick response time, often less than 1 ms [45].

The contrast ratio increased with doping as depicted from Fig. 3. The device demonstrated high stability, maintaining consistent spectral transmittance across states even after 20 cycles of switching the electric field on and off over 170 s.

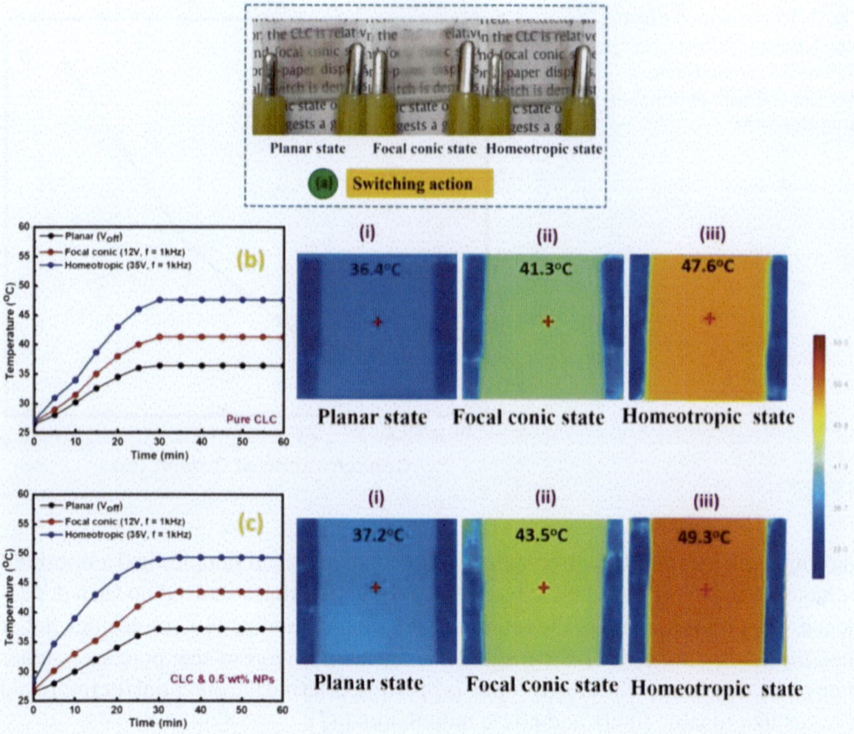

Fig. 2 [a] Optical switching of the device, [**b** and **c**—(**i, ii, iii**)] are the Photographs of infrared thermal pictures of samples illumination by simulated sunlight (100 mW cm^{-2}). Reproduced with permission from Ref. [45]

The thermal regulation capabilities of both CLC and ZnO NPs-doped CLC systems were evaluated under a solar simulator. Infrared thermal imaging revealed temperatures of 37.2 °C in the P state, 43.5 °C in the FC state, and 49.3 °C in the H state, as shown in Fig. 2[b, c]. By switching these states, the device can tune light and heat. Moreover, it can significantly reduce energy consumption by minimizing the use of heating and cooling systems in winter and summer, respectively [45].

Pathinti et al. investigated CLC systems doped with green synthesized (using *Syzygium cumini*, jamun leaf extract) Titanium dioxide (TiO_2) NPs and observed a reduction in both absorbance and bandgap, while photoluminescence (PL) was enhanced with TiO_2 doping. Broadband reflection analysis showed a blue shift with increased temperature, and homogeneous alignment was seen in the TiO_2 doped CLCs in Polarizing optical microscope (POM) textures. Additionally, dielectric studies indicated that the TiO_2-doped CLC systems exhibited a lower relaxation frequency than pure CLCs [46].

Kim et al. studied polymer/CLC nano-composites, which exhibit fast, deformation-free electro-optic tuning, unlike conventional CLCs. They also investigated the relationship between helical pitch and threshold voltage in five samples with

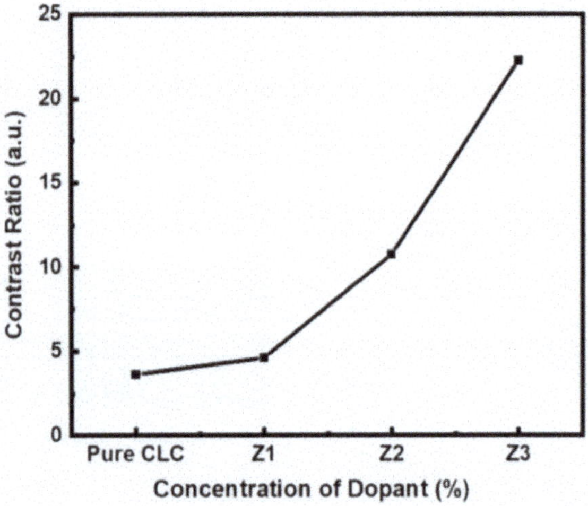

Fig. 3 Contrast ratio versus pure CLC and NPs doped CLC composites. Reproduced with permission from Ref. [45]

varying pitch lengths. They found that while unpolymerized samples had a threshold voltage inversely proportional to pitch, polymerized samples showed no such dependence. This was ascribed to the reorientation of nano-confined LC molecules rather than the unwinding of the helical structure. The study suggests that pore size, rather than pitch, determines the electro-optic properties, offering the potential for improved devices like tunable filters and phase modulators [47].

Shivaraja et al. have shown the electro-optic switching property of functionlized silver NPs (f-AgNPs) of varying concentrations distributed into 4-*trans*-pentyl cyclohexyl cyanobenzene (5PCH) in twisted nematic (TN) cells. The room temperature calamatic NLC, 5PCH, and functionalized (hexane thiol capped) f-AgNPs with an average diameter of 3–5 nm were employed in this work. The POM images verified the electro-optic switching of nematic in TN cells under applied voltage and also verified the uniform dispersion of NPs. The dielectric anisotropy decreased after dispersing f- AgNPs into 5PCH in the TN cells. The threshold voltage (V_{th}) was dropped upto 12.7% with 0.05 wt% of f-AgNPs dispersed 5PCH. With 0.025 wt% of f-AgNPs dispersed 5PCH, when 0.025 weight percent of f-AgNPs doped 5PCH was compared to pure 5PCH, the switching voltage—that is, the voltage needed to change the transmission intensity from 90 to 10% of highest intensity decreased by 69%. When compared to pure 5PCH, the V-T graph's steepness increased by almost 230% in the 0.025 weight percent of f-AgNPs dispersed 5PCH, which is beneficial for displays. However, a higher concentration of f-AgNPs doped nanocomposites exhibits slower switching compared to a lower concentration of f-AgNPs but faster switching than the pure 5PCH due to the aggregation of NPs in NLC. When comparing the f-AgNPs doped nanocomposites to pure 5PCH, the ion transport test reveals a decrease in the maximum ion current, suggesting that impurity ions are trapped on the f-AgNPs surface [48].

Chauhan et al. have prepared pure polymer-stabilized cholesteric liquid crystals (PSCLCs), 0.1 wt.% and 0.5wt.% TiO_2 NPs doped PSCLC composites in planar configuration, the dielectric and electro-optic properties of prepared samples were observed. In this work, NLC material (BL036) (alkyl cyanobiphenyl), chiral dopant CB15 with helical twisting power (HTP) 6.7 were used to induce a chirality into host NLC. Norland optical adhesive (NOA65) polymer of refractive index 1.52 was utilized to prepare the composites. TiO_2 NPs: a mixture of rutile and anatase was used as a doping material of average size 50–70 nm. The phase transition of TiO_2 NPs doped PSCLC composites enhanced (3 °C) for 0.5 wt.% than pure PSCLC system. The morphological analysis suggested that in the PSCLC system doped with TiO_2 NPs, the LC molecules operate at low voltage. For the 0.5 weight percent TiO_2 NPs doped PSCLC system, the permittivity decreased from 9.62 to 7.75, however, the dielectric anisotropy of the composites raised to 27.83%.

The saturation voltage decreased from 10.5 to 7.5 V, and the V_{th} decreased from 3.2 to 1.4 V, with 0.5 wt.% doped TiO_2 NPs represented by Fig. 4. As the concentration of TiO_2 NPs increased, PL intensity increased as well. TiO_2 NPs' ability to capture ions greatly improved the PSCLC system's physical characteristics as shown in Fig. 5. The findings indicate that TiO_2 NPs can regulate the ionic influence on PSCLC. The enhanced performance could aid in developing PSCLC devices with reduced energy usage [38].

Middha et al. investigated the effects of chirality on the behavior of NLC molecules doped with silver nanoparticles (Ag NPs). After adding chirality to doped NLCs, significant enhancement in photoluminescence, electro-optical, and thermoelectric behavior has been observed, as shown in Fig. 6 and Table 1 [49].

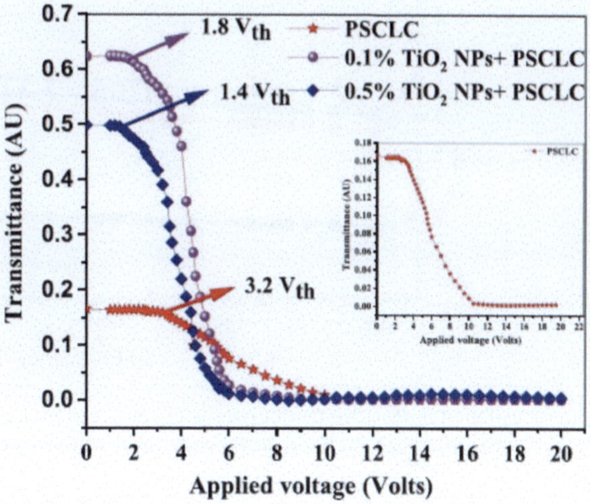

Fig. 4 Voltage versus Transmittance curve for pure and doped samples. Reproduced with permission from Ref. [38]

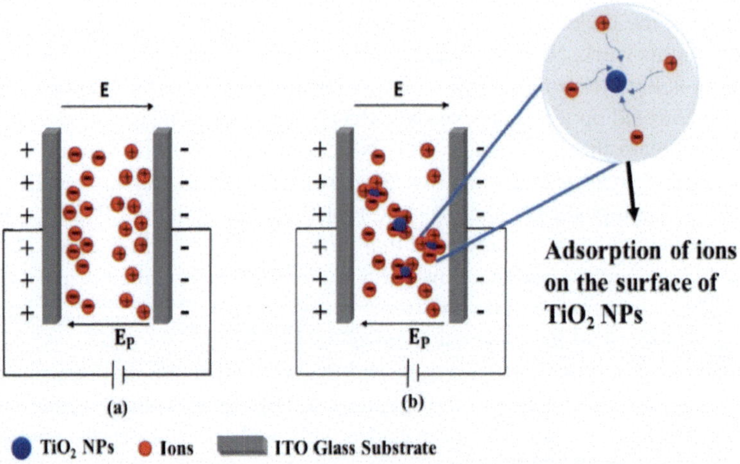

Fig. 5 Adsorption of TiO_2 nanoparticles in PSCLC matrix. Reproduced with permission from Ref. [38]

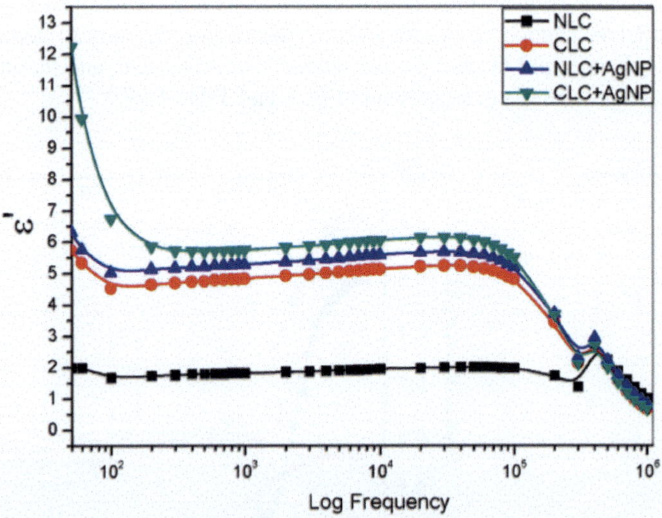

Fig. 6 Variation of Relative permittivity with respect to frequency between pure NLC, CLC and their mixtures with AgNPs. Reproduced with permission from Ref. [49]

4.2 *Effect of Quantum Dots (QDs)*

Advances in nanotechnology have made it possible to produce semiconducting QDs, which have been studied [50]. QDs are especially useful for observing size distribution challenges due to their size-dependent quantum confinement properties [51].

Table 1 Summarized results with the effect of doping of AgNPs in NLC and CLC matrix

Sample	Planar → iso (°C)	Conductivity at 200 kHz ($S \cdot cm^{-1}$)	ϵ' (60 Hz, 30 °C)	Threshold voltage (V)	Peak PL intensity (a. u.)	λ_{max} (nm)
NLC	95–121	6.902×10^{-6}	1.97	10	140	384
CLC	95–117	2.544×10^{-5}	5.46	8	190	407
NLC + 0.002 wt.% AgNPs	122–154	3.159×10^{-5}	5.75	7	290	410
CLC + 0.002 wt.% AgNPs	76–107	2.792×10^{-5}	10.02	6.3	617	413

Reproduced with permission from Ref. [49]

They possess unique optical characteristics, including high quantum yields, fluorescence emission tunability based on size, a large absorption cross-section, broad excitation spectra, better color purity, and high stability. These features have generated significant interest in QDs for applications in photodetectors, QD displays, lasers, next-generation photonic devices and biomedical images [52]. Numerous studies have explored circularly polarized fluorescence emission of QDs in CLCs and their tunable emission properties [53, 54]. Additionally, CdSe/ZnS QDs have been utilized to modify the electro-optical characteristics of NLCs, decreasing the effect of switching times [55] and V_{th} for the Fredericks transition [56]. Additionally, QDs also used for tuning the electro-optical and dielectric properties of ferroelectric LCs [57–59]. Here, we explored the effects of QDs on the physical properties of CLCs, such as electro-optical, dielectric, absorbance, and photoluminescence (PL).

Due to their higher photostability and interaction with LC molecules, QDs not only act as a nanoscale source of fluorescence but also act as structural stabilizers for CLCs, as proved by Chen et al. [56]. Pathak et al. have determined PL, Ultraviolet-visible (UV-Vis) absorption spectra, Full width at half maximum (FWHM), etc. This study used Cadmium Selenide/Zinc Sulfide (CdSe/ZnS) QDs of diameter ~6 nm as a dopant for pure CLC and QD dispersed CLC composites. The optical absorption and emission peaks of these QDs are located in the 530–550 nm and 545–565 nm ranges, respectively. It has been discovered that when the number of QDs in pure CLC grows, so does the PL intensity. This enhanced PL intensity is also supported by decrement in FWHM parameter. A study on UV absorbance revealed that for QD-dispersed CLC systems, the intensity of absorption decreases. Following QDs dispersion into CLC, a reduction in the band gap is also seen. Materials with a smaller band gap are highly advantageous for optoelectronic devices. As a result, they provided insight into how QDs affect CLC optical characteristic [57].

Gulsum Kocakulah et al. have utilized chiral dopant S-811, with helical twisting power (HTP) of ~11 μm^{-1} and Indium Phosphide/Zinc Sulfide (InP/ZnS) QDs, a

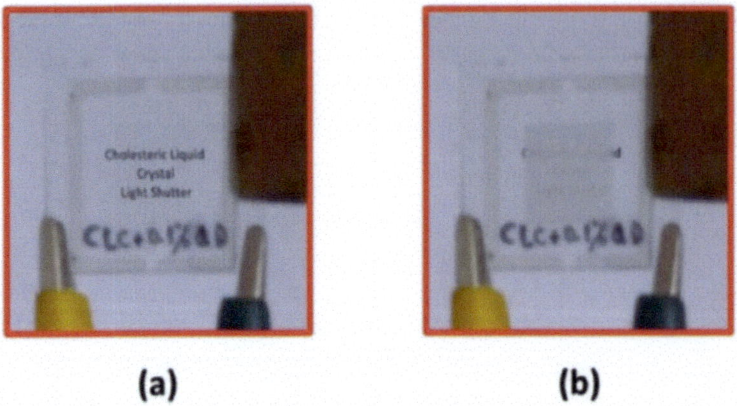

Fig. 7 Operating mechanism of CLC + 0.1wt.% QDs (**a**) at 0 V (**b**) at 7.5 V and 50 Hz frequency. Reproduced with permission from Ref. [58]

left-handed chiral dopant and investigated Pristine and InP/ZnS QDs dispersed CLC mixtures produced transparent and opaque appearance in the absence of voltage (P-state) and in the presence of certain voltage (FC-state) as shown in Fig. 7 respectively. InP/ZnS QDs contribution resulted in a decrease in the VP-FC, τR, and Eg parameters of CLC. In addition, the values of ε', ε'' and σac parameters of CLC increased on dispersion of InP/ZnS QDs at lower frequencies. Specifically, it was found that the CLC + 0.1wt.% QDs sample's optical, electro-optical, and dielectric properties produced the best results out of all the samples that were analyzed and were helpful in the creation of light shutters [58].

Chauhan et al. have fabricated PSCLC composite by using CB15 for chirality, polymer; poly methyl methacrylate (PMMA). Cadmium Selenide (CdSe) QDs with an average size of 3.8 nm were used as a doping material in PSCLCs. They studied the CdSe QDs effect in different concentrations on the morphological, optical, dielectric and photoluminescence properties. According to voltage-dependent morphological tests, LC molecules change from homogeneous to homeotropic alignment at a comparatively lower voltage than undoped and other CdSe QDs-doped PSCLC samples when the optimal concentration of CdSe QDs is reached (0.1 wt.%). For samples of PSCLCs doped with 0.25 and 0.1 wt.% CdSe QDs, the dielectric permittivity rise by 37.4% and 36.25%, respectively. At 0.1 wt.% CdSe QDs-doped PSCLCs sample, dielectric anisotropy is enhanced by approximately 68.32% while dielectric loss decreases. Low threshold and saturation voltages are also evident in the electro-optical measurements. The findings suggested that the LC molecules might be oriented in the direction of the applied electric field with a lower voltage. PL intensity was increased for PSCLCs doped with 0.25 wt.% QDs. These results suggested that an energy-efficient shutter might be created with an ideal concentration of CdSe QDs in PSCLC [59].

Kocakulah et al. prepared CLC with 5CB NLC, S-811 chiral dopant and other composites also prepared with azo dye Methyl red (MR) and CdSeS/ZnS QDs-doped CLC composites. The thermal, morphological, electro-optical and dielectric properties of these structures named as CLC and CLC/MR/QDs were examined. The phase transition temperature decreased due to the rising molecular weight and number of impurity ions in the samples when compared to used 5CB NLC. The phase transition temperature of the CLC/MR/QDs sample at lower temperatures is very important for the creation of temperature-controlled CLC light shutter devices that can function in the low-temperature range. The presence of azo dye MR and the impacts of CdSeS/ZnS QDs caused the drop in transition voltage for transition from P state to FC state and FC state to H state by 10.2 and 9.8%, respectively, in comparison to the pure CLC sample. POM images and photographs concluded that the samples showed opaque and transparent properties. However, frequency and voltage- dependent dielectric characteristics of the samples were examined. It was found that the ε' value in the P-state of the CLC/MR/QDs was higher than that of the CLC because of the enhanced electrical polarization in the applied electric field [60].

Kocakulah et al. have studied the impact of CdSeS/ZnS QD on the electro-optical and dielectric properties of CLC. The transmittance values of CLC-QDs composites in the FC state reduced compared to pure CLC which is advantageous for light shutter device applications. As QDs concentration increased, the composites' response time and transition voltage values similarly dropped. Furthermore, when the voltage increased, so did the real and imaginary components of the complex dielectric constant. The obtained results indicate that the dielectric and electro-optical properties of the CLC are greatly impacted by the CdSeS/ZnS QDs [61].

The study by Chemingui et al., observed that the introduction of Cu/In/ZnSe-QDs forms clusters that distort the CLC helix, causing an imperfect planar texture and broadening the photonic band gap. These broadening shifts the central and band-edge wavelengths as QDs concentration increased. The addition of QDs significantly enhances PL intensity, supported by a reduction in the full width at half maximum (FWHM). UV absorbance increased, with a bit red shift, and the optical band gap narrowed after QDs dispersion. Textural analysis revealed a slight reduction in the clearing temperature and a disturbance in the CLC director field due to QDs self-aggregation. The increased PL intensity from QDs dispersion suggests the potential for improving PL-CLC display devices [62].

Bobrovsky et al. made significant advancements in the development of new optical materials by doping CLCs with fluorescent CdSe/ZnS QDs. These materials exhibited both optically and electrically controlled circularly polarized emission, showcasing the ability to manipulate light in CLC systems with semiconductor QDs. This has potential for applications in advanced display technologies and optoelectronics, where circular polarization and emission control are crucial [63].

Cao et al. (2017) further explored the use of QDs in CLCs by demonstrating tunable amplified spontaneous emission in graphene QDs-doped CLCs. This work highlighted the versatility of graphene QDs in enhancing light emission properties, especially through the unique photonic properties of CLCs, enabling the tuning of light output through external stimuli [64].

Rodarte et al. (2012) explored the role of the CLC matrix's selective photonic bandgap in modulating the polarization of QDs photoluminescence. By leveraging the one-dimensional photonic cavity formed by the CLCs, the study revealed a way to control the spectral and polarization characteristics of QDs emission, a valuable property for optical filters, lasers, and sensors [65].

Lukishova et al. (2012) reported that QDs, when placed in a CLC micro-resonator, exhibited amplified fluorescence under continuous wave excitation, particularly in the near-infrared (IR) region. This amplification occurs due to the resonance in the photonic bandgap of the CLC host, making the system promising for IR emission devices and optical amplifiers [66].

Additionally, Lee et al. (2014) demonstrated an optically stable and tunable QDs nanocrystal-embedded CLC composite laser. This innovation combined the tunability of QDs nanocrystals with the optical properties of CLCs to produce a highly adaptable laser system, with potential applications in tunable laser sources and optical communications [67]. Table 2 represents the outcomes of doping of different materials into CLCs composites.

Table 2 Effect of doping materials in CLCs

Author	Materials	Outcomes	Year	References
Kocakulah et al.	5CB LC, Chiral dopant S811, MR azo dye	Observed Planer, focal conic, and homeotropic states	2020	[68]
Liang et al.	LC, Chiral dopant S811, polymer RM257	CR and driving voltage increased with chiral dopant	2012	[69]
Wu et al.	LC E7, Chiral agent R5011, Graphene nanoplatelets (GNPs)	Switched from the planar to focal conic state at lower operating voltage	2015	[70]
Middha et al.	LC ZLI-4151, Chiral dopant CB15, carboxylic (COOH) functionalized Multi-walled carbon nanotubes (fMWCNT)	Improved the ac conductivity and optical anisotropy and significantly reduced threshold	2016	[71]
Middha et al.	NLC ZLI-4151, Chiral dopant CB15, 0.002 wt.% Ag NPs	Permittivity increased from 5.46 to 10.02, threshold reduced from 8 to 6.3 V, and PL intensity enhanced from 190 to 617	2016	[72]
Singh et al.	Chiral LC (2CD6HBN)	Enhanced activation energy (2.57 eV)	2013	[73]
Jessy et al.	LC 5CB, Cholesteryl - oleyl carbonate	Higher chirality in 3 N + 7C mixture	2018	[74]

4.3 Effects of Different Chiral Molecules and Polymer Doping

Moheghi et al. presented a bistable switchable light shutter utilizing CLC doped with tetrabutyl ammonium bromide. This shutter displays two stable states at zero voltage: a transparent P and a scattering FC, as shown in Fig. 8. The results conclude that the transition from the transparent P state to the opaque FC state is achieved by applying a low-frequency voltage pulse (60 Hz), while the return to the transparent state is triggered by a high-frequency voltage pulse (2 kHz) [75].

Kim et al. demonstrated the effect of terminal chain length on the HTP in achiral bent-core molecules doped in a CLC; longer terminal chains showed larger HTPs. The behaviors of distribution changes for n = 8 to 12 and n = 12 to 16 are quite different despite the linear relationship because of excessive bulkiness [76].

Sun et al. prepared a CLC system by using NLCs (SLC-1717), chiral dopant (S811) and then prepared polymer dispersed cholesteric liquid crystal (PDCLC) with acrylate monomer was a mixture of 3,5,5-trimethelhexyl acrylate and 1,4-butanedioldiacrylate. The mercaptan monomers were 2, 2-(ethylenedioxy) diethanethiol (95%), trimethylopropane tris(3-mercaptopropionate) (95%) and pentaerythritol tetra(3-mercaptopropionate) PETMP (95%). They analyzed PDCLC films based on thiol/acrylate monomers. The addition of varying amounts of S811 along the network bone changed the LC domain size distribution, which in turn affected the electro-optical characteristics of the PDCLC films. The chiral dopant significantly impacted the morphology of the PDCLC film. The impact of the co-polymerized thiol monomer's content and functionality in the PDCLC films has been investigated. As previously stated, the electro-optical characteristics of the PDCLC film created with 7 wt.% PETMP and 3 wt.% S811 were superior to those prepared with different amounts of PETMP. According to the findings of this investigation, the chiral dopant is crucial to the development of the polymer network [77].

Kocakulah et al. fabricated CLCs cells which were prepared by using NLCs E7 (E7 is the mixture of 5CB (51 wt.%), 7CB (25 wt.%), 8OCB (16 wt.%), and 5CT (8 wt.%)); with left- handed chiral dopant S-811 under planar condition with 7.7 μm

Fig. 8 Photographs of the bistable CLC light shutter under 0 V. (**a**) Transparent state and (**b**) Opaque state. Reproduced with permission from Ref. [75]

space thickness. They investigated and compared the effect of cobalt (II) phthalocyanine [CoPc] concentration on the electro-optical and dielectric parameter of CLC. For this purpose, CLC, CLC + 0.1wt.% CoPc, CLC + 0.2wt.% CoPc, and CLC + 0.3wt.% CoPc samples were made. Electro-optical analysis and POM pictures revealed that all samples were opaque in the FC-state and transparent in the P-state. The voltage at FC-state was shown to drop as the concentration of CoPc in CLC increased, with the CLC + 0.3% CoPc sample exhibiting the lowest voltage. Dielectric measurements revealed that, at low frequencies, ε' rose as the concentration of CoPc in CLC increased. The frequency-dependent variation of the dielectric strength ($\Delta\varepsilon$) graphs showed that FC state of CoPc doped CLC composites rises in comparison to pure CLC, but these values fell as the concentration of CoPc in CLC increased. Therefore, it was determined that the dielectric properties were likewise impacted by the change in CoPc concentration. In conclusion, it is believed that the electro-optical and dielectric properties of CoPc doped CLC composites can be applied to the technology of optoelectronic devices, particularly light shutters [78].

Chauhan et al. prepared two sample by using LC BL036 with a chiral dopant CB15 and polarization is done by using polymer NOA65 under UV-interaction. The one cell (S_1) is prepared without any surface treatment while the other one (S_2) is treated with nylon. The morphological investigation discovered the phase transition alteration. According to the dielectric experiments, the surface-treated sample's permittivity and ac conductivity rise. At 40 °C, the values of ε' for samples S_2 and S_1 are 6.3 and 3.17, respectively. The surface-treated sample also exhibits an improvement in intensity in the photoluminescence spectrum. According to the electro-optical studies, the surface-treated PSCLC sample has a higher threshold and saturation voltage [38].

Yu-Kai Chang have investigated the cationic surfactant cetyltrimethylammonium bromide (CTAB) which was dispersed into CLC was prepared with high-HTP chiral additive R-5011 (with helical twisting power 115 μm) at 2.75 wt.% mixed into nematic host E7. Dielectric and electro-optical studies were investigating of these samples. The diffusion of doped ions within the CLC was indicated by a noticeable blue shift in the complex dielectric spectrum caused by the doping concentrations of CTAB (ranging from 0 to 0.20 wt.%). The diffusion coefficient increased from $0.34 \times 10–6\ cm^2{\cdot}s\ 1$ at c = 0 wt.% to $49.38 \times 10–6\ cm^2{\cdot}s\ 1$ at c = 0.30 wt.% at room temperature, and the mobile ion density increased from 2.17×10^{13} to $4.05 \times 10^{13}\ cm^3$, respectively. The increment in these values occurred at concentrations of CTAB from 0 to 0.50 wt.%. When an appropriate concentration (0.05 wt.%) of CTAB is doped, the electrohydrodynamic (EHD) effect can be successfully induced to subsequently generate the high- transmission ULH state between the two low-transparity FC and dynamic scattering (DS) states. It was discovered that the inclusion of CTAB significantly increases the operating frequency of the DS state in the expanded frequency range up to 300 Hz in addition to effectively utilizing the EHD effect to generate the ULH state [79].

4.4 Effect of Different Dyes

The conventional CLC films used in displays have shown limited optical performance for switchable windows, primarily due to poor shielding.

Oh et al. (2016) proposed a CLC light shutter combining dye-doping with a PDLC film, showing how dichroic dyes enhance optical control in switchable windows [80].

Kim et al. (2017) demonstrated bistable switching between homeotropic and focal-conic states in ion-doped CLC cells, suggesting that doping enhanced bistability and switching performance [81]. Additionally, Yu et al. (2015) showcased the simultaneous control of haze and transmittance in dye-doped CLC cells, further highlighting the ability to fine-tune optical properties for diverse applications [82].

Dichroic dyes exhibit shape anisotropy, resulting in directionally dependent absorption (anisotropic absorption). When these rod-like dye molecules are doped in LCs, they align with the LC director and can be electrically switched, enabling adjustable absorption based on the applied electric field. Oh et al. (2017) explored this further by integrating azo dyes into CLCs, achieving optical and electrical switching capabilities, enhancing CLC films' functionality for light modulation. This interaction between the dye's anisotropic properties and the LC's reorientational behavior under electric fields makes such systems highly effective for applications in switchable windows [83].

Gahrotra et al. developed a CLC mixture by combining chiral dopant R1011 (4 wt.%) with LC, which has dual frequency properties. Research has been done on the impact of azo orange 3 dichroic dye (0.4 wt.% of LC). They showed how to apply a low frequency (1 kHz) voltage at 30 V to pristine and dye-doped CLC light shutters, which can be readily converted from P state to FC state. Once the voltage is withdrawn, a stable FC state is maintained. In comparison to pure CLC, POM textural investigations demonstrated dark state in dye-doped CLC and verified the perfect opaque (FC) condition. Additionally, electro-optical experiments showed that dye-doped CLC with enhanced CR had lower threshold and operating voltages by approximately 10.9% and 6%, respectively, compared to pristine CLC. Additionally, POM textures revealed the onset of a temperature-dependent isotropic phase transition, which was observed to be decreased by around 8.4 °C for dye-doped CLC in comparison to pure CLC. As a result, adding dye to CLC greatly improves its electro-optical qualities and CR, making it more suitable for light shutter applications [84].

In order to prepare the liquid crystalline material CLC, a mixture composed of 5CB coded nematic LC of 96 wt% and the left-handed chiral dopant S811 of 4 wt% was used. Azo dye MR was also used as the dispersal agent. Kocakulah et al. have observed frequency-dependent dielectric properties of the CLC and CLC/MR composites in VIS and under different UV light intensities. Higher dielectric constant values were acquired in the low-frequency region by increasing UV light intensity in the CLC/MR composite structure. In the presence of voltage, the maximum CLC/MR sample growth and the highest relaxation frequency f_R value for CLC/MR were noted at 90 mW/cm^2 UV light intensity. Furthermore, De0 decreases with MR integration and even more as UV light intensity rises. For CLC samples, exposure to UV light also increased conductivity and raised the fc value. However, after 60 mW/cm^2 of UV light intensity, the dual-frequency LC behavior of the CLC/MR composite structure vanished, and it was linked to the *trans–cis* photo-isomerization of MR molecules. The findings indicate that adding MR improves the CLC composite's electrical and dielectric characteristics [68].

The CLC mixture of 1 μm pitch length (p) was prepared using the nematic LC (MDA-00-3969) of dielectric anisotropies + 3.2 and 2.7 at low frequency ($\leq$1 kHz) and high frequency ($\geq$50 kHz), respectively and 4 wt% of chiral dopant (R1011). The dye doped CLC mixture was formed using 0.5 wt% of azo orange 3 dichroic dye dispersed to pure CLC. Gahrotra et al. used pure and dye- doped CLC mixes to create bistable CLC light shutters, which exhibit good planar and FC textures for the transmission and scattering modes, respectively. The planar and FC states shown good bistability at zero voltage and were found to be immediately switchable between transparent and scattering states by applying high (50 kHz) and low (1 kHz) frequency voltages, respectively. The operating voltage in the direct switching between planar and FC states has decreased due to dye doping, and vice versa. In dye-doped CLC, the direct switching voltages were 27 V, while in pure CLC light shutters, they were 29 V. Additionally, the dye-doped CLC light shutter had a higher CR. The planar states achieved with high frequency (50 kHz) were found more stable with CR values of 99.57 and 115.45 for pure and dye-doped CLCs, respectively. However, the isotropic temperature of dye-doped CLC was shown to have significantly dropped in both POM and Differential scanning calorimetry (DSC) investigations. Therefore, these materials are very suitable for energy-efficient bistable light shutters and display devices due to their enhanced electro-optical performance with reduced switching voltage, higher contrast, and increased stability of CLC light shutter with the addition of dichroic dye [85].

Kumar et al. addressed this by doping the films with dichroic dyes and increasing their thickness. While these adjustments individually yielded minimal improvements, their combination produced a significant enhancement in screening performance. The thicker films enhanced scattering in the focal conic state and facilitated a multi-domain planar state, leading to colored reflection and turbid scattering. Dichroic dyes efficiently absorb transmitted and scattered light in the voltage-off state, while the film maintains good transmission in the voltage-on state. This improved the spectral properties, making CLC films suitable for electrically switchable windows, with potential applications in privacy, UV, and heat control by tuning the cholesteric pitch and dye absorption. [86]

5 Conclusion and Future Scope

In conclusion, the doping of various materials such as nanoparticles, QDS, dyes, and polymers has significantly broadened the scope of CLCs system. The physical properties such as optical, electro-optical, thermal, and dielectric studies of CLCs are altered differently by each doping substance. Nanoparticles have been found to enhance electro-optic performance and strengthen the photonic band gap, leading to more precise control of reflection wavelengths and polarization characteristics. QDs, owing to their size-dependent quantum confinement effects, provide adjustable photoluminescence, rendering them especially appropriate for use in display technologies and sensors. The incorporation of dyes into CLCs enhances light absorption and emission, which is advantageous for tunable laser systems and smart window technologies. Polymers augment the structural integrity and durability of CLCs, broadening their applicability in flexible electronics and display technologies.

Future prospects for exploration and innovation in this field are substantial. Future research may concentrate on hybrid doping strategies that integrate various dopants, such as nanoparticles and QDs, to attain synergistic effects, potentially resulting in enhanced control of optical and electro-optic properties. The incorporation of advanced nanomaterials, including graphene and perovskite nanoparticles, offers promising prospects for the advancement of next-generation energy-efficient displays, sensors, and tunable photonic devices. The investigation of sustainable and eco-friendly doping agents could accelerate the advancement of environmentally conscious CLC-based technologies for sustainable energy efficient devices. The other promising avenue for future research exists in the biomedical field, where CLCs infused with biocompatible agents may provide novel opportunities in medical diagnostics, bioimaging, and drug delivery. Furthermore, the development of more intelligent and responsive CLC-based systems, capable of dynamically adapting to external stimuli such as light, temperature, or electric fields, may result in advancements in smart windows, adaptive lenses, and wearable technologies. Finally, this research that combines LC science with nanotechnology, optics, and materials engineering has made it possible for big steps forward in both basic understanding and practical uses. This means that CLC- based technologies have a bright future ahead of them.

References

1. Mitov, M., Dessaud, N.: Going beyond the reflectance limit of cholesteric liquid crystals. Nat. Mater. **5**, 361 (2006). https://doi.org/10.1038/nmat1619
2. Mc Conney, M.E., White, T.J., Tondiglia, V.P., Natarajan, L.V., Yang, D., Bunning, T.J.: Dynamic high contrast reflective coloration from responsive polymer/cholesteric liquid crystal architectures. Soft Matter **8**, 318 (2012). https://doi.org/10.1039/c1sm05980g
3. Pindak, R.S., Huang, C.-C., Ho, J.T.: Divergence of cholesteric pitch near a smectic-A transition. Phys. Rev. Lett. **32**, 43 (1974). https://doi.org/10.1103/PhysRevLett.32.43
4. Talukder, J.R., Lin, H.-Y., Wu, S.-T.: Photo- and electrical-responsive liquid crystal smart dimmer for augmented reality displays. Opt. Express **27**, 18169 (2019). https://doi.org/10.1364/OE.27.018169
5. Lee, K.M., et al.: A different perspective on cholesteric liquid crystals reveals unique color and polarization changes. ACS Appl. Mater. Interfaces **12**, 37400 (2020). https://doi.org/10.1021/acsami.0c09845

6. Dierking, I.: Chiral liquid crystals: structures, phases, effects. Symmetry **6**, 444 (2014). https://doi.org/10.3390/sym6020444
7. Pan, Y., et al.: Microfluidic construction of responsive photonic microcapsules of cholesteric liquid crystal for colorimetric temperature microsensors. Adv. Opt. Mater. **11**, 2202141 (2023). https://doi.org/10.1002/adom.202202141
8. Goodby, J.W.: Chirality in liquid crystals. J. Mater. Chem. **1**, 307 (1991). https://doi.org/10.1039/JM9910100307
9. Coates, D.: Development and applications of cholesteric liquid crystals. Liq. Cryst. **42**, 653 (2015). https://doi.org/10.1080/02678292.2015.1020454
10. Yang, D.-K., Huang, X.-Y., Zhu, Y.-M.: Bistable cholesteric reflective displays: materials and drive schemes. Annu. Rev. Mater. Sci. **27**, 117 (1997). https://doi.org/10.1146/annurev.matsci.27.1.117
11. Lu, H., et al.: Electrically switchable multi-stable cholesteric liquid crystal based on chiral ionic liquid. Opt. Lett. **39**, 6795 (2014). https://doi.org/10.1364/OL.39.006795
12. Ponti, S., Becchi, M., Oldano, C., Taverna, P., Trossi, L.: Optical properties of short pitch cholesteric liquid crystals. Liq. Cryst. **28**, 591 (2001). https://doi.org/10.1080/02678290010020201
13. Moheghi, A., Nemati, H., Li, Y., Li, Q., Yang, D.-K.: Bistable salt doped cholesteric liquid crystals light shutter. Opt. Mater. **52**, 219 (2016). https://doi.org/10.1016/j.optmat.2015.12.033
14. Oh, S.-W., Ji, S.-M., Han, C.-H., Yoon, T.-H.: A cholesteric liquid crystal smart window with a low operating voltage. Dyes Pigm. **197**, 109843 (2022). https://doi.org/10.1016/j.dyepig.2021.109843
15. Li, C.-C., et al.: Enhanced image quality of OLED transparent display by cholesteric liquid crystal back-panel. Opt. Express **25**, 29199 (2017). https://doi.org/10.1364/OE.25.029199
16. Zhang, H., et al.: Electrically induced coloration of polymer-stabilized cholesteric liquid crystal films with broadband reflection capability for smart windows. Dyes Pigm. **203**, 110316 (2022). https://doi.org/10.1016/j.dyepig.2022.110316
17. Huang, X.-Y., et al.: Full-color reflective cholesteric liquid crystal display. In: Liquid Crystal Materials, Devices, and Applications VII, vol. 3635, p. 120 (1999). https://doi.org/10.1117/12.343863
18. Chilaya, G.: Always cholesterics… but sometimes chiral smectic C, TGB and blue phases. Mol. Cryst. Liq. Cryst. **561**, 8 (2012). https://doi.org/10.1080/15421406.2012.686692
19. Zhang, X., Zhou, Z., Halder, S., Hu, L., Yang, D.-K.: Mechanism of switching between planar and focal conic states in a bistable cholesteric-liquid-crystal reflective display. Phys. Rev. Appl. **22**, 044020 (2024). https://doi.org/10.1103/PhysRevApplied.22.044020
20. Huang, Y., Zhou, Y., Doyle, C., Wu, S.-T.: Tuning the photonic band gap in cholesteric liquid crystals by temperature-dependent dopant solubility. Opt. Express **14**, 1236 (2006). https://doi.org/10.1364/OE.14.001236
21. Shibaev, P.V., Kopp, V.I., Genack, A.Z.: Photonic materials based on mixtures of cholesteric liquid crystals with polymers. J. Phys. Chem. B **107**, 6961 (2003). https://doi.org/10.1021/jp0222189
22. Yang, Y.-C., Kee, C.-S., Kim, J.-E., Park, H.Y., Lee, J.-C., Jeon, Y.-J.: Photonic defect modes of cholesteric liquid crystals. Phys. Rev. E **60**, 6852 (1999). https://doi.org/10.1103/PhysRevE.60.6852
23. Arsenault, A.C., Puzzo, D.P., Manners, I., Ozin, G.A.: Photonic-crystal full-colour displays. Nat. Photonics **1**, 468 (2007). https://doi.org/10.1038/nphoton.2007.140
24. Natarajan, L.V., et al.: Electro-thermal tuning in a negative dielectric cholesteric liquid crystal material. J. Appl. Phys. **103**, 094906 (2008). https://doi.org/10.1063/1.2913326
25. Lin, T.-H., et al.: Electrically controllable laser based on cholesteric liquid crystal with negative dielectric anisotropy. Appl. Phys. Lett. **88**, 061122 (2006). https://doi.org/10.1063/1.2168259
26. Kumar, P., Kang, S.-W., Lee, S.H.: Advanced bistable cholesteric light shutter with dual frequency nematic liquid crystal. Opt. Mater. Express **2**, 1121 (2012). https://doi.org/10.1364/OME.2.001121

27. McConney, M.E., Tondiglia, V.P., Hurtubise, J.M., Natarajan, L.V., White, T.J., Bunning, T.J.: Thermally induced, multicolored hyper-reflective cholesteric liquid crystals. Adv. Mater. **12**, 1453 (2011). https://doi.org/10.1002/adma.201003552
28. Balenko, N., Shibaev, V., Bobrovsky, A.: Polymer dispersed cholesteric liquid crystals with combined photo- and mechanochromic response. J. Mol. Liq. **401**, 124637 (2024). https://doi.org/10.1016/j.molliq.2024.124637
29. Froyen, A.A.F., Debije, M.G., Schenning, A.P.H.J.: Polymer dispersed cholesteric liquid crystal mixtures for optical time-temperature integrators. Adv. Opt. Mater. **10**, 2201648 (2022). https://doi.org/10.1002/adom.202201648
30. Gardymova, A.P., Krakhalev, M.N., Zyryanov, V.Y., Gruzdenko, A.A., Alekseev, A.A., Rudyak, V.Y.: Polymer dispersed cholesteric liquid crystals with a toroidal director configuration under an electric field. Polymers **13**, 732 (2021). https://doi.org/10.3390/polym13050732
31. Xu, Z., Liu, M., Liu, Y., Pan, Y., Yang, L., Ge, D.: Mechano-optical response behavior of polymer-dispersed cholesteric liquid crystals for reversible and highly sensitive force recorders. ACS Appl. Mater. Interfaces **15**, 3673 (2023). https://doi.org/10.1021/acsami.2c20959
32. Manaila-Maximean, D., et al.: Polymer-dispersed cholesteric liquid crystals reflecting in the infrared region. Mol. Cryst. Liq. Cryst. **417**, 199 (2004). https://doi.org/10.1080/15421400490478830
33. Xing, H., Wang, X., Xu, J., Wei, J., Guo, J.: Polymer network microstructures and electro-optical properties of a pressure-sensitive cholesteric liquid crystal device. RSC Adv. **3**, 17822 (2013). https://doi.org/10.1039/c3ra41673a
34. Matsuyama, A.: Theory of polymer-dispersed cholesteric liquid crystals. J. Chem. Phys. **139**, 174905 (2013). https://doi.org/10.1063/1.4828940
35. Li, C.-Y., et al.: Electro-optical properties of a polymer dispersed and stabilized cholesteric liquid crystals system constructed by a stepwise UV-initiated radical/cationic polymerization. Crystals **9**, 282 (2019). https://doi.org/10.3390/cryst9060282
36. Ma, J., Zheng, Z.-G., Liu, Y.-G., Xuan, L.: Electro-optical properties of polymer stabilized cholesteric liquid crystal film. Chin. Phys. B **20**, 024212 (2011). https://doi.org/10.1088/1674-1056/20/2/024212
37. Lee, K.M., et al.: Recent advances in electro-optic response of polymer-stabilized cholesteric liquid crystals. Mol. Cryst. Liq. Cryst. **50**, 2540 (2023). https://doi.org/10.3390/ma16062248
38. Chauhan, G., Malik, P., Deep, A.: Morphological, dielectric, electro-optic and photoluminescence properties of titanium oxide nanoparticles enriched polymer stabilized cholesteric liquid crystal composites. J. Mol. Liq. **376**, 121406 (2023). https://doi.org/10.1016/j.molliq.2023.121406
39. Chauhan, G., Malik, P., Deep, A.: Improved performance of cadmium selenide quantum dots-doped polymer stabilized cholesteric liquid crystals for light shutter. Liq. Cryst. **50**, 2540 (2023). https://doi.org/10.1080/02678292.2023.2264249
40. Manhas, N., Adimule, V., Lakshminarayana, P., Keri, R., Kumar, L.S., Bathula, C.: Utilization of CO^{2+} doped Eu_2O_3 nanostructures and their effect on the dielectric and photoisomerization of liquid crystals. Phys. Fluids **35** (2023). https://doi.org/10.1063/5.0171578
41. Adimule, V., Yallur, B.C., Gowda, A.H.J.: Advanced sensors based on carbon nanomaterials. In: Carbon Nanomaterials-Based Sensors, pp. 259–268. Elsevier (2022). https://doi.org/10.1016/B978-0-323-91174-0.00004-4
42. Adimule, V., Sharma, V., Kumar, P., Keri, R., Sharma, K.: Perylene-3, 4, 9, 10-tetracarboxylic dianhydride dye intercalated liquid crystal molecules: synthesis, spectroscopic characterizations and photoluminescence dynamics. J. Mol. Liq. **413** 125990 (2024). https://doi.org/10.1016/j.molliq.2024.125990
43. Adimule, V., Bathula, C., Sharma, K., Keri, R.: Study of structural, optical, dielectric properties poly-3-butyl thiophene reinforced with CO_3O_4 nanoparticles. Nano-Struct. Nano-Objects **38**, 101150 (2024). https://doi.org/10.1016/j.nanoso.2024.101150
44. Hsiao, Y.-C., Lee, W.: Lower operation voltage in dual-frequency cholesteric liquid crystals based on the thermodielectric effect. Opt. Express **21**, 23927 (2013). https://doi.org/10.1364/OE.21.023927

45. Pathinti, R.S., Tatipamula, A.K., Vallamkondu, J.: ZnO nanoparticles dispersed cholesteric liquid crystal based smart window for energy saving application. J. Alloys Compd. **963**, 171198 (2023). https://doi.org/10.1016/j.jallcom.2023.171198
46. Pathinti, R.S., Gollapelli, B., Jakka, S.K., Vallamkondu, J.: Green synthesized TiO_2 nanoparticles dispersed cholesteric liquid crystal systems for enhanced optical and dielectric properties. J. Mol. Liq. **336** (2021). https://doi.org/10.1016/j.molliq.2021.116877
47. Kim, H., Kobashi, J., Maeda, Y., Yoshida, H., Ozaki, M.: Pitch-length independent threshold voltage of polymer/cholesteric liquid crystal nano-composites. Crystals **5**, 302 (2015). https://doi.org/10.3390/cryst5030302
48. Shivaraja, S.J., Gupta, R.K., Kumar, S., Manjuladevi, V.: Enhanced electro-optical response of nematic liquid crystal doped with functionalised silver nanoparticles in twisted nematic configuration. Liq. Cryst. **47**, 1678 (2020). https://doi.org/10.1080/02678292.2020.1755901
49. Middha, M., Kumar, R., Raina, K.K.: Effects of chirality on optical and electro-optic behavior of nematic liquid crystals doped with functionalized silver nanoparticles. J. Mol. Liq. **219**, 631 (2016)
50. Mody, V., Siwale, R., Singh, A., Mody, H.: Introduction to metallic nanoparticles. J. Pharm. Bioallied Sci. **2**, 282 (2010). https://doi.org/10.4103/0975-7406.72127
51. Vasudevan, D., Gaddam, R.R., Trinchi, A., Cole, I.: Core-shell quantum dots: properties and applications. J. Alloys Compd. **636**, 395 (2015). https://doi.org/10.1016/j.jallcom.2015.02.102
52. Singh, G., Fisch, M., Kumar, S.: Emissivity and electrooptical properties of semiconducting quantum dots/rods and liquid crystal composites: a review. Reports Prog. Phys. **79**, 56502 (2016). https://doi.org/10.1088/0034-4885/79/5/056502
53. Shurpo, N.A., Vakshtein, M.S., Kamanina, N.V.: Effect of CdSe/ZnS semiconductor quantum dots on the dynamic properties of nematic liquid-crystalline medium. Tech. Phys. Lett. **36**, 319 (2010). https://doi.org/10.1134/S1063785010040097
54. Konshina, E.A., Gavrish, E.O., Orlova, A.O., Artem'ev, M.V.: Effect of dispersed CdSe/ZnS quantum dots on optical and electrical characteristics of nematic liquid crystal cells. Tech. Phys. Lett. **37**, 1011 (2011). https://doi.org/10.1134/S1063785011110071
55. Kumar, A., Prakash, J., Khan, M.T., Dhawan, S.K., Biradar, A.M.: Memory effect in cadmium telluride quantum dots doped ferroelectric liquid crystals. Appl. Phys. Lett. **97**, 2012 (2010). https://doi.org/10.1063/1.3485298
56. Chen, L.J., De Lin, J., Huang, S.Y., Mo, T.S., Lee, C.R.: Thermally and electrically tunable lasing emission and amplified spontaneous emission in a composite of inorganic quantum dot nanocrystals and organic cholesteric liquid crystals. Adv. Opt. Mater. **1**, 637 (2013). https://doi.org/10.1002/adom.201300124
57. Pathak, G., Hegde, G., Prasad, V.: Octadecylamine-capped CdSe/ZnS quantum dot dispersed cholesteric liquid crystal for potential display application: Investigation on photoluminescence and UV absorbance. Liq. Cryst. **48**, 579 (2021). https://doi.org/10.1080/02678292.2020.1799085
58. Kocakülah, G., Köysal, O.: Determination of optical band gap, transmittance, and dielectric constant parameters of Indium Phosphide/Zinc Sulfide quantum dot doped cholesteric liquid crystals. Opt. Mater. (Amst). **149**, (2024). https://doi.org/10.1016/j.optmat.2024.115101
59. Chauhan, G., Malik, P., Malik, P., Deep, A.: Improved performance of cadmium selenide quantum dots-doped polymer stabilized cholesteric liquid crystals for light shutter. Liq. Cryst. **50**, 2540 (2023). https://doi.org/10.1080/02678292.2023.2264249
60. Kocakülah, G., Köysal, O.: Electro-optical, dielectric and morphological properties of a cholesteric liquid crystal light shutter: the influence of azo dye and quantum dot nanoparticles. J. Electron. Mater. **51**, 6864 (2022). https://doi.org/10.1007/s11664-022-09913-z
61. Kocakülah, G., Köysal, O.: Electro-optical and dielectric response of quantum dot- doped cholesteric liquid crystal composites. J. Mater. Sci. Mater. Electron. **33**, 5489 (2022). https://doi.org/10.1007/s10854-022-07739-0
62. Chemingui, M., Yang, D., Wang, Y., Ayeb, H., Cai, W., Tang, A., Drevensek-Olenik, I., Zhang, X., Xu, J.: The influence of quantum dots on the optical properties of a room temperature cholesteric liquid crystal. Liq. Cryst. **49**, 2095 (2022). https://doi.org/10.1080/02678292.2022.2103850

63. Bobrovsky, A., Mochalov, K., Oleinikov, V., Sukhanova, A., Prudnikau, A., Artemyev, M., Shibaev, V., Nabiev, I.: Optically and electrically controlled circularly polarized emission from cholesteric liquid crystal materials doped with semiconductor quantum dots. Adv. Mater. **24**, 6216 (2012). https://doi.org/10.1002/adma.201202227
64. Cao, M., Yang, S., Zhang, Y., Song, X.: Tunable amplified spontaneous emission in graphene quantum dots doped cholesteric liquid crystals, (n.d.). https://doi.org/10.1088/1361-6528/aa67ca
65. Rodarte, A.L., Gray, C., Hirst, L.S., Ghosh, S.: Spectral and polarization modulation of quantum dot emission in a one-dimensional liquid crystal photonic cavity. Phys. Rev. B Condensed Matter Mater. Phys. **85**, 1 (2012). https://doi.org/10.1103/PhysRevB.85.035430
66. Lukishova, S.G., Bissell, L.J., Winkler, J., Stroud, C.R.: Resonance in quantum dot fluorescence in a photonic bandgap liquid crystal host. Opt. Lett. **37**, 1259 (2012). https://doi.org/10.1364/OL.37.001259
67. Chen, L.J., De Lin, J., Lee, C.R.: An optically stable and tunable quantum dot nanocrystal-embedded cholesteric liquid crystal composite laser. J. Mater. Chem. C **2**, 4388 (2014). https://doi.org/10.1039/C4TC00128A
68. Kocakülah, G., Yıldırım, M., Köysal, O., Ercan, İ: Influence of UV light intensity on dielectric behaviours of pure and dye-doped cholesteric liquid crystals. J. Mater. Sci. Mater. Electron. **31**, 22385 (2020). https://doi.org/10.1007/s10854-020-04740-6
69. Liang, H.H., Wang, P.H., Wu, C.C., Hsu, S.C., Lee, J.Y.: Polymer network morphology and electro-optical properties of electrothermal switchable bistable polymer- stabilized cholesteric texture light shutters with various chiral dopant concentrations. Mol. Cryst. Liq. Cryst. **552**, 111 (2012). https://doi.org/10.1080/15421406.2011.606012
70. Wu, P.-C., Lisetski, L.N., Lee, W.: Suppressed ionic effect and low-frequency texture transitions in a cholesteric liquid crystal doped with graphene nanoplatelets. Opt. Express **23**, 11195 (2015). https://doi.org/10.1364/OE.23.011195
71. Middha, M., Kumar, R., Raina, K.K.: Improved electro-optical response of induced chiral nematic liquid crystal doped with multi-walled carbon nanotubes. Ferroelectrics **495**, 75 (2016). https://doi.org/10.1080/00150193.2016.1136765
72. Middha, M., Kumar, R., Raina, K.K.: Effects of chirality on optical and electro-optic behavior of nematic liquid crystals doped with functionalized silver nanoparticles. J. Mol. Liq. **219**, 631 (2016). https://doi.org/10.1016/j.molliq.2016.03.087
73. Singh, D.P., Varia, M.C., Gupta, S.K., Sagar, L.K., Kumar, S., Yadav, S.P., Manohar, R.: High-temperature chiral nematic phase in naphthalene and cholesterol derivative liquid crystal: characterisation and dielectric relaxation study. Phys. Chem. Liq. **51**, 663 (2013). https://doi.org/10.1080/00319104.2013.771263
74. Jessy, P.J., Radha, S., Patel, N.: Morphological, optical and dielectric behavior of chiral nematic liquid crystal mixture: Study on effect of different amount of chirality. J. Mol. Liq. **255**, 215 (2018). https://doi.org/10.1016/j.molliq.2018.01.160
75. Moheghi, A., Nemati, H., Li, Y., Li, Q., Yang, D.-K.: Bistable salt doped cholesteric liquid crystals light shutter. Opt. Mater. (Amst). **52**, 219 (2016). https://doi.org/10.1016/j.optmat.2015.12.033
76. Kim, B.C., Walker, M., Jo, S.Y., Wilson, M.R., Takezoe, H., Choi, S.W.: Effect of terminal chain length on the helical twisting power in achiral bent-core molecules doped in a cholesteric liquid crystal. RSC Adv. **8**, 1292 (2018). https://doi.org/10.1039/C7RA11589J
77. Sun, Y., et al.: A study on the electro-optical properties of thiol-ene polymer dispersed cholesteric liquid crystal (PDCLC) films. Molecules **22** (2017). https://doi.org/10.3390/molecules22020317
78. G. Kocakülah, Comparison of electro-optical and dielectric properties of cobalt (II) phthalocyanine doped cholesteric liquid crystals, Opt. Mater. (Amst). 141, 113949 (2023). https://doi.org/10.1016/j.optmat.2023.113949
79. Chang, Y.K., Wu, P.C., Lee, W.: Effects of ion doping on the dielectric and electro- optical properties of chiral nematic liquid crystal. J. Mol. Liq. **405**, 125084 (2024). https://doi.org/10.1016/j.molliq.2024.125084

80. Oh, S.W., Baek, J.M., Heo, J., Yoon, T.H.: Dye-doped cholesteric liquid crystal light shutter with a polymer-dispersed liquid crystal film. Dye. Pigment. **134**, 36 (2016). https://doi.org/10.1016/j.dyepig.2016.06.045
81. Kim, J.-H., Huh, J.-W., Oh, S.-W., Ji, S.-M., Jo, Y.-S., Yu, B.-H., Yoon, T.-H.: Bistable switching between homeotropic and focal-conic states in an ion-doped chiral nematic liquid crystal cell. Opt. Express **25**, 29180 (2017). https://doi.org/10.1364/OE.25.029180
82. Yu, B.H., Huh, J.W., Heo, J., Yoon, T.H.: Simultaneous control of haze and transmittance using a dye-doped cholesteric liquid crystal cell. Liq. Cryst. **42**, 1460 (2015). https://doi.org/10.1080/02678292.2015.1061149
83. Oh, S.W., Baek, J.M., Kim, S.H., Yoon, T.H.: Optical and electrical switching of cholesteric liquid crystals containing azo dye. RSC Adv. **7**, 19497 (2017). https://doi.org/10.1039/C7RA01507K
84. Gahrotra, R., Sharma, V., Malik, P., Kumar, P.: Textural and electro-optical study of cholesteric liquid crystal for light shutter. AIP Conf. Proc. **2220**, 5 (2020). https://doi.org/10.1063/5.0001809
85. Gahrotra, R., Sharma, V., Dogra, A.R., Malik, P., Kumar, P.: Performance augmentation of bistable cholesteric liquid crystal light shutter-effect of dichroic dye on morphological and electro-optical characteristics, Opt. Mater. (Amst). **127**, 112243 (2022). https://doi.org/10.1016/j.optmat.2022.112243
86. Kumar, V., Jeong, K., Kang, S.: Dyes and pigments dichroic-dye-doped short pitch cholesteric liquid crystals for the application of electrically switchable smart windows. Dye. Pigment. **166**, 403 (2019). https://doi.org/10.1016/j.dyepig.2019.03.045

Bistable Cholesteric Liquid Crystals: Electro-optic Applications

Ridhima Gahrotra, Pankaj Kumar, and Vandna Sharma

Abstract This chapter provides a comprehensive overview of bistable cholesteric liquid crystals (CLCs), exploring its unique electro-optic characteristics and recent advancement in technological field. The chapter commences with a detailed introduction to CLCs, highlighting its helical structure and their associated thermal and optical properties. Furthermore, an analysis on bistability within CLCs is provided, that focusses on the existence of two distinct stable states and its implications for both natural and technological innovations. A substantial portion of the chapter emphasises on the role of chiral dopants in the formation of CLCs. Moreover, it discusses that how these chiral dopants influences the helical pitch and alignment of cholesteric molecules, thereby influencing the electro-optical characteristics and potential applications. It further delves into the optical properties of bistable CLCs, highlighting their potential for novel display devices. It also covers various techniques for inducing and controlling bistability in CLCs, including certain methods for tuning the pitch and alignment in CLCs. Some fabrication and characterization techniques are also discussed in detail to provide an insight of latest tools and techniques for analysing bistable CLCs system. By integrating theoretical insights with practical applications, this chapter serves as a critical resource for understanding the dynamic behaviour of bistable cholesteric liquid crystals and their impact on modern technology and material science.

Keywords Liquid crystals · Cholesteric · Bistability · Electro-optic

R. Gahrotra (✉)
Department of Applied Sciences, Chitkara University Institute of Engineering and Technology, Chitkara University, Punjab, India
e-mail: ridhima.gahrotra@chitkara.edu.in

P. Kumar · V. Sharma
Centre for Liquid Crystal Research, Chitkara University Institute of Engineering and Technology, Chitkara University, Punjab, India
e-mail: pankaj.kumar@chitkara.edu.in

V. Sharma
e-mail: vandnasharma@chitkara.edu.in

P. Kumar et al. (eds.), *Cholesteric Liquid Crystals*, Engineering Materials,
https://doi.org/10.1007/978-981-96-3821-5_10

1 Introduction to Cholesteric Liquid Crystals (CLCs)

Liquid crystals (LCs) have been a topic of great interest among both academicians and researchers across the globe and have been interesting class of materials since its inception. Cholesteric liquid crystals (CLCs)—a type of liquid crystal (LC) also termed as chiral nematic LCs have attracted profound interest due to their molecular arrangement, unique helical super-structure, electro-optical properties, recent advancements in technological aspects [1, 2]. Additionally, the quest to obtain new materials with novel applications has tremendously enhanced the Cholesteric liquid crystal (CLC) research. The increasing demand for technological advancements gives rise to a new approach called Guest–Host approach. This approach includes the doping of LC (Host) with various substances (Guest) i.e. dyes, polymers, and nanomaterials to enhance the inherent properties of host and significantly fulfilling the technological needs. However, while this approach is promising, it is essential to explore other avenues, such as the development of new LC structures that can further enhance the sensitivity and functionality of optical devices, potentially leading to breakthroughs in smart materials and applications [3, 4].

1.1 Structure of Cholesteric Molecule

The term 'Cholesteric' was invented by George Friedel in 1922 [5]. This phase of LC is made of rod-like molecules and exhibits long-range orientational order but no positional order. The long-range CLC molecules are oriented along the common director ($\hat{n}$) in each layer, with each plane rotated at a certain angle relative to the neighbouring planes. The chiral centers added to nematic LCs to form CLCs exert intermolecular forces between the molecules and align them at a small angle with one another in each layer. This chirality causes the molecules to align in a helicoidal structure, with each layer twisted slightly to the layer below to it as shown in Fig. 1. The distance over which the molecules complete one full twist of 2π can be termed as "pitch" of the helix [6, 7]. Another significant property of CLCs is selectively reflect the circularly polarized light through the helix, a phenomenon known as Bragg's reflection [8]. This selective reflection is dependent on mainly two factors: one is the pitch of the helix (p) and the other is average refractive indices (Δn) of the LC material [9, 10] such that

$$\lambda = \Delta n.p \tag{1}$$

where λ denotes specific wavelength of incident light. In any case, if λ is greater than the helical pitch, the medium is treated as optically homogeneous (planar) and exhibits Grandjean texture and if p is greater than the resolution limit of microscope, fingerprint texture is observed in the focal conic state. Additionally, for any CLC molecule, the magnitude of the p depends much on the factors including temperature,

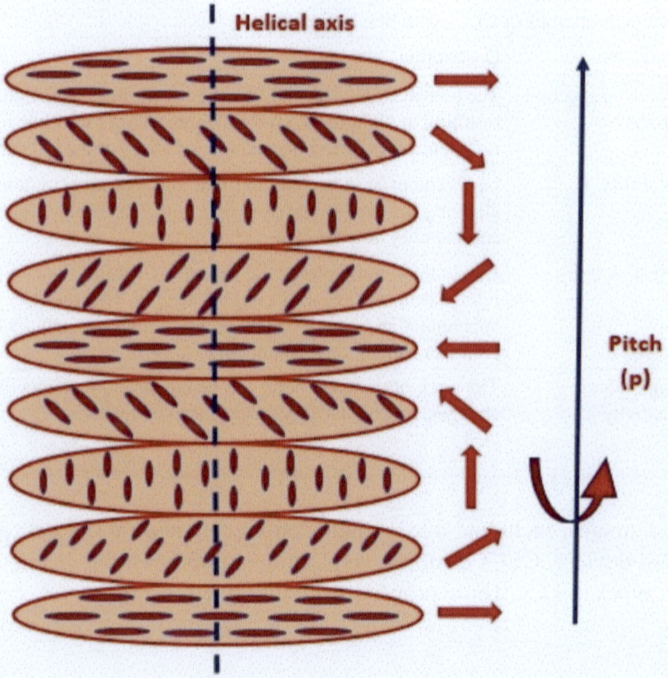

Fig. 1 Schematic representation of the molecules in the cholesteric phase

structure, and amount of the chiral dopant added [11]. Also, the concentration of chiral dopant (c) is dependent on helical twisting power (HTP) and p as

$$c = (p \times HTP)^{-1} \tag{2}$$

The HTP value mainly helps in determining the possibility of CLC helical pitch controllability such as degree of p switching and its handedness (either left or right), thereby results in limiting its display applications. Higher the value of HTP can largely reduce the concentration of the chiral dopant; resulting in no undesirable change in the physical properties (e.g. viscosity, clearing point), that form the basis of optical display addressing.

1.2 Properties of CLCs

Over the years, some comprehensive research has been directed to explore the properties and behaviour of CLCs, leading to significant developments in this field. The ability of CLCs to selectively reflect light and exhibit bistability made them ideal candidates for display technologies. Since then, CLCs have created applications

Table 1 Unique properties of CLCs with description

S. no	Properties	Description
1	Selective reflection of light	The helical structure of CLCs reflects specific wavelengths of light, resulting in vivid colors and iridescence. This selective reflection is dependent on p and the refractive indices of the LC material [10]
2	Bistability	CLCs can exist in two stable states under certain conditions, allowing for the development of low-power display technologies and memory devices [12]
3	Optical activity	Due to their chiral nature, CLCs can switch the plane of polarized light, a property known as optical activity. This property is appreciated in various optical applications, that includes circular polarizers and optical isolators [13, 14]
4	Temperature sensitivity	The pitch of the CLC helix can be influenced by temperature changes, making CLCs useful in temperature sensing applications [15]

in various domains, including telecommunications, biomedical devices, and smart architectural devices. CLCs exhibit several unique properties that distinguish them from other types of LCs. These properties are described in Table 1.

2 Fundamentals of Bistability in CLCs

Bistability in CLCs discusses the ability of the materials to exist in two stable and distinct states under the external applied conditions. This characteristic is significantly remarkable as it allows the formation of such devices that can switchable between the two states without the need for any continuous supply to maintain them. This means that if a device is switched to a particular state once, it remains in that state even if the external stimulus (such as an electric field) is removed, leading to energy-efficient applications [16, 17]. The significance of bistability lies in its potential for developing low-power and stable state optical and electronic devices. In display technologies, for instance, bistable CLCs can be used to create screens that do not require constant refreshing to maintain an image, thereby reducing power consumption. Additionally, bistability is crucial for memory devices where information needs to be stored without continuous power supply to enhance the data storage efficiency and reliability [18, 19].

CLCs cells prepared conventionally can adopt two different states-planar (P) and focal conic (FC) states. The schematic is shown in Fig. 2. In the P state, the director in each layer is nearly perpendicular to the polarized incident light [20]. In the FC state, the CLC domains are randomly oriented and the incident light gets scattered due to differences in refractive index [21, 22]. On applying low-frequency voltage

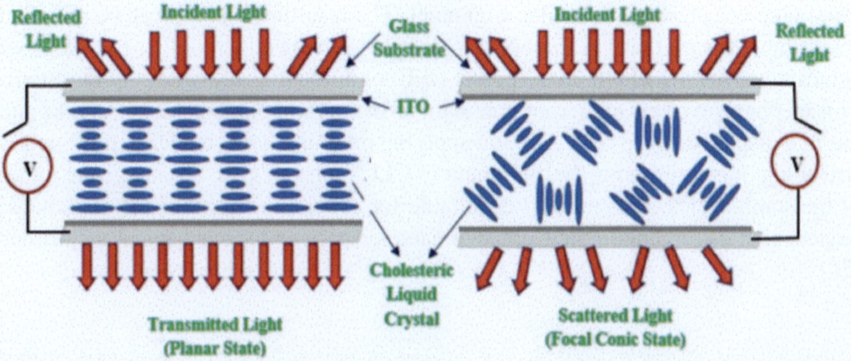

Fig. 2 Schematic diagram of configuration of CLC and operational principle of two stable states: planar (P) and focal conic (FC) states after switching OFF the applied voltages

to the P state, the FC state is obtained and this state is maintained unless some high-frequency voltage is applied. Further, on removing the applied voltage to attain FC and P states, still sustained in the same states.

2.1 *Mechanisms Leading to Bistable States*

Bistable states in CLCs can be induced through several mechanisms, including the application of electric or magnetic fields, thermal effects, and mechanical stress. Understanding these mechanisms is essential for designing and controlling bistable CLC devices.

Electric Fields: When electric field is applied to CLC filled cell, it can cause the helical structure to unwind that leads to different stable configurations. The application of external electric field at high frequency unwinds the helical structure causing the molecules to align in random orientation and when low-frequency is applied at the same time the CLC molecules attain the actual orientation (director aligned perpendicular to the glass substrate) will be obtained. Further on removing the external field at any stage, the CLCs can remain in either of the stable states (planar or focal conic) due to the energy difference between the two states. Due to this property, such materials have exhibited attraction in bistable display technologies, where the images can be sustained with no continuous power [23]. Similar to electric fields, magnetic fields can also be employed to align the helix in CLCs. Direct switching between planar and focal conic states is feasible by controlling the strength and direction of the magnetic field. The method can be used in those applications that require non-contact control of the LC states [24].

Additionally, the temperature change can also affect the pitch of the CLC helix, that leads to the switching between two stable states. On providing continuous heat or cool to the sample-filled cell placed in a temperature controller, the material can also induce changes in molecular alignment. The resulting changes can be recorded using the software that has the options to control the heating/cooling rates and set values saved in the memory in the power OFF condition in both stable configurations. This mechanism offers numerous applications in the form of temperature sensors and thermal display devices [25]. Also, on applying pressure or mechanical stress such as stretching or compressing, the alignment of CLC molecules can be changed to either of the stable states. This method is suitable for developing flexible and stretchable devices that can maintain their optical characteristics under mechanical distortions [26].

2.2 Applications of Bistable CLCs in Nature and Technology

Several natural phenomena and technological applications have been prepared using bistable CLCs. Some remarkable examples are: The presence of CLCs in exoskeletons of some beetles, butterflies, and other insects' results in iridescent colours. Because of their bistability, these naturally occurring structures enable organisms to change hues or retain a specific coloration without consuming energy. For instance, the green weevil exhibits vibrant colors which results from the selective reflection by CLCs in its cuticle [27]. Another most prominent application of bistable CLC is in display technologies. Bistable displays, such as electronic paper (e-paper), utilize CLCs to create high-contrast images that remain stable without continuous power. CLCs are used in bistable displays, including electronic paper (e-paper), to provide high-contrast images that hold up without constant power. These screens are found in low-power devices including e-readers and signage [28].

Furthermore, bistable CLCs are employed in smart windows that have the ability to switch between transparent and opaque states. On applying an external electric field, the orientation of the LC molecules can be altered, which enables the window to control light transmission and offers privacy or shading on demand. Once the desired state is achieved, it can be maintained without additional power [29]. Furthermore, the unique optical properties of bistable CLCs make them suitable for use in tunable optical filters and reflectors. These devices can selectively reflect specific wavelengths of light, making them useful in telecommunications, imaging systems, and laser technologies. By adjusting the pitch of the CLC helix, the reflected wavelength can be tuned to the desired range [30, 31]. Bistable CLCs are also being explored for use in biomedical sensors and diagnostic devices. Their ability to change optical properties in response to external stimuli, such as temperature or pH changes, makes them ideal for detecting and monitoring various physiological parameters. For example, CLC-based sensors can be used to monitor body temperature or detect the presence of specific biomolecules [32]. In conclusion, the fundamentals of bistability in CLCs encompass the definition, mechanisms, and examples of bistable states in both nature and technology. The unique properties of bistable CLCs make them valuable for a wide range of applications, from low-power displays and smart windows to advanced optical devices and biomedical sensors. Ongoing research and development in this field continue to expand the potential uses of bistable CLCs, promising further innovations and advancements.

3 Role of Chiral Dopants in Inducing Cholesteric Phases

Chiral dopants play a vital role in inducing and stabilizing the cholesteric phase in LCs. These dopants introduce chirality into the LC mixture, causing the molecules to twist and form a helical structure. The pitch of the cholesteric helix is directly influenced by the concentration and type of chiral dopants used [33]. Typically, the chiral dopants are being added to nematic LCs to induce the cholesteric phase. The dopant molecules have a twisted structure, which imparts a chiral influence on the surrounding LC molecules, leading to the formation of a helical structure. In CLC, the helical superstructure has specific handedness: negative corresponds to left-handed helix and positive corresponds to right-handed helix. This periodic helical superstructure selectively reflects the circularly polarized light at specific wavelengths that correspond to P [34]. The strength of this chiral influence is also quantified by the helical twisting power (HTP) as given in Eq. (2), which measures the ability of a chiral dopant to induce a helical twist in the LC [34].

Recent research has focused on developing new chiral dopants with higher HTP values and better thermal stability. These advances have enabled the creation of CLCs with finely tunable pitches, allowing for precise control over their optical properties. Additionally, the use of mixtures of different chiral dopants can provide further flexibility in tuning the pitch and enhancing the stability of the cholesteric phase [35]. Below are some key findings of the researchers that highlight the advancements. Hung et al. fabricated high transmittance and very responsive bistable CLC optical device using a new synthesized acrylic chiral dopant added to dual frequency nematic LC to form CLC. The incorporation of this chiral dopant resulted in 97.6% transmittance and 1.5 ms response time with bistable display performance. Each state was also maintained either planar "**ON**" and focal conic "**OFF**" states without providing any continuous electric field [36]. Additionally, Shaban et al. demonstrated two enantiomeric chiral dopants R5011 (right-handed) and S5011 (left-handed) at different concentrations (0–4 wt./wt.%) added to nematic E44 to prepare CLCs. The prepared CLC cells were used to study the ionic properties (electrical conductivity, ion density, and ion diffusivity) of chiral dopants in order to form CLCs. Phase-Transition temperature-dependent study was performed using transmission and dielectric spectroscopy. According to thermal studies, the clearing temperature was reduced by 8 °C as the dopant concentration goes on increasing up to 4.0 wt.% for both right-handed and left-handed CLCs. After thoroughly examining the role of each chiral dopant to the ionic characteristics of all the resulting CLCs highlights a novel area of investigation. It was found that the ionic properties proportional to the concentration of the chiral dopant was not directly linked to the handedness of the chiral molecules. Also, the dielectric studies and its related parameters showcased that the R5011 doped CLCs were significantly more ion-rich when compared to the host nematic (E44) and other chiral agent (S5011) at other different concentrations.

An equal amount of chiral dopants R5011 and S5011 were mixed with the nematic host E44 to form a CLC. The ionic properties of both the samples prepared with either both chiral dopants or singly doped samples comprising R5011 were found quite similar. The higher density of ions detected in the E44/R5011 mixtures maybe due to the degradation of dopant with time. The surface alignment of the cell either antiparallel or twisted by 90° had no significant effect on the observed ionic behaviour. The temperature dependent studies show that, the elongation of helical pitch (i.e., weakening of HTP) was monitored above $T > \sim 60$ °C. Also, dielectric properties of E44/R5011, the optical data of both E44/R5011 and E44/S5011 were found identical. The experimental findings emphasize on the potentially overlooked ionic contribution of a chiral additive in a constituted CLC [37].

4 Electro-optical Properties of Bistable CLCs

4.1 Bragg Reflection and Selective Reflection of Light

CLCs exhibit a unique optical property known as Bragg reflection. This phenomenon arises from their helical molecular structure, which acts as a one-dimensional photonic crystal. The periodic variation in the refractive index along the helical axis causes constructive interference for certain wavelengths of light, leading to selective reflection [38]. The wavelength (λ) of the reflected light is given by Bragg's law:

$$\lambda = nP \sin\theta \tag{3}$$

where, n is the average refractive index of the LC, p is the pitch of the helix, and θ is the angle of incidence [39]. This selective reflection results in vivid colors and iridescence, with the reflected wavelength being highly dependent on the pitch of the helical structure.

By adjusting the pitch, CLCs can reflect different wavelengths of light, making them suitable for various optical applications such as displays, filters, and sensors [40]. The helical pitch varies for a wider range in CLC system from 0.1 μm to several thousand μm. For the case of high chirality (short pitch), wavelength and pitch are approximately equal, the propagation of light in this case is described by the eigen values that becomes elliptical. However, the eigen waves are circular for limiting case. Here the limiting case refers to the selective reflection that occurs at a wavelength λ (B) due to Bragg's reflection. Additionally, this diffraction leads to some remarkable optical properties [11].

$$m\,\lambda(B) = nP\sin\theta \tag{4}$$

where m represents the order of diffraction. For the case of low chirality (long pitch), the wavelength is very small than p, the propagation here is described by the superposition of two eigen waves $E_{\parallel}$ *and* $E_{\perp}$ to the director. This type of configuration of CLCs have applications in conventional twisted nematic displays.

4.2 Birefringence and Orientational Order Parameter

Birefringence is an intrinsic and fundamental optical property of any optical material. Usually, the CLCs are birefringent which means that both the refractive indices are different along ($n_{\parallel}$) and perpendicular ($n_{\perp}$) to the optic axis [42]. The birefringence of the material decreases with increase in temperature and becomes zero when isotropic phase is achieved. This also helps in determining another most important parameter of CLCs, called 'Orientational Order Parameter' and is given by:

$$S = \frac{1}{2} < 3cos^2\theta - 1 > \quad (5)$$

Here, <> represents the thermal averaging and θ represents the angle between the director and the molecular axis. For crystalline solid, S = 1 (perfect alignment of molecules), for LCs, S ranges between 0.3 and 0.9 (higher the value more ordered the phase) and for isotropic liquids, S = 0 (complete random orientation of the molecules). Another importance of S lies in the fact that all the properties of LC depend on it and changes in S with temperature can be considered remarkable from technological point of view. Higher the order parameter and lower the birefringence values are best for practical applications [43, 44].

4.3 Bistablity, Colour Changes and Wavelength Tuning

Optical bistability in CLCs refers to their ability to exist in two stable optical states under the same set of external conditions. This bistability is a result of the energy barriers between different molecular alignments, which can be manipulated using external stimuli such as electric fields, magnetic fields, temperature changes, or light exposure [24]. The two stable states typically correspond to different reflectance or transmission properties, allowing for the creation of devices that can be switchable between these states without continuous power. This property is particularly useful in the development of low-power display technologies. Bistable CLC displays can maintain an image without the need for constant refreshing, significantly reducing power consumption. Figure 3 shows the generalized voltage-transmission graph of dual frequency CLC showing the bistability of opaque and transparent states with low and high frequency states. This makes them ideal for applications such as e-readers, digital signage, and other portable electronic devices [45–47]. Additionally, optical

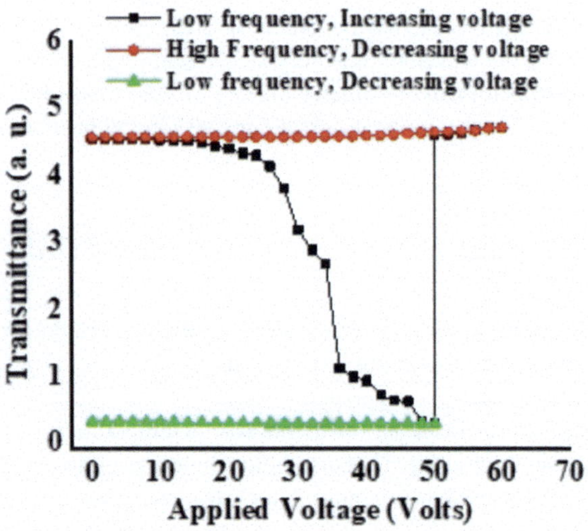

Fig. 3 Transmittance versus applied voltage curve showing the bistability of CLCs

bistability is leveraged in memory devices where the stable states can represent binary information (0 and 1). These CLC-based memory devices offer advantages in terms of energy efficiency and non-volatility, meaning the stored information is retained even when the power is turned off [48].

Furthermore, the helical super-structure of CLCs enables dynamic color changes and wavelength tuning through various external stimuli. This tunability is a key feature that makes CLCs versatile over a wide range of applications. As on applying external electric field, the pitch of the CLC helix can be altered, resulting in changes in the reflected wavelength. This allows for the dynamic adjustment of color in display technologies and tunable optical filters. The electric field can reorient the LC molecules, leading to changes in the helical structure and thereby shifting the reflected color [49]. Also, temperature variations can affect the helical pitch of the CLC, causing changes in the reflected wavelength. As the temperature increases, the pitch typically lengthens, leading to a shift towards longer wavelengths (redshift). On the contrary, cooling can shorten the pitch, resulting in a shift towards shorter wavelengths (blueshift). This property is utilized in temperature sensors and thermally responsive materials [50].

Furthermore, on applying mechanical stress, such as stretching or compressing the CLC, the molecular alignment and pitch of the helix can be modified. This mechanical tuning enables the development of stretchable and flexible optical devices that can change color or optical properties in response to deformation. Such devices are useful in wearable technology and adaptive optics [51]. Additionally, Photo-responsive chiral dopants can be used to induce changes in the helical structure upon exposure to specific wavelengths of light. This photonic tuning allows for the control of optical properties using light, enabling the creation of light-sensitive devices and sensors [52].

The ability to tune the optical properties of CLCs through various stimuli makes them highly adaptable for applications in displays, smart windows, tunable lasers, and sensing technologies. The ongoing research and development in this area continue to expand the potential uses of CLCs, leading to innovative solutions in photonics and optoelectronics [53, 54].

4.4 Contrast Ratio and Stabilization Rate

The bistability of the CLCs also depends on the two bistable factors termed as contrast ratio (CR) and stabilization rate (SR) in terms of transmittance. In order to obtain opaque and transparent state at low and high frequency, CR signifies the ratio of maximum transmittance in the transparent state to the minimum transmittance in opaque state as represented in Eq. 6. Additionally, the SR represents the stability of the opaque and transparent states by measuring the ratio of transmittance under applied voltage to the difference of transmittances with and without applied voltage. The detailed formula is depicted in Eq. (7) [55, 56]:

$$CR = \frac{T_{max}}{T_{min}} \; or \; \frac{T_{transparent}(V > 0, high\,frequency)}{T_{opaque}(V > 0, low\,frequency)} \; or \; \frac{T_{transparent}(V = 0, low\,frquency)}{T_{opaque}(V > 0, low\,frequency)} \tag{6}$$

$$SR = \frac{T_{opaque(V>0,\,low\,frequency)}}{\left|T_{opaque(V>0,\,low\,frequency)} - T_{opaque(V=0,\,low\,frequency)}\right|} \; or \; \frac{T_{transparent(V>0,\,high\,frequency)}}{\left|T_{transparent(V>0,\,high\,frequency)} - T_{planar(V=0,\,high\,frequency)}\right|} \tag{7}$$

where, T_{max} and T_{min}, $T_{transparent}$ and T_{opaque} represent the maximum, minimum, transparent and opaque transmittances, respectively.

5 Characterization Techniques for Bistable CLCs

Several tools and apparatuses are used for characterization of the bistable CLC sample cells including polarized optical microscopy (POM), electron microscopy, UV–Vis spectrophotometer, differential scanning calorimetry (DSC), fourier transform infrared spectroscopy (FT-IR), X-ray diffraction (XRD), photoluminescence spectroscopy, dielectric analysis, birefringence method etc. [52, 56–66]. The in-depth analysis on the practice of few of these apparatuses is provided below.

5.1 Polarized Optical Microscopy (POM)

In order to study the morphology of the CLC filled cells with the application of external field at varied frequencies, a polarized optical microscope (POM) arranged with cross-polarizers is used, the schematic is shown in Fig. 4. To obtain the information related to surface morphology of the CLCs, POM uses a technique that employs the use of polarized light. Herein, two polarizers made of polymeric films are placed perpendicular to one another, so that material appears bright under crossed polarizers [57]. The polarizers were enclosed with iodoquinine sulphate crystallites oriented in the same direction absorbs the ordinary ray and allows the extraordinary ray to pass through, thereby limiting the transmitted ray of light to single plane. When the polarised light is being incident on the sample filled cell and upon passing through the analyser, combines the ordinary and extra-ordinary ray results in a vector sum of the two rays. When the sample cell is rotated at 45° angle to the polarizers, it shows maximum brightness, however it shows complete darkness at 90° due to destructive interference of the light rays.

Although the sample placement with appropriate lighting is extremely important to study the optical properties. However, it is also crucial to have the microscope camera settings done accurately so that any image taken reflects true condition. The white balance must be correct, or it will completely throw off the colors seen in the

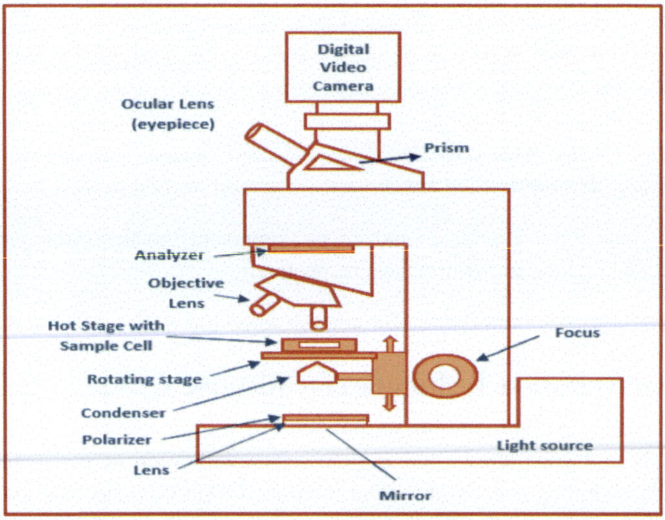

Fig. 4 A detailed schematic of polarized optical microscopy (POM)

pictures. This can be calibrated by bringi ng part of the slide that does not contain the sample into view and using the auto-white balance function on the software. It is highly preferable to have the correct colors showing when the picture is taken as what is seen live from the eyepiece. It helps with ensuring fidelity of the taken pictures.

5.2 X-Ray Diffraction (XRD)

It is a common tool used for the structural investigation of LCs. The complete analysis of XRD provides the direct information related to the positional order in LC. It also helps in analysing the structure of a particular LC phase. XRD works on the principle of Bragg's reflection $2d \sin \theta = n\lambda$ where 'd' represents the distance of adjacent atomic planes, 'θ' is the path difference and 'n' is an integer which is a multiple of wavelength 'λ' that gives the order of reflection [58]. The incident and reflected beam make an angle of 2θ with each other. XRD setup includes—X-ray tube, Gobel mirror, soller slit, sample chamber for controlling the temperature and alignment of the sample with the help of magnetic or electric field, mounted on a goniometer and the output is obtained through a detector as shown in Fig. 5.

The goniometer holds the crystal in the beam of X-rays and rotates the sample to precise positions and different orientations to obtain enough diffraction data for good quality analysis. Nowadays, the goniometer has four circles, and provides three angles of rotation for the crystal sample. The incident beam emerging from X-ray tube is always polychromatic. So, Gobel mirror is used as a monochromator to select a single wavelength. The sample chamber is mounted at the centre of the goniometer which allows the rotation around the vertical axis. The detector is attached to soller slit to enhance the focus and accuracy of measurements.

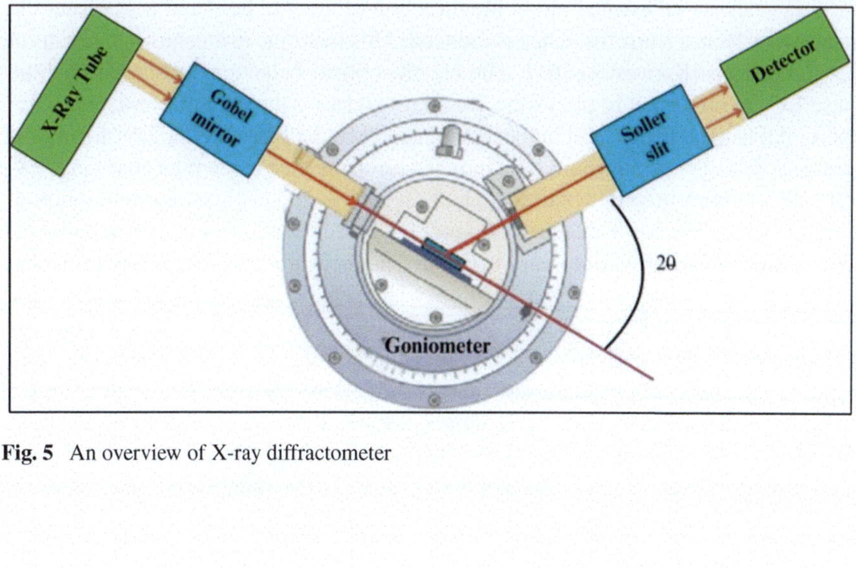

Fig. 5 An overview of X-ray diffractometer

5.3 UV–Visible Spectrometer

The UV–visible spectroscopy is a powerful tool to measure the optical properties of LC sample including transmittance, absorption, reflectance, etc. If a material possesses higher transmittance in the visible region, it clearly indicates lower absorption and reflectance. If the sample does not absorb the incident light for a given wavelength range, then the incident light intensity (I) is equal to transmitted light intensity (Io) (I = Io). However, in practice all the materials do not possess complete transmittance, Io is always lower than I (Io < I). The relation between absorption (A), and transmittance (T) is given according to Beer-lambert's law [60], and the absorption of the material is expressed as follows,

$$A = -\log(T) = -\log(\mathrm{Io}/\mathrm{I}) \tag{8}$$

The absorption of the material and optical band gap depends on the film thickness and absorption coefficient of the materials where absorption coefficient is one of the intrinsic properties of the material.

5.4 Birefringence Method

The difference in refractive index along and perpendicular to the electric field is called induced birefringence. The Kerr effect or the quadratic electro-optic effect (QEO effect) corresponds to a change in the refractive index of a material in response to an applied electric field. The induced index change is directly proportional to the square of the electric field. The blue phases are supposed to be optically isotropic since the molecules in this are randomly distributed. On application of electric field, the molecules are reoriented along the direction of electric field and hence become anisotropic. Hence a birefringence is induced. This isotropic to anisotropic transition can be called as Kerr effect [61]. The electro-optical behaviour exhibited by blue phase LCs approximately obeys the Kerr law given by the equation $\Delta n = \lambda K E^2$ where, Δn is the electric field induced birefringence, λ is the wavelength of the light used, and K is the Kerr constant. The induced birefringence in the blue phase of CE8-LCP cell was measured by obtaining the phase retardation using Senarmont method. A yellow filter (589 nm) is placed in front of the source to allow only monochromatic light at wavelength 589 nm to pass through. The measurement setup consists of the polarizer with the transmission axis of 45°. The sample is placed with no tilt. This is followed by a quarter wave plate placed parallel to the polarizer. The analyzer is placed perpendicular to the polarizer. In the voltage-off state, the sample is optically isotropic and does not introduce any phase retardation. The fast axis of the $\lambda/4$ plate coincides with the transmission axis of the polarizer; thus, the polarization state does not change. Since the polarizer and analyzer are crossed, the linearly polarized light is absorbed by the analyzer. As the voltage is applied to the sample, phase retardation is introduced and this can be obtained by noting the angle through which the analyzer has to be rotated to bring back the position where the polarizer and analyzer is in crossed position and the analyzer completely blocks out the light.

5.5 *Differential Scanning Calorimetry (DSC)*

DSC is a method of thermal analysis that is commonly used to study thermal transitions. We have used DSC to better understand the temperatures at which the LC phases undergo phase transitions. DSC measures the heat absorbed or released by a material as a function of temperature or time (isothermally). The enthalpy changes at a transition can also be used to identify the nature of the phase transitions and can serve as a complimentary tool to the POM by means of which distinct textures of the phases can be observed. Using DSC, it is possible to understand what materials go through during energy gain or loss [56].

5.6 *Dielectric Analysis*

The measurement of dielectric parameters relating to materials as they change state provides considerable information relating to molecular structure and change in the electrical properties of the system. Dielectric anisotropy (ε) is one of the most important physical properties of liquid crystalline compounds, which in essence determines the lower threshold voltages of LC displays. The complex dielectric constant of the LC is described as

$$\varepsilon^* = \varepsilon' - i\varepsilon'' \tag{9}$$

where ε' is the real and ε" is the imaginary parts (dielectric loss) of the dielectric constant. Frequency dependent cole–cole form of this equation given following form:

$$\varepsilon^*(\omega) = \varepsilon_\infty + \frac{(\varepsilon_s - \varepsilon_\infty)}{1 + (i\omega\,\tau_0)^{1-\alpha}} \tag{10}$$

where ε_s is the low frequency dielectric constant and ε_∞ is the high frequency dielectric constant, τ_0 is the average relaxation time, ω and α are the angular frequency and distribution parameter coefficient, respectively [64].

6 Advancements in Bistable CLCs

Bistable technology plays a pivotal role in the development of optical elements as it requires no power unless any stable state needs to be changed. This technological advancement in the field has enabled new approaches in smart cities, health-care sector, communication system, sensors etc. [65]. Junior et al. (2023) reviewed CLC sensors for quality human health based on nano-cellulose derivatives. The application of this sensors includes numerous parameters ranging from strain sensing to DNA and chirality assessment. The author has also provided a SWOT analysis, where the strengths relate to the sustainability, cost efficiency and flexibility in preparation. These strengths lead to widespread opportunities that is related to the use of new green technology regulations, innovation in signal processing and different method for the functioning and fabrication of these CLC sensors. However, the weakness of this technology comes with the necessity to overcome some challenges on the shape, size, large scale producibility of the sensors. Also, the threats are the standards and regulations set for sensor technology which requires accuracy and expected competition of the sensor industry [66]. Some detailed analysis on recent advancements in the field of bistable CLC is discussed below in Table 2.

Some detailed analysis with supporting data of the advancements in CLC technology provided in the above table is discussed here. Zhang et.al reported the influence of HTP on the optical bistable properties of polymer morphology in samples A1–A4. Figure 6a shows the SEM images of prepared samples. It was observed that size of the mesogenic epoxy polymer microspheres decreased from A1 to A4 with increased anchoring effect. Due to the large anchoring effect, the transmittance gets lowered at 50 Hz, 90 V from sample A1 to A4 and shows FC state (Fig. 6b) and on removing the electric field after 24 h, the state was maintained (Fig. 6c). The transmittance curves at 15 kHz, 90 V exhibits P state for samples A1–A4 that results in high transmittance and on removing the electric field after 24 h, negligible changes were observed as shown in Fig. 6d and Fig. 6e, respectively. This phenomenon confirmed that the parallel alignment of N-CLC in the samples had good stability and the existence of mesogenic epoxy polymer microspheres has no effect on alignment of molecules.

Also, transmission spectra and photographs were observed for sample A4 for 30 after being placed in both FC and P state as shown in Fig. 7a. The transmittance in visible range was large in P state and <30% in FC state. Besides this, A4 sample displayed the infrared shielding function, especially in the FC state. This makes A4 sample suitable for bistable electro-optical window and exhibits potential applications in energy efficient devices. Then, three simulated houses were prepared, as shown in Fig. 7b, where a sunlamp of 500–3000 nm was placed to simulate sunlight and sample A4 was placed in two states for 30 days respectively, and an empty LC cell was prepared as control sample that can be used as the window. The recorded temperature with electronic thermometer was shown in Fig. 7c. The sun lamp and thermometer is placed 42 cm to each other. When stimulated house was exposed to sun lamp for 100 min, the temperature was stable at 30.1 °C which uses control

Table 2 Comparison of recent advancements of bistable CLCs with details of used materials, dopants/cell configuration, and outcomes

Reference(s)	Materials	Dopants/cell configuration	Outcomes
Zhang et al. [67]	Negative nematic LC: HNG762600-100 Chiral dopants: R811, R1011, R2011 and R5011 LC epoxide: E6M Curing agent: Capcure 3–800 Base catalyst: DMP-30	Cell thickness: 19.0 ± 1.0 μm Sample cells: A mixture of 6.0 wt.% E6M, 6.0 wt.% Capcure 3–800, 87.0 wt.% N-ChLCs and 1.0 wt.% DMP-30 were mixed with different chiral dopants and filled into the cell by thermal curing at 120 °C, resulting in four sample cells (A1, A2, A3, A4)	• Studied, influence of HTP on polymer morphology and optical bistable properties of prepared sample cells • The sample A4 possessing the chiral dopant R5011 with a large HTP exhibited good bistability, and its large- and low-transmittance states can be maintained for a long time (>30 days) in the absence of electric fields • Detailed outcome is provided in Figs. 7 and 8

(continued)

Table 2 (continued)

Reference(s)	Materials	Dopants/cell configuration	Outcomes
Zhao et al. [52]	Negative nematic LC: SLC 3 MV–8173 Chiral dopant: R811 UV-polymerizable monomers mixture: polyethylene glycol diacrylate (PEGDA 600), 1,6-hexanediol, iso-bornyl methacrylate (IBMA) and hydroxypropyl methacrylate (HPMA) Photo-initiator: RG651 Commercial PDLC: D001	Transparent PET film: 110 μm Cell thickness: 20 μm Sample cells: prepared using two PET conductive transparent electrodes with mixture evenly filled in it to form films. The films were covered with photomask to form polymer spacer columns	• This paper showcases the comparative analysis of polymer stabilized cholesteric textures with and without polymer spacer column • Electro-optic properties were investigated: one for different monomer mixture concentration (A1-A5) and other for different mask times (B1-B5) • Maximum transmittance in planar state and minimum transmittance in focal conic state was found for the cell with 5% monomer mixture (A3) and 300 s mask polymerisation time (B5) that leads to enhanced contrast ratio • Also, the bistable states were maintained for more than 7 days without applying any external field

(continued)

Table 2 (continued)

Reference(s)	Materials	Dopants/cell configuration	Outcomes
Hung et al. [36]	Dual frequency LC: HEF951800–100 Photo-initiator: IRG651 Synthesized chiral dopant: 4-(((R)–1-butoxy-1-oxopropan-2-yl)oxy)phenyl 4′-(((S)–3,7-dimethyloct-6-en-1-yl)oxy)-(1,1′-biphenyl)–4-carboxylate, 4′-((6-hydroxyhexyl)oxy)-(1,1′-biphenyl)–4-carboxylic acid, 4-((6-(acryloyloxy)hexyl)oxy)-(1,1′-biphenyl)–4-carboxylic acid, (R)–4-((1-butoxy-1-oxopropan-2-yl)oxy)phenyl 4′-((6-(acryloyloxy)hexyl)oxy)-(1,1′-biphenyl)–4-carboxylate	Cell gap: 16 μm	• This paper presents the synthesis of novel liquid crystalline acrylic chiral dopant • Bistable display performance achieved high transmittance of 97.6% and 1.5 ms response time which shows reliable electro-optical characteristics for applications in smart window
Zhang et al. [68]	Nematic LC: E7 and X1 Dimer: CB7CB Chiral dopant: R5011 (120 μm^{-1}) and CB15 (8 μm^{-1})	Cell thickness: 5 μm	• Authors reported a new mechanism in order to switch bistable reflective displays at low operating voltage • The addition of dimer reduces the driving voltage of the switching states leading to heliconical state formation

(continued)

Table 2 (continued)

Reference(s)	Materials	Dopants/cell configuration	Outcomes
Zhao et al. [69]	Bistable CLC: DYE10900 Chiral dopant: R5011(right-handed chiral) Photo initiator: IRG651 Liquid crystalline acrylate monomer: C6M Non-liquid crystalline acrylate monomer: HDDA Fluorescent dyes: Dye 1 (perylene) and Dye 2 (di-isobutylperylenedicarboxylate)	Sample cells: 4 pre-polymer mixtures were prepared with varied concentration of CLC, photo-initiator and monomer Also, 5 sample cells were prepared with dye 1 and dye 2 individually in varied concentrations	• This paper presents dye coated polymer stabilized bistable CLC films for display applications • Fluorescent dye concentration of 4.0 mg/ml yields optimised performance with high brightness and contrast ratio (7.44 for dye 1 and 5.56 for dye 2)
He et al. [70]	Negative nematic LC: HNG726200-100 Photo initiator: Irg651 Chiral dopant: S811 Monomer: C6M	Cell thickness: 20 μm Sample cells: Two sample groups were prepared-Group A (1–5 wt/wt.% varied concentration of C6M + 5 wt./wt.% S811), Group B (3 wt/wt.% C6M + {1,3,5,7,9} wt./wt.% S811)	• Bistable polymer stablised LC based light shutter was developed using different polymer concentration and chiral dopant • Optimised energy efficient electro-optical properties were achieved when polymer concentration is set to 3wt./wt.% and 5 wt./wt.% for chiral dopant • Detailed outcome is provided in Figs. 9 and 10

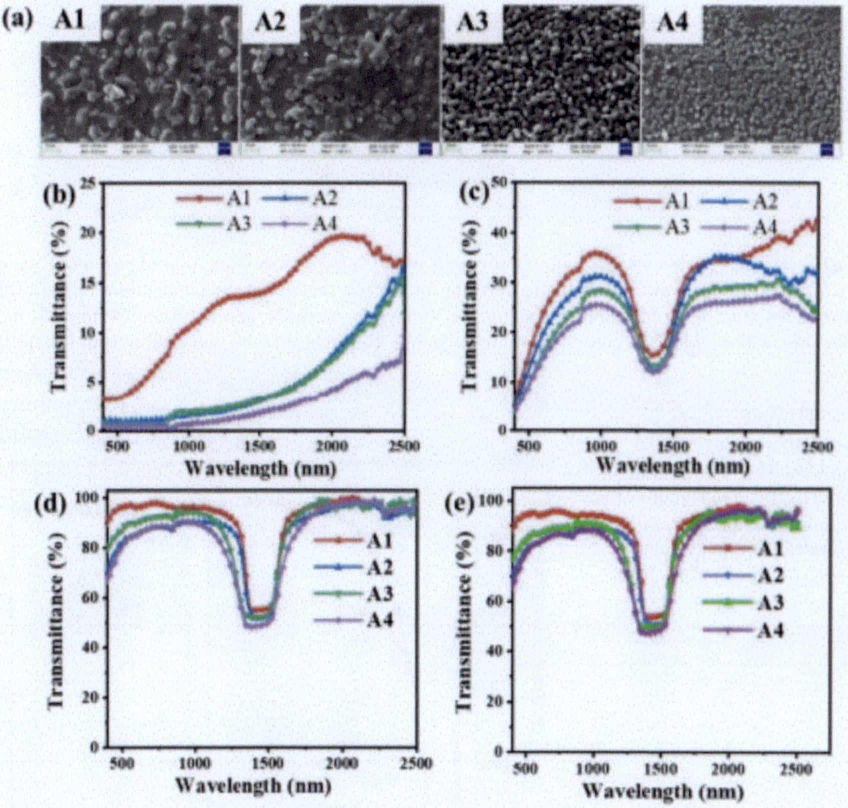

Fig. 6 **a** SEM images of samples A1–A4. Transmission spectra: **b** at low frequency electric field (50 Hz, 90 V); **c** after removing electric field for 24 h; **d** at high frequency electric field (15 kHz, 90 V); **e** after removing electric field for 24 h; of samples A1–A4. Reproduced with permission from Ref. [67]

sample as the window. However, it was 26.4 °C when sample A4 was placed in the FC state and 28.7 °C in the P state. This shows that FC state can reduce the thermal effect of sunlight. In general, the polymer microspheres filled N-CLC composite films can be used in the visible and infrared regulating windows to achieve both the privacy protection and the thermal management effect of buildings.

A study by He et al. aimed to compare the stability of prepared sample cells in long duration through transmittance versus time curves as plotted in Fig. 8. Initially the samples were switched to P state and then low frequency voltage was applied at “ON1” stage. After the pulse, the transmittance of A1 and A2 increases sharply at “OFF1” stage, resulting in unstable FC state. However, remaining three samples remains in the FC state for quite long duration, while samples A4~A5 attained low-transmittance P state. Thus, sample A3 was tested for following stages. Afterwards, high frequency voltage was applied at “ON2” stage to obtain P state and “OFF2”

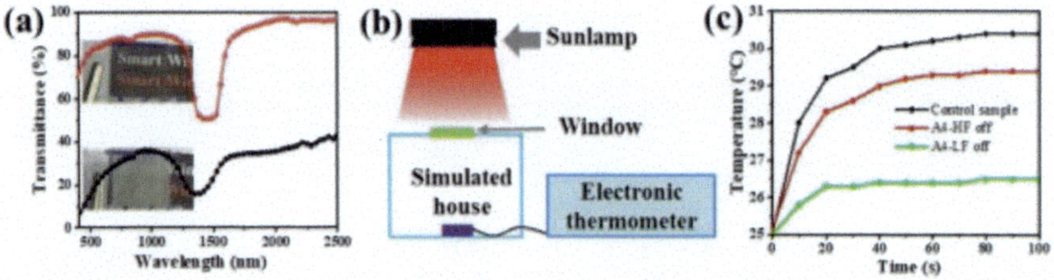

Fig. 7 **a** Transmission spectra with photographs of A4 sample after being placed in F and P state for 30 days, respectively; **b** schematic diagram for the test method of the temperature utilized for sample A4 after being placed in two states for 30 days respectively, and the control sample as the windows. **c** Temperature recorded in simulated house. Reproduced with permission from Ref. [67]

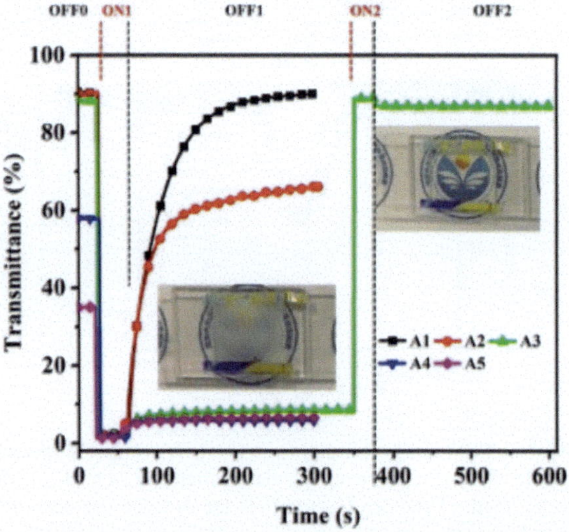

Fig. 8 Transmittance versus time variations of A1~A5 samples. The inset photos show the FC and P states of sample A3 with a picture below it. Reproduced with permission from Ref. [70]

stage transmittance was recorded. This exhibits A3 has good bistability for both P and FC state.

Further, morphological characteristics were observed for sample A3 using POM as shown in Fig. 9. Initially, Fig. 9a represents P state at no applied voltage. When low-frequency voltage is applied, the helix of the CLC molecules starts unwinding and some oily streaks were observed (Fig. 9b–f). At 60 V, 100 Hz the molecules gets distorted completely and resulted in FC state (Fig. 9g). Subsequently, when high-frequency voltage 60 V, 5 kHz was applied the sample switches back to P state as depicted in Fig. 9h. On considering all the electro-optical characteristics, A3 sample with 3 wt./wt.% polymer concentration provides the optimised results with best performances.

In 2022, Ridhima et al. demonstrated an energy efficient device for light shutter applications. The textural study of the sample cells (pure and dye doped CLC) using

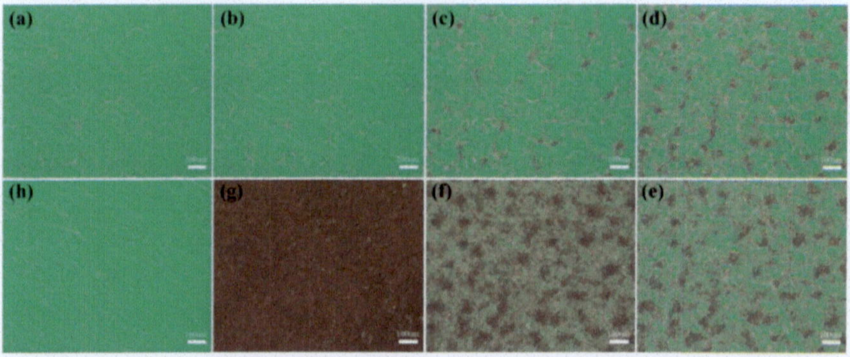

Fig. 9 POM textures of sample A3 under different electric fields with crossed polarizers. **a** 0 V 100 Hz, **b** 10 V 100 Hz, **c** 20 V 100 Hz, **d** 30 V 100 Hz, **e** 40 V 100 Hz, **f** 50 V 100 Hz, **g** 60 V 100 Hz and **h** 60 V 5 kHz. Reproduced with permission from Ref. [70]

POM was demonstrated in Fig. 10. Initially both the sample cells were in P state, under low frequency voltage, minor perturbations in the cells were observed. However, at 29 V, 1 kHz (pure CLC) and 27 V, 1 kHz (dye doped CLC) considerable distortions were observed in the cells and FC state was achieved 1 kHz). Subsequently, when high frequency voltage was applied, the cells were switched to P state. Although fine-tuned P and FC states were observed for both samples but dye molecules absorb the leakage of light in FC state, which serves as an advantage to obtain minimum transmittance in the present state of CLC molecules [56].

Further, Fig. 11 reported the bistability of the prepared sample cells. Figure 11a1 and b1 represents the perfect FC states of pure and dye doped cells (as reported earlier) and in switch OFF condition negligible changes were observed (Fig. 11a2 and b2). Subsequently, Fig. 11c1 and c2 exhibited good P states at 29 V and 27 V, 50 kHz for pure and dye doped CLCs, respectively. Although some oily streaks were observed in P state of both cells, but is found lesser in dye doped cell. This is due to the inherent property of the dye to provide alignment to CLC molecules. This resulted into excellent stability for the cells and each state and can only be switched to other state when low (1 kHz) or high (50 kHz) frequency voltage was applied.

Herein the electro-optical characteristics were found in conjunction with the textural properties as shown in Fig. 12. Initially the cells in P state showed maximum transmittances at 4.564 a.u. (pure CLC) and 4.609 a.u. (dye-doped CLC). When voltage is applied at 1 kHz, no perturbations in the alignment of the molecules were observed upto 18 V (pure CLC) and 16 V (dye-doped CLC). However, at 22.3 V and 20.71 V in pure and dye doped CLCs, respectively, the uniformity of P state gets distorted and FC state with smaller domains was obtained. On again switching the cells under frequency voltage, reduction in the transmittance was found to be 0.33277 a.u. and 0.2801 a.u. respectively, FC state was achieved. Both the states were found to be stable electro-optically as well, when the voltage was reduced to 0 V.

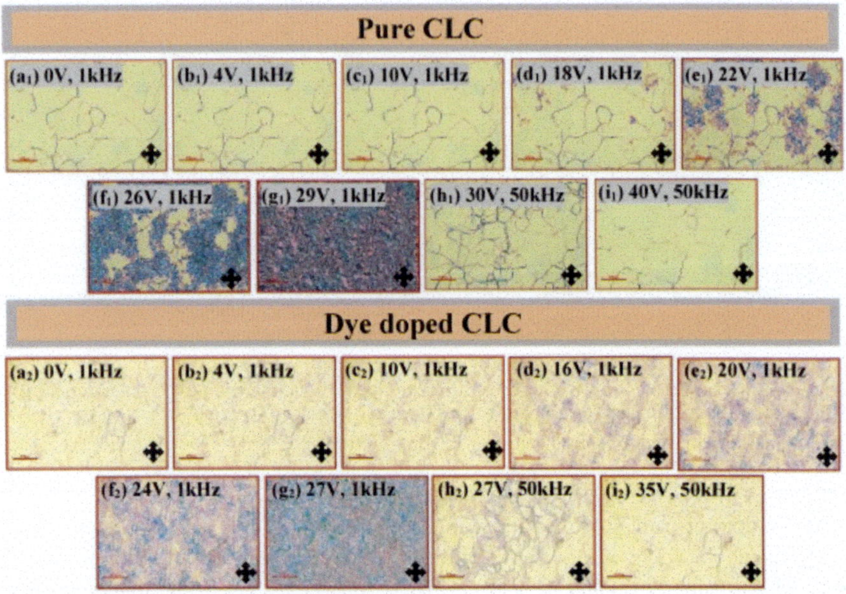

Fig. 10 The POM textures of pure and dye doped CLCs under crossed polarizers at different applied voltages for low (1 kHz) and high (50 kHz) frequency. Reproduced with permission from Ref. [56]

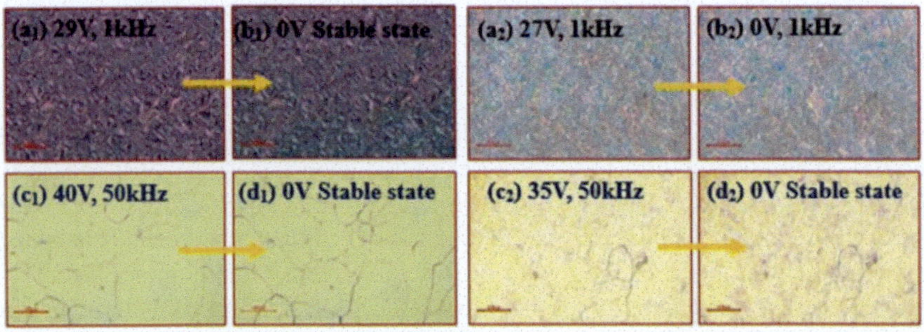

Fig. 11 POM textures confirming bistability for both pure (a1 → b1), (a2 → b2) and dye doped (c1 → d1), (c2 → d2) CLCs in FC and P states, respectively. Reproduced with permission from Ref. [56]

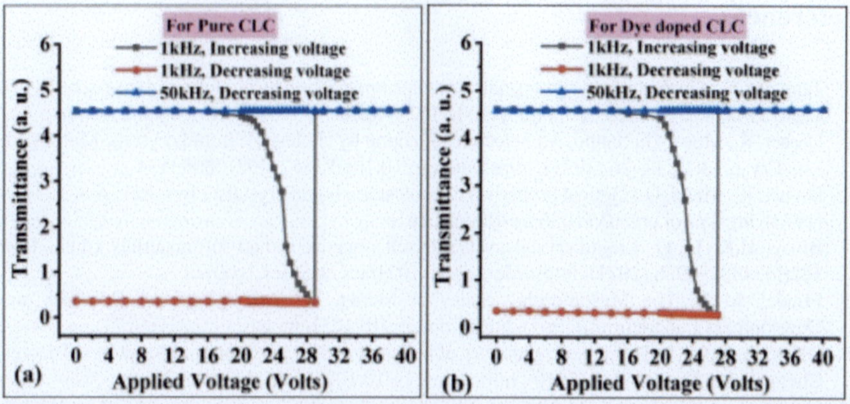

Fig. 12 The electro-optical characteristics for pure and dye doped CLC, from P to FC state at 1 kHz and FC to planar state at 50 kHz. Both the states show negligible transmittance variation on reducing voltages to zero at low and high frequencies. Reproduced with permission from Ref. [56]

7 Conclusion

Bistable CLCs serve as next generation materials as they offer numerous applications. They can have two stable states-P and FC state even in the absence of applied electric field. The helical super-structure of CLCs makes them ideal for electro-optical systems. Chiral dopant serves as ideal in regulating the helical alignment and form of the cholesteric molecules. As a result, the optical characteristics are altered, creating further opportunities for real-world applications in the field of smart materials, sensors, architectural windows and displays.

This chapter covers recent advances in sample preparation, characterization etc. that allow improvement and our understanding of their dynamic behaviour. However, it opens the door for novel applications in sensors, adaptive materials, and display technologies. Bistability in CLCs have significant implications for energy-efficient displays and smart optical systems, highlighting their importance in recent technological aspects. This research connects the gap between fundamental science and practical application by merging theoretical concepts with real-world methodologies. As researchers continue to investigate and create in this area, the future of bistable CLCs is bright, particularly for developing sustainable, high-performance materials to solve new problems in display technology.

References

1. Tamaoki, N.: Cholesteric liquid crystals for color information technology. Adv. Mater. **13**(15), 1135–1147 (2001). https://doi.org/10.1002/1521-4095
2. Dreher, R., Meier, G., Saupe, A.: Selective reflection by cholesteric liquid crystals. Mol. Cryst. Liq. Cryst. **13**(1), 17–26 (1971). https://doi.org/10.1080/15421407108083534
3. Dreher, R., Meier, G.: Optical properties of cholesteric liquid crystals. Phys. Rev. A **8**(3), 1616 (1973). https://doi.org/10.1103/PhysRevA.8.1616
4. Bisoyi, H.K., Li, Q.: Liquid crystals: versatile self-organized smart soft materials. Chem. Rev. **122**(5), 4887–4926 (2021). https://doi.org/10.1021/acs.chemrev.1c00761
5. Friedel, M.G.: The Mesomorphic States of Matter. US Army Engineer Research and Development Laboratories (1967). https://doi.org/10.1051/anphys/192209180273
6. Olsen, K., Bohr, J.: The generic geometry of helices and their close-packed structures. Theoret. Chem. Acc. **125**, 207–215 (2010). https://doi.org/10.1007/s00214-009-0639-4
7. Duraes, A.D., Gezelter, J.D.: A theory of pitch for the hydrodynamic properties of molecules, helices, and achiral swimmers at low Reynolds number. J. Chem. Phys. **159**(13) (2023). https://doi.org/10.1063/5.0152546
8. Zhou, H., Wang, H., He, W., Yang, Z., Cao, H., Wang, D., Li, Y.: Research progress of cholesteric liquid crystals with broadband reflection. Molecules **27**(14), 4427 (2022). https://doi.org/10.3390/molecules27144427
9. Golicha, H.S., Omar, M.H., Mbithi, N.M.: Optical techniques in the determination of pitch lengths in the cholesteric and chiral smectic C phases. J. Mater. Sci. Eng. **10**(3) (2021)
10. Wang, L., Luo, S., Liu, S., Yang, P., He, Z., Miao, Z.: Progress in the study of broad-wave reflection of cholesteric liquid crystals. Liq. Cryst. 1–17 (2024). https://doi.org/10.1080/02678292.2024.2406881
11. Min, L.K., Marsh, Z.M., Crenshaw, E.P., Tohgha, U.N., Ambulo, C.P., Wolf, S.M., Carothers, K.J., Limburg, H.N., McConney, M.E., Godman, N.P.: Recent advances in electro-optic response of polymer-stabilized cholesteric liquid crystals. Materials **16**(6), 2248 (2023). https://doi.org/10.3390/ma16062248
12. Wang, Y., Nie, H., Han, J., An, Y., Zhang, Y.M., Zhang, S.X.A.: Green revolution in electronic displays expected to ease energy and health crises. Light: Sci. Appl. **10**(1), 33. https://doi.org/10.1038/s41377-020-00455-9
13. Vetrov, S.Y., Timofeev, I.V., Shabanov, V.F.: Localized modes in chiral photonic structures. Phys. Usp. **63**(1), 33 (2020). https://doi.org/10.3367/UFNe.2018.11.038490
14. Golovina, T.G., Konstantinova, A.F., Timofeev, V.I.: Features of optical activity in inorganic and organic materials. Crystallogr. Rep. **65**, 653–677 (2020). https://doi.org/10.1134/S1063774520050077
15. Zhang, W., Froyen, A.A., Schenning, A.P., Zhou, G., Debije, M.G., de Haan, L.T.: Temperature-responsive photonic devices based on cholesteric liquid crystals. Adv. Photonics Res. **2**(7), 2100016 (2021). https://doi.org/10.1002/adpr.202100016
16. Shen, W., Li, G.: Recent progress in liquid crystal-based smart windows: materials, structures, and design. Laser Photonics Rev. **17**(1), 2200207 (2023). https://doi.org/10.1002/lpor.202200207
17. Miao, Z., Chen, X., Zhang, Y., Wang, D., Wang, L.: Bistable cholesteric liquid crystal films with excellent electro-optical performance and spacing stability for reflective displays. ACS Appl. Polym. Mater. **5**(1), 476–484 (2022). https://doi.org/10.1021/acsapm.2c01594
18. Yang, D.K., Shiu, J.W., Yang, M.H., Che, J.: Flexible reflective display based on cholesteric liquid crystals. Flexible Flat Panel Displays 167–206 (2023). https://doi.org/10.1002/9781118751077.ch10
19. Janaki, V.P., Moorthi, S., Mohan, M.M.: Dynamic memory tuned by frequency in a homologous thermotropic liquid crystal. Eur. Phys. J. E **46**(6), 38 (2023). https://doi.org/10.1140/epje/s10189-023-00300-y

20. Oh, S.W., Baek, J.M., Heo, J., Yoon, T.H.: Dye-doped cholesteric liquid crystal light shutter with a polymer-dispersed liquid crystal film. Dyes Pigm. **134**, 36–40 (2016). https://doi.org/10.1016/j.dyepig.2016.06.045
21. Baek, J.M., Oh, S.W., Kim, S.H., Yoon, T.H.: Fabrication of an initially-focal-conic cholesteric liquid crystal cell without polymer stabilization. Displays **52**, 55–58 (2018). https://doi.org/10.1016/j.displa.2017.10.002
22. Kim, S.H., Oh, S.W., Yoon, T.H.: Enhancement of absorption and haze with hybrid anchoring of dye-doped cholesteric liquid crystals. Opt. Express **26**(11), 14259–14266 (2018). https://doi.org/10.1364/OE.26.014259
23. Cao, Y., Derakhshani, M., Fang, Y., Huang, G., Cao, C.: Bistable structures for advanced functional systems. Adv. Func. Mater. **31**(45), 2106231 (2021). https://doi.org/10.1002/adfm.202106231
24. Zhang, J., Zhang, Y., Yang, J., Wang, X.: Beyond color boundaries: pioneering developments in cholesteric liquid crystal photonic actuators. Micromachines **15**(6), 808 (2024). https://doi.org/10.3390/mi15060808
25. Yang, T., Xie, D., Li, Z., Zhu, H.: Recent advances in wearable tactile sensors: materials, sensing mechanisms, and device performance. Mater. Sci. Eng. R. Rep. **115**, 1–37 (2017). https://doi.org/10.1016/j.mser.2017.02.001
26. Wu, C., Zeng, S., Wang, Z., Wang, F., Zhou, H., Zhang, J., Ci, Z., Sun, L.: Efficient mechanoluminescent elastomers for dual-responsive anticounterfeiting device and stretching/strain sensor with multimode sensibility. Adv. Func. Mater. **28**(34), 1803168 (2018). https://doi.org/10.1002/adfm.201803168
27. Sharma, V., Crne, M., Park, J.O., Srinivasarao, M.: Bouligand structures underlie circularly polarized iridescence of scarab beetles: a closer view. Mater. Today: Proc. **1**, 161–171 (2014). https://doi.org/10.1016/j.matpr.2014.09.019
28. Herle, D., Martin, O.J., Villanueva, L.G., Quack, N.: Emulating paper: a review of reflective display technologies. J. Opt. Microsyst. **4**(2), 020901–020901 (2024). https://doi.org/10.1117/1.JOM.4.2.020901
29. Gahrotra, R., Sharma, V., Dogra, A.R., Malik, P., Kumar, P.: Temperature dependent morphological and electro-optical characteristics of dye doped cholesteric liquid crystal. ECS Trans. **107**(1), 5481 (2022). https://doi.org/10.1149/10701.5481ecst
30. Manhas, N., Adimule, V., Lakshminarayana, P., Keri, R., Kumar, L.S., Bathula, C.: Utilization of CO^{2+} doped Eu_2O_3 nanostructures and their effect on the dielectric and photoisomerization of liquid crystals. Phys. Fluids **35**(10) (2023). https://doi.org/10.1063/5.0171578
31. Ahn, S., Ko, M.O., Kim, J.H., Chen, Z., Jeon, M.Y.: Characterization of second-order reflection bands from a cholesteric liquid crystal cell based on a wavelength-swept laser. Sensors **20**(16), 4643 (2020). https://doi.org/10.3390/s20164643
32. Hussain, S., Zourob, M.: Solid-state cholesteric liquid crystals as an emerging platform for the development of optical photonic sensors. Small **20**(7), 2304590 (2024). https://doi.org/10.1002/smll.202304590
33. Kurochkin, O., Nazarenko, K., Tereshchenko, O., Golub, P., Nazarenko, V.: The helical twisting power of chiral dopants in lyotropic chromonic liquid crystals. Liq. Cryst. **50**(1), 110–120 (2023). https://doi.org/10.1080/02678292.2022.2114030
34. Kim, Y., Tamaoki, N.: Photoresponsive chiral dopants: light-driven helicity manipulation in cholesteric liquid crystals for optical and mechanical functions. Chem. Photo. Chem. **3**(6), 284–303 (2019). https://doi.org/10.1002/cptc.201900034
35. Dörfler, H.D.: Chirality, twist and structures of micellar lyotropic cholesteric liquid crystals in comparison to the properties of chiralic thermotropic phases. Adv. Coll. Interface. Sci. **98**(3), 285–340 (2002). https://doi.org/10.1016/S0001-8686(01)00095-1
36. Hung, Y.H., Chang, K.T., Chiu, Z.Y., Chang, J.Y., Yen, K.T., Liu, C.Y.: Fabrication of high-transmittance and fast-response bistable dual-frequency CLC cells. J. Taiwan Inst. Chem. Eng. **146**, 104866 (2023). https://doi.org/10.1016/j.jtice.2023.104866
37. Shaban, H., Wu, P.C., Jia, Y.F., Lee, W.: Overlooked ionic contribution of a chiral dopant in cholesteric liquid crystals. Materials **17**(20), 5080 (2024). https://doi.org/10.3390/ma17205080

38. Valyukh, S., Arwin, H., Birch, J., Järrendahl, K.: Bragg reflection from periodic helicoidal media with laterally graded refractive index. Opt. Mater. **72**, 334–340 (2017). https://doi.org/10.1016/j.optmat.2017.06.018
39. Kumar, P., Kang, S.W., Lee, S.H.: Advanced bistable cholesteric light shutter with dual frequency nematic liquid crystal. Opt. Mater. Express **2**(8), 1121–1134 (2012). https://doi.org/10.1364/OME.2.001121
40. Pooja, P., Takkar, R., Malik, P., Sharma, V., Kumar, P.: Morphological, electro-optical and UV-visible spectroscopic analysis of pure cholesteric liquid crystal. In: AIP Conference Proceedings, vol. 3209, no. 1. AIP Publishing (2024). https://doi.org/10.1063/5.0228638
41. Kipfer, P., Klappert, R., Herzig, H.P., Grupp, J., Dandliker, R.: Improved red color with cholesteric liquid crystals in Bragg reflection mode. Opt. Eng. **41**(3), 638–646 (2002). https://doi.org/10.1117/1.1447229
42. Dong, X., Huang, L., Zeng, H., Lin, Z., Ok, K.M., Zou, G.: High-performance sulfate optical materials exhibiting giant second harmonic generation and large birefringence. Angew. Chem. **134**(10), e202116790 (2022). https://doi.org/10.1002/ange.202116790
43. Raja, S.J., Rao, S.S., Charlcedony, R.: Design and analysis of dispersion-compensating chalcogenide photonic crystal fiber with high birefringence. SN Appl. Sci. **2**, 1–10 (2020). https://doi.org/10.1007/s42452-020-2308-0
44. Farezi, N., Khoshsima, H., Zakerhamidi, M.S., Ranjkesh, A., Yoon, T.H.: Temperature-dependent characterization of optical parameters in novel high birefringence nematic liquid crystal mixtures: an empirical investigation. Phys. Scr. **99**(2), 025980 (2024). https://doi.org/10.1088/1402-4896/ad1eac
45. George, N., Nair, A.B., Simi, P.K., Joseph, R.:. Polymers in display devices. In: Specialty Polymers, pp. 445–457. CRC Press (2023). https://doi.org/10.1201/9781003278269
46. Lin, K.W., Lu, J.Y., Tseng, H.Y., Lin, C.Y., Huang, L.Y., Lin, Y.L., Chang, H.H., Lin, T.H.: Study of polymer-stabilized cholesteric liquid crystals for reflective displays in active matrix driving. J. Inform. Display 1–8. https://doi.org/10.1080/15980316.2024.2372334
47. Selvaraj, P., Tsai, Y.C., Wu, C.T., Huang, C.T., Liu, C.K., Cheng, K.T.: Electro-optically addressable and rewritable transparent liquid crystal bistable waveguide display devices. Opt. Laser Technol. **182**, 112165 (2025). https://doi.org/10.1016/j.optlastec.2024.112165
48. Faneca, J.F.: Tunable Silicon Integrated Photonics Based on Functional Materials. University of Exeter (United Kingdom) (2020)
49. Shen, W., Zhang, H., Miao, Z., Ye, Z.: Recent progress in functional dye-doped liquid crystal devices. Adv. Func. Mater. **33**(6), 2210664 (2023). https://doi.org/10.1002/adfm.202210664
50. Dou, S., Xu, H., Zhao, J., Zhang, K., Li, N., Lin, Y., Li, Y.: Bioinspired microstructured materials for optical and thermal regulation. Adv. Mater. **33**(6), 2000697 (2021). https://doi.org/10.1002/adma.202000697
51. Chen, X., Chen, J., Song, X., Du, T., Deng, X., Deng, Z., Lan, R.: Bioinspired mechanochromic liquid crystal materials: from fundamentals to functionalities and applications. Adv. Mater. (2024). https://doi.org/10.1002/adma.202403766
52. Adimule, V., Keri, R., Rajendrachari, S., Kendrekar, P., Kulkarni, C.V.: Highly photoluminescence and wide band gap insulating metal hybrid nanoparticles array of samarium-doped SrO: CoO: synthesis, characterizations and sensor characteristics. J. Mater. Sci.: Mater. Electron. **34**(5), 442 (2023). https://doi.org/10.1007/s10854-023-09899-2
53. Mysliwiec, J., Szukalska, A., Szukalski, A., Sznitko, L.: Liquid crystal lasers: the last decade and the future. Nanophotonics **10**(9), 2309–2346 (2021). https://doi.org/10.1515/nanoph-2021-0096
54. Zia, A., Saeed, S., Man, T., Liu, H., Chen, C.X., Wan, Y.: Next-generation interfaces: integrating liquid crystal technologies in augmented and virtual reality–a review. Liq. Cryst. Rev. **12**(1), 30–56 (2024). https://doi.org/10.1080/21680396.2024.2401786
55. Sharma, V., Kumar, P., Chinky, C., Malik, P., Raina, K.K.: Preparation and electrooptic study of reverse mode polymer dispersed liquid crystal: Performance augmentation with the doping of nanoparticles and dichroic dye. J. Appl. Polym. Sci. **137**(22), 48745 (2020). https://doi.org/10.1002/app.48745

56. Gahrotra, R., Sharma, V., Dogra, A.R., Malik, P., Kumar, P.: Performance augmentation of bistable cholesteric liquid crystal light shutter-effect of dichroic dye on morphological and electro-optical characteristics. Opt. Mater. **127**, 112243 (2022). https://doi.org/10.1016/j.optmat.2022.112243
57. Cowling, S.J.: Optical Microscopy Studies of Liquid Crystals. Handbook of Liquid Crystals (2014)
58. Azároff, L.V.: X-ray diffraction by liquid crystals. Mol. Cryst. Liq. Cryst. **60**(1–2), 73–97 (1980). https://doi.org/10.1080/00268948008072426
59. Pathinti, R.S., Dasari, S.G., Gollapelli, B., Gogula, S., Reddy, R., Vallamkondu, J.: Enhanced ethanol gas detection using TiO_2 nanorods dispersed in cholesteric liquid crystal: synthesis, characterization, and sensing performance. J. Alloy. Compd. **994**, 174731 (2024). https://doi.org/10.1016/j.jallcom.2024.174731
60. Chemingui, M., Yang, D., Wang, Y., Ayeb, H., Cai, W., Tang, A., Drevensek-Olenik, I., Zhang, X., Xu, J.: The influence of quantum dots on the optical properties of a room temperature cholesteric liquid crystal. Liq. Cryst. **49**(15), 2095–2107 (2022). https://doi.org/10.1080/02678292.2022.2103850
61. Fridkin, V.M., Golovina, T.G., Konstantinova, A.F., Evdishchenko, E.A.: Peculiarities of electric properties of various materials. Crystallogr. Rep. **67**(4), 494–520 (2022). https://doi.org/10.1134/S1063774522040083
62. Kim, S., Jeong, T.I., Park, J., Ciappina, M.F., Kim, S.: Recent advances in ultrafast plasmonics: from strong field physics to ultraprecision spectroscopy. Nanophotonics **11**(11), 2393–2431 (2022). https://doi.org/10.1515/nanoph-2021-0694
63. Adimule, V., Sharma, V., Kumar, P., Keri, R., Sharma, K.: Perylene-3, 4, 9, 10-tetracarboxylic dianhydride dye intercalated liquid crystal molecules: synthesis, spectroscopic characterizations and photoluminescence dynamics. J. Mol. Liq. (2024). https://doi.org/10.1016/j.molliq.2024.125990
64. Pathinti, R.S., Gollapelli, B., Jakka, S.K., Vallamkondu, J.: Green synthesized TiO_2 nanoparticles dispersed cholesteric liquid crystal systems for enhanced optical and dielectric properties. J. Mol. Liq. **336**, 116877 (2021). https://doi.org/10.1016/j.molliq.2021.116877
65. Alhamarneh, R.A., Mahinderjit Singh, M.: Strengthening internet of things security: surveying physical unclonable functions for authentication, communication protocols, challenges, and applications. Appl. Sci. **14**(5), 1700 (2024). https://doi.org/10.3390/app14051700
66. Leal-Junior, A., Soares, M.S., de Almeida, P.M., Marques, C.: Cholesteric liquid crystals sensors based on nanocellulose derivatives for improvement of quality of human life: a review. Adv. Sens. Res. **2**(10), 2300022 (2023). https://doi.org/10.1002/adsr.202300022
67. Zhang, H., Gao, J., Guo, Z., Wang, X., Li, Q., Miao, Z.: The influence of the helical twisting power on the bistable electro-optical performance of the liquid crystalline epoxide/mercaptan/negative-dielectric-anisotropy cholesteric liquid crystal composite films. Opt. Mater. **142**, 114091 (2023). https://doi.org/10.1016/j.optmat.2023.114091
68. Zhang, X., Zhou, Z., Halder, S., Hu, L., Yang, D.K.: Mechanism of switching between planar and focal conic states in a bistable cholesteric-liquid-crystal reflective display. Phys. Rev. Appl. **22**(4), 044020 (2024). https://doi.org/10.1103/PhysRevApplied.22.044020
69. Zhao, Y., Zhu, M., Zhao, Y., He, Z., Zhang, H., Ma, C., Miao, Z.: Dye-coated polymer stabilized cholesteric liquid crystal films for bistable reflective displays. Opt. Mater. **134**, 112991 (2022). https://doi.org/10.1016/j.optmat.2022.112991
70. He, Z., Zeng, J., Zhu, S., Zhang, D., Ma, C., Zhang, C., Miao, Z.: A bistable light shutter based on polymer stabilized cholesteric liquid crystals. Opt. Mater. **136**, 113426 (2023). https://doi.org/10.1016/j.optmat.2022.113426

Effect of Nanomaterial's Dispersion on the Dynamic Responses of Cholesteric Liquid Crystals

Neeraj Puri

Abstract Cholesteric liquid crystals (CLCs) are a class of liquid crystals that have wide range of applications in various fields due to their unique electrical and optical properties such as low viscosity, capability to change colour with variation of electric field, temperature and light etc. This chapter describes the effect of dispersion of nanomaterials on the dynamic responses of CLCs for dielectric and optical applications in liquid crystal displays, sensors, smart materials, luminescent displays, UV modulators, dye lasers, UV detectors etc. The addition of nanomaterials in the CLCs alters their dynamic responses in terms of electro-optical behaviours and mark them promising materials for advanced technology applications. Through a comprehensive review of studies of various researchers, this chapter provides a review that the nature, type, size, shape, concentration, dispersion techniques etc. influence the mesogenic behaviour and alignment within the cholesteric liquid crystal (CLC) matrix for enhanced performance for display and non-display applications.

Keywords Dynamic · Nanomaterials · Optical · Dielectric · Dispersion · And Cholesteric

1 Introduction

Liquid crystals (LCs) are organic compounds that manifest a phase of matter with combined properties of crystalline solids and isotropic liquids. Such phases are known as mesomorphic or liquid crystal (LC) phases. Based on the shape, chemical structure and molecular dynamics, LCs are classified as nematic, chiral nematic/cholesteric and smectic LCs [1]. The LC molecules typically feature a rod-shaped structure, exhibit rigidity in their lengthy axis direction and readily polarizable. These LCs have a defining character of alignment along a common axis n, termed as director [2]. The anisotropic properties are result of this alignment, where characteristics of material vary depending on the measurement direction. This unique characteristic

N. Puri (✉)
School of Natural Sciences, GNA University, Phagwara, Punjab, India
e-mail: neerajvenus@gmail.com

P. Kumar et al. (eds.), *Cholesteric Liquid Crystals*, Engineering Materials,
https://doi.org/10.1007/978-981-96-3821-5_11

underlines the distinctive optical properties used by the scientists across numerous applications [3].

LCs are used mainly in number of digital devices including displays. As the properties of LCs are affected after doping with the other materials like polymers, dyes and nanomaterials, it makes LCs as promising materials for future smart devices. Nanomaterials are latest and widely experimenting materials now days in LCs. A large number of research works are getting reported in diverse phases of LCs.

2 Chirality in Liquid Crystals (LCs)

Chiral molecules are molecules with no mirror image symmetry and optical active i.e. that they rotate the plane of the polarized light. Molecular chirality significantly influences the LC properties. A variety of new phases are formed when the chiral molecules are induced into LC phases that have very unusual properties. Such new phases are called chiral phases. Chirality swaps nematic and smectic phases by their chiral versions [2].

3 Cholesteric Liquid Crystals (CLCs) or Chiral Nematic LCs

The cholesteric liquid crystals (CLCs) or chiral nematic LCs are typically comprised of nematic mesogenic molecules holding a chiral center. It creates intermolecular forces which favor alignment between molecules at a minor angle to one another that results to the formation of an assembly like an arrangement of very thin 2-D nematic kind layers. The director in each layer is twisted as shown in Fig. 1. In such an assembly, the director move in a constant helical pattern around the layer normal. Pitch is one of the significant features of the cholesteric mesophase and defined as the distance taken by the director to rotate one complete turn in the helix. The helical pitch controls the selective reflections of the cholesteric phase and hence makes them appropriate for the electro-optic display devices. This pitch can be fine-tuned to any required value by the addition of a chiral compound to a nematic structure in a suitable amount.

The essential condition for the cholesteric mesophase construction is that the molecules are optically active i.e. chiral. Due to this cause cholesteric phase is also acknowledged as the chiral nematic phase, represented by the asterisk N*. The asterisk signifies the existence of chiral molecules. Once a chiral agent is doped with a nematic LC, the below relationship provides the pitch of the mixture [2]

$$\mathrm{p} = \frac{1}{HTP.X_C} \tag{1}$$

Fig. 1 Director orientation in cholesteric liquid crystals

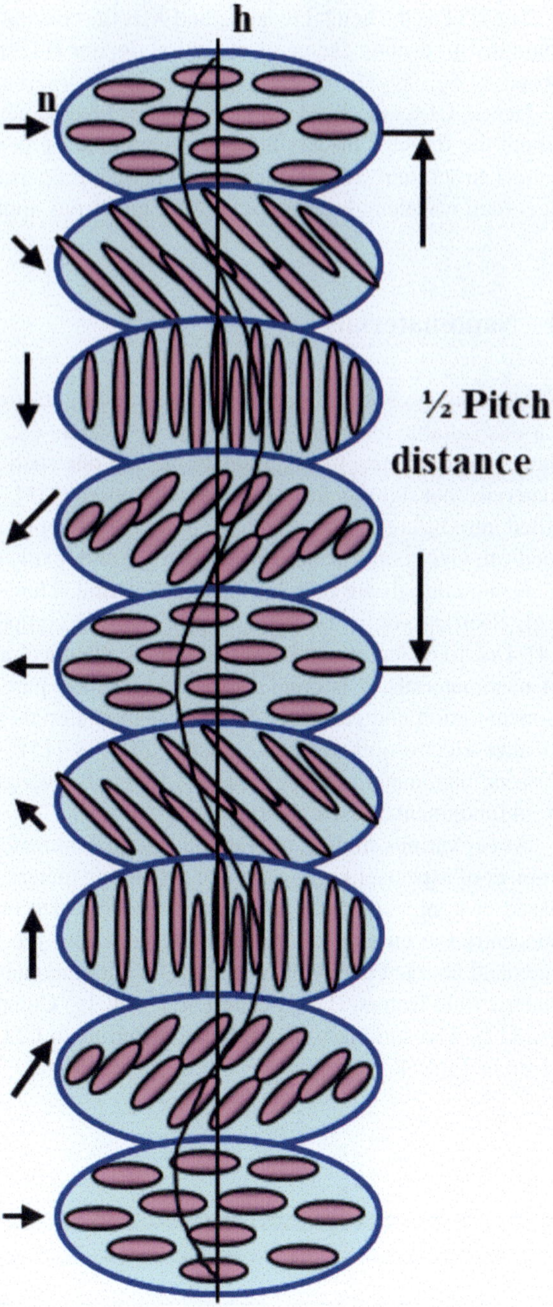

The HTP is the helical twisting and $X_c$is the concentration of the chiral agent. The main distinguishing factor of the chiral dopant HTP is determined by the nematic host.

Hence, CLCs or chiral nematic phase is layered nematic LC along with chirality where the director orients helically remaining perpendicular to helical axis. This helical structure results in exceptional optical properties such as selective reflection of certain wavelengths that are vital for numerous applications.

4 Nanomaterials

There is always scope for a new research and interesting results with the addition of various suitable materials such as dyes/polymers etc. into LCs as it can alter their molecular dynamics [4]. Nanomaterials are one such kind of promising dispersing materials that significantly alter the LC properties [5]. These nanomaterials are classified into different types and have tremendous applications in the various fields viz medical, electronics, food, pharmaceutical and textile etc. [6–10].

Nanomaterials are both artificial and natural. These have unique physical, chemical, electrical, optical, magnetic properties in comparison to bulk matter [7, 11–14]. Due to different synthesis techniques, chemical nature, size, shape, origin etc. of nanomaterials, it is crucial to recognize their part in modifying LC properties. Concentration and chemical nature of nanomaterials again significantly contribute to alter electro-optical characteristics of LCs [15]. In addition to technological aspects, nanomaterial dispersion in LCs is also very encouraging and stimulating from fundamental research approach [16].

As per various studies, interaction among nanomaterials and LCs depends upon a number of aspects, like (a) nature of dispersing materials (organic, inorganic, carbon based or composite type), (b) dimensionality, size or shape of nanomaterials, (c) concentration amount of the nanomaterials into the LC matrix, (d) physical and chemical characteristics of both the LCs and nanomaterials, (e) amount of applied electric field factors, (f) agglomeration state, (g) Composition, and (h) Morphology etc. [17]. The summary of used nanomaterials in CLCs with their applications are given in Table 1.

Table 1 Summary of applications of nanomaterials dispersed cholesteric liquid crystals

Nanomaterials	Results	Applications
Platinum nanoparticles	Formation of periodic structure with inter-distance correlated to the helical pitch and molecular chirality	Liquid crystal displays
Polyhedral oligomeric silsesquioxane (POSS) nanoparticles	Cluster formation of nanoparticles/multi domain textures	Bistable CLC devices, multi-wavelength laser
Gold nanoparticles	Film thickness depending self-assembly formation effecting selective reflection	Soft technology and optical materials applications, smart materials for engineering
Silver nanoflakes	Enhanced birefringence and lasing intensity, time-luminance stability and enhanced thermal resistance	High performance lasers and electro-optic devices
Single-walled carbon nanotubes	Enhanced switching time, dielectric anisotropy, conductivity	Opto-electronic devices. Liquid crystal displays, filters etc
Multi wall carbon nanotubes	Enhanced memory	Memory devices
Graphene nanoplatelets	Reduction in ionic density, diffusivity, relaxation frequency and operating voltage	Electro-optic and photonic devices
Carbon nanoparticles/ shungite carbon	Mesomorphic and rheological characteristics like viscosity, elastic modulus, loss modulus and complex viscosity alters	Tribological applications
$BaTiO_3$ ferroelectric nanopowder	Increase in sensitivity and ultrasonic velocities microelectronics	Microelectronic devices cost-effective photovoltaic devices, and custom-shaped containers that may serve as coatings
Silica nano-particles	Surface anchoring/ high density complex ordered surface patterned nanoparticle-decorated emulsions formation with cross-linkable controlled via the underlying elastic field	Liquid crystal displays
Cellulose nanoparticle	Enhanced transparency and fast response time	Liquid crystal displays
Tin oxide nano-particles and nanorods	Reduction in relaxation frequency, Enhanced photoluminescence, Enhanced sensitivity and fast gas detection	Fast electro-optic devices, luminescent displays, highly sensitive gas sensors
Zinc oxide nanoparticles	Decrease in the response time and transmittance changes	Electrically tunable smart windows

5 Influence of Different Nanomaterials on Dynamic Responses of CLCs for Different Applications

5.1 Platinum Nanoparticles

Platinum nanoparticles are metallic nanoparticles and due to their biocompatibility nature, they have wide range of applications in the fields of biomedical, optical and catalysis. Platinum nanoparticle dispersions in the CLCs results in a long range ordering of nanoparticle assemblies.

Transmission Electron Microscopy (TEM) studies of 1.5 wt% of platinum nanoparticles dispersed mechanically in Oligomers with two different side-chains attached to a siloxane cyclic chain via a non-chiral and a chiral mesogen spacers depict the fingerprint patterns and inter-distance ribbon like structures of nanoparticle assemblies at the interface [18]. By varying the molar fraction of cholesterol-containing mesogen in the LC host, it was observed that the distance between ribbons is correlated to the helical pitch which thereby evolves into a simple control parameter to control the distance between the nanoparticle ribbons. An unusual segregation occurs at the boundaries of the anisotropic film with particle ordering localized near the film-air interface. The spacing between ribbons is directly linked to the helical pitch which in turn relates to the molecular chirality [19]. The potential application of such self-assembly formation is appealing alternative to ridge-and-valley structured surfaces that require the presence of periodic relief of pre-chosen substrates by depositing hybrid fluid on various surfaces provided that the film presents a cholesteric organization. The study illustrates that by choosing suitable materials with various pitches, nanoparticle architectures with controlled periodicity could be produced the creation of varied periodic structures. As optical properties are associated with the helix pitch, these new material systems could have potential optoelectronic applications in LC displays [20].

5.2 Polyhedral Oligomeric Silsesquioxane (POSS) Nanoparticles

The polyhedral oligomeric silsesquioxane (POSS) nanoparticles are hybrid inorganic–organic materials with cage-like structures. These have tremendous bio medical and optical high-tech applications in industries. When POSS nanoparticles with diameter of few nanometers are dispersed in CLCs, they get adsorbed on the inner surfaces of the Indium Tin Oxide (ITO) coated glass substrates and lower the corresponding anchoring energy. As a result the cholesteric crystal molecules are aligned vertically. Furthermore, there is cluster formation due to hydrogen bond formed between hydroxyl compounds in the side chains of POSS nanoparticles. The properties of these POSS clusters are monitored by the nanoparticle concentration and the cooling rate of the POSS-CLC mixture from a higher temperature than the clearing point. It is observed that the POSS cluster steady the focal-conic texture by disrupting the planar texture into multi-domain assembly resulting into an extensive distribution of alignment of the helical axes of CLC.

Also, reflectance is one important electro-optic parameter that decreases with increase in voltage from 10 to 40 V and achieves original value as soon as the driving voltage is greater than 45 V. A grayscale switching is visible between 40 and 45 V applied fields. The higher POSS concentration and higher cooling rate consequences in the better appearance of black and white CLC based devices. Therefore, POSS Clusters may be used to broaden the planar texture reflection spectrum to produce a white appearance and stabilize the bistable states of the CLCs [21–23]. The dispersion of different concentrations of POSS nanoparticles (NPs) in dye doped CLCs results in multi-wavelength lasing peaks due to formation of multiple planar domains. The vertical alignment of POSS NPs clusters adsorbed on the surface of substrate through a rapid thermal annealing process strongly influences the formation of these multiple planar domains. The multi-wavelength lasing emission is produced due to resultant effect of multiple longitudinal resonant modes of a Fabry–Perot etalon between the cell substrates and the fluorescence photons amplification with the resonant wavelengths within the broadening long-wavelength edge of the reflection band of multi-domain CLCs [24].

Chen et al. (2016) [24] found that by changing the cooling rate of the cell and concentration of POSS NPs, it is possible to adjust the amount of multi-wavelength lasing peaks.

Figure 2 shows the schematic of experiment setup used in the study. In addition, during heating/cooling process of the cell, thermo-reversible control of multi-wavelength lasing emission can be achieved by the control of thermally induced phase separation procedure of the cell. As there is no unstable diffusion for the oversaturated chiral dopant and room temperature operation of lasing output, more convenient, reliable, thermo-reversible and multi-wavelength tunable laser devices can be fabricated by dispersing self-assembled POSS NPs in the dye doped CLCs for the displays and laser industries.

5.3 Gold Nanoparticles

Due to unique opto-electronic properties of gold NPs of producing vibrant colors after interacting with visible light, these have been extensively used by the artists of various countries for centuries. They are being valuable in optoelectronics, chemical, engineering, biological and medical applications up to large extent. The opto-electronic features of gold NPs depend upon their size, shape, aggregation etc.

In CLCs, gold NPs of 3 nm average size get self-aggregated to form two and three-dimensional arrangements that depend entirely upon the cell film thickness. In order to avoid isotropic aggregation of NPs that leads to unstable films, film thickness plays a vital role [25]. Textural studies reveal that NPs patterns occur in the areas

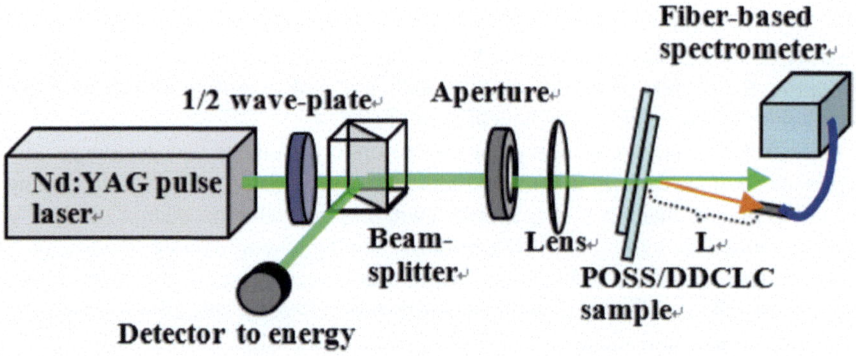

Fig. 2 Experimental arrangement in POSS/CLC Laser. (Reproduced with permission from Ref. [24])

near to air-material interface where there are strongest distortions in the molecular-director distribution. The optical properties like reflection are also altered as the film thickness varies. Gold NPs change the nature of air-material interfaces that results a lag in blue-shift produced from helical-axis orientation distribution disturbance. Such mixtures have are useful in optical materials and soft technology [26].

A textural modulation in the gold NPs dispersed CLC film is observed near air-material interface to control the formation of NPs aggregation and their localization. This modulation strongly depends upon NPs concentration. At low concentrations, it is found that NPs have weak interaction and form homeotropic texture near film surface. However, at high concentrations, gold NPs penetrate within the LC texture and electromagnetic interactions takes place between them resulting into a strongly red shifted localized surface plasmon resonance. This resonance is sensitive towards light polarization due to the alignment of the anisotropic NP aggregates formed by the textural modulation of the CLC. Molecular dynamics of such gold NP-CLC mixtures are promising for new smart materials in engineering [27].

Another study of gold NP dispersion in cholesteric liquids demonstrates that LC films exhibit linear traps for NPs when subject to competing anchoring conditions at the interface, particularly planar unidirectional versus homeotropic [28]. These traps may be either regions with significant elastic distortion or oriented linear defects with distorted cores of few nanometers narrow. Such phenomenon is related to helical rotation near air-material interface that suggests the stability of the gold NPs–CLC hybrid system favors the latter scenario. The traps facilitate the NP needles formation of less than 5 nm wide and exceed 50 nm in length showcasing a strong and well defined anisotropic localized surface plasmon resonance response as shown in the Fig. 3.

NPs spacing within these needles is around 50% smaller compared to same NPs deposited on a substrate without LC or confined within smectic dislocation cores. These studies establish a limit on defect size in distorted CLC films and highlights how the nature of defects influences the crystallographic properties of the resulting

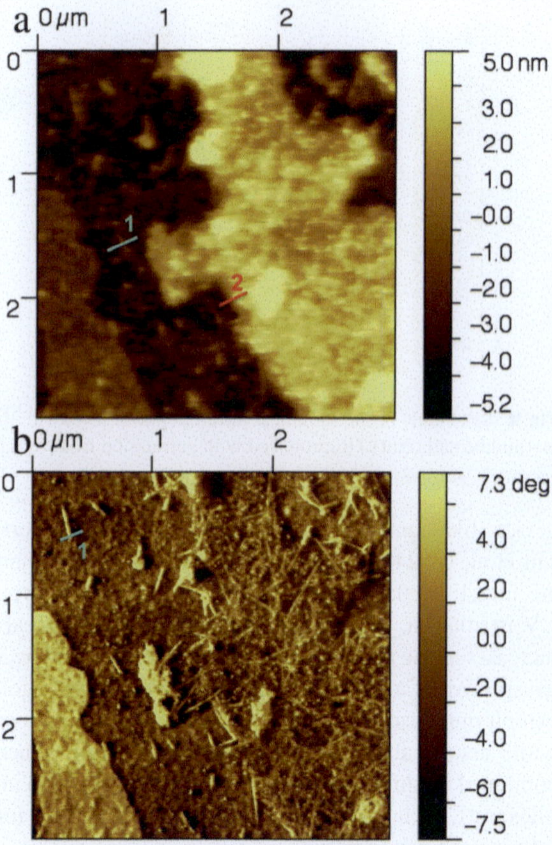

Fig. 3 **a** AFM topography and **b** Phase Data for an area with a plasmon resonance peak. (Reproduced with permission from Ref. [28])

NPs self-assembly. Finally, these hybrid films confine NPs to specific areas within CLC film's thickness paving the way for a variety of hybrid systems with anisotropic optical properties. Investigating the localized plasmon resonance of trapped NPs also provides valuable insights into the characteristics of highly deformed regions in the CLC films [28].

The photonic and plasmonic properties of CLC systems in the substrates coated with photo alignment layer (PAL) of commercially available azo dye and gold NPs array are greatly influenced by optical radiations due to increase in refractive index transition from air to photo-alignment layer and then to CLC as depicted in the Fig. 4. In context of Mie theory, there is a correlation between the refractive index of the medium around gold NPs and localized plasmon resonance frequency.

When exposed to resonant radiation, the local heating of the CLC medium triggered changes in its optical properties, transitioning from a liquid crystalline phase to an isotropic phase, which resulted in a corresponding blue shift in localized plasmon resonance. These findings are helpful for the development of plasmonic architectures with tunable LPR frequencies, as well as integration into modern photonic

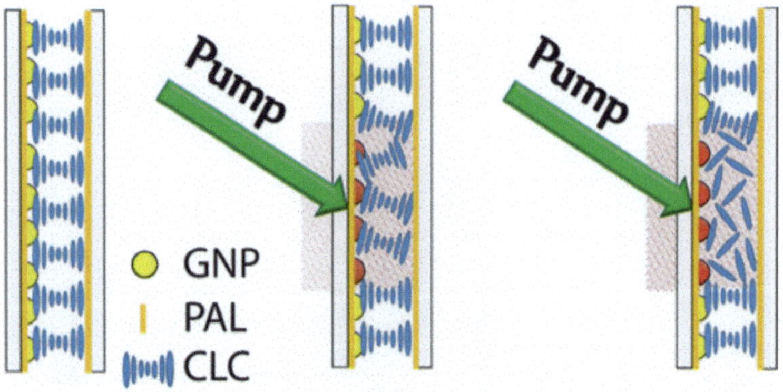

Fig. 4 Schematic of the sample at different power level with the pump beam turned off (left) and on (middle and right). (Reproduced with permission from Ref. [29])

devices like controllable light filters and smart mirrors [29]. In the recent studies by Liu et al. (2024) [30], helical assemblies of gold nanorods in photo-tunable CLC are investigated, followed by modulating their morphological asymmetry through UV irradiation. The optical asymmetry of the helical assemblies of gold nanorods increases as the pitch decreases and the distance between the gold nanorods reduces, irrespective of azimuthal variations. Moreover, the gold nanorods mixed with CLCs exhibit impressive and stable reflective properties compared to conventional CLCs, while also enabling modulation of light flux at near-infrared wavelengths. The combined modulation of reflection and transmission in gold nanorods helical assemblies and CLC hosts represents a promising direction for future research in multi-level information encryption.

5.4 *Silver Nanoflakes*

Recently, in other interesting studies of LC laser devices by Liu et al. (2024) [31], it is found that dispersing silver nanoflakes in CLCs composites, birefringence is enhanced, contributing to the dipole moment and reducing pump energy consumption. Silver nanoflakes cause helical twisting in CLC phase. Thus, addition of small concentration of silver nanoflakes not only improved the reflection of the hybrid but also optimized pump energy utilization, resulting in brighter lasing due to their excellent localized surface polarization resonance. The LC laser created by blending 0.10 wt% silver nanoflakes into polymer-stabilized CLCs containing 1 wt% reactive mesogens demonstrated lower pump energy consumption at 0.291 μJ/pulse. Additionally, the laser showed a 16.2% increase in lasing intensity, improved time-luminance stability, and greater thermal resistance in the 26–38 °C range. These findings advance random laser technology across various fields and support the development of high-performance optoelectronic devices.

5.5 Carbon Nanotubes

Carbon nanotubes are basic structures comprising of carbon atoms and members of the fullerene structural family including buckyballs. While buckyballs are spherical in shape, a nanotube is cylindrical with a diameter in the range of nanometers and at least one end typically capped with a hemisphere of the buckyball structure. The diameter of a nanotube is of the order of a few nanometers and length up to several millimeters. The nanotube structures comprise of one or more coaxial cylinders of graphene sheet placed between two halves of a fullerene molecule (buckyball). They are produced using three basic techniques i.e. chemical vapour deposition, arc-discharge and laser ablation. Carbon nanotubes are categorized into single-walled carbon nanotubes and multiwalled carbon nanotubes.

A single-walled carbon nanotube (SWCNT) is a one-atom thick sheet of graphite known as graphene that is rolled up into a seamless cylinder. It has diameter of the order of one nanometer and a tube length of many thousands of times longer. Multi-walled carbon nanotubes (MWCNTs) are made of several layers of graphite rolled to a tube shape. Two models are used to describe the structures of multi-walled nanotubes. First one is Russian Doll model, where sheets of graphite are arranged in concentric cylinders and the other one is Parchment model, where a single sheet of graphite is rolled in around itself. The interlayer distance in multi-walled nanotubes is approximately 3.3 Å i.e. the distance between graphene layers in graphite. The MWCNTs are less expensive and more chemical resistant than SWCNTs. So, they are chosen over SWCNTs especially when functionalization is required. When it comes to applications part, carbon nanotubes are one of the best among nanomaterials as they are extensively used by physicists, chemists, material scientists, biologists etc. Due to their unique electrical, mechanical and thermal properties, like high strength to weight ratio, higher resistant to chemical attack, high surface area, high thermal conductivity etc. carbon nanotubes are promising materials in the mechanical and electronics applications. They are majorly used in the sensors, actuators, field emission displays, batteries, scanning probes microscopic tips for metrology, field effect transistors, nano-electronics, electrodes, polymer matrix composites etc.

Carbon nanotubes have also transported many changes and new opportunities in the field of LC research as they have high aspect ratio i.e. length to thickness ratio [32]. They offer improved self-alignment to the LC molecules and thus alter their physical properties like dielectric properties, switching response and morphological behavior. Research in this area is throwing up numerous surprises because LCs provides an appropriate medium to control their orientations and field induced reorientations [17].

Dispersion of the single-walled carbon nanotubes in the cholesteric phase result in an increase in the anchoring energy due to π–π electron stacking between carbon nanotubes, CLC molecules and the alignment layers. Thus, surface property is modified and a significant improvement in response time for controlled birefringence cells is observed [33] through continuous waveform and one-pulse switching methods for potential applications in LC displays. A jump in the optical transmission is reported in three different types of cholesteric matrices when dispersed with 0.05 and 0.1 wt% of multi-walled carbon nanotubes as carbon nanotubes get aligned perpendicular to the helical axis. However, no effect on helical pitch was observed after dispersion of multiwall carbon nanotubes as there is no change in selective transmission properties like position of maximum wavelength or shape of selective reflection bands [34]. Koysal (2010) [35] depicts that 0.5 wt% of single-walled carbon nanotubes can change the dielectric behaviour and electrical conductivity of CLCs. An increase in dielectric anisotropy parameters and electrical conductivity at high voltage was observed due to reorientation of nanotubes along director field, caused by elastic interaction of LC molecules. The possible reason may be that after the electric field is applied, electric torque tends to align LC molecules parallel to the electric field that in turn reorient long axis of carbon nanotubes. Such materials promise new design electro-optic switches like electro-optical switches and optical filters.

In another interesting study, it is found that CLC-carbon nanotubes composites have application in gas sensors with a high dynamic range. This composite possesses a change in electrical resistance and reflected color different gas diffusion stages due to absorption of gas vapors. The CLC molecules form a multi-domain structure after absorbing gas vapors that enables carbon nanotubes to form conducting networks. The application of such composite materials is quite useful in medical devices to detect low acetone concentration and environment protection for detection of high acetone [36]. In another comparative optical transmission and electric conductivity studies of carbon nanotubes aggregation in nematic and cholesteric phases, Lisetski et al. (2013) [37] found that chiral dopants slow down the formation of aggregation in cholesteric phase. This is again due to the reason that carbon nanotubes try to locate in the planes of quasi-nematic layers. Such composites are promising useful materials with more time stability.

Memory efficiency of 0.02 wt% of multi-walled carbon nanotubes dispersed in photosensitive CLC is calculated using below relation by Yaroshchuk et al. (2013) [38]:

$$\frac{T_m - T_0}{T_{max} - T_0} \tag{2}$$

where T_m, T_0 and T_{max} are optical transmission values at initial zero state, final zero state and maximum of optical transmission versus applied voltage curve.

A dual-mode operation with optical switching between reversible electro-optic mode and optically settable memory mode was observed as carbon nanotubes stabilize the cholesteric filamentary domains. In contrast to the pure cholesteric, carbon nanotubes dispersed CLC composite exhibit slower structural changes and maintain a stable network of oily streaks within the unwound cholesteric homeotropic structure. This texture results in an irreversible (memory) electro optic response as the network of cholesteric domains preserves the planar orientation of cholesteric phase established by the electric field. These composites may have applications for advanced LC displays based on the new concept.

Samoilov et al. (2014) [39] examined the potential optoelectronic applications of single-walled carbon nanotubes in a cholesteric matrix subjected to UV irradiation under comparable conditions. Their research on selective reflection spectra and optical transmission at various carbon nanotubes concentrations, temperatures, and LC matrix compositions revealed that there is a nonlinear relationship between optical density and carbon nanotubes concentration. The formation of aggregated carbon nanotubes was associated with a minimum in optical density that is sensitive to partially reversible UV-induced *trans–cis–trans* isomerization effects.

Based on the analysis of tribotechnical investigations, [40] conclude that cholesteric additive X-7, combined with multi-walled CNT micro additives, exhibits strong anti-friction properties. Findings indicate that incorporating cholesteric ether of stearic acid -X-7 into industrial oil I-20A in various concentrations significantly enhances the wear resistance of friction pairs. A slight increase in the friction ratio is also observed. The X-7 additive at a concentration of 20% demonstrated the best thickening properties.

Rheological studies show that the addition of nanotubes alters the balance between the elastic and viscous components, favouring an increase in the elastic properties of the sample. This change in turn affects the tribological characteristics of industrial oil. The ability to modify the compositions and weight ratios of components in the lubrication materials, along with the flexibility to adjust the conditions of tribotechnical investigations, underscores the promising potential of further research in this area. We believe that exploring similar binary systems of additives in moldable lubrication media also holds significant potential for oil I-20A.

In a similar manner, primary transducer of sulphur dioxide (SO_2) optical sensor could be designed using single, double and multi-walled carbon nanotubes dispersed cholesteric-nematic mixture LCs composite materials [41]. From the spectral characteristics of the composites, the sensitivity is calculated as a function of composition using below formula. The results depict that highest sensitivity coefficient is for 0.3 wt% of carbon nanotubes and sulphur dioxide interacts with nanocomposites for around 10 s.

$$S = \frac{\Delta\lambda}{\Delta C} \tag{3}$$

Again, data on optical transmission and selective reflection is presented for suspensions of single-walled carbon nanotubes (CNTs) in a photoactive cholesteric matrix. At low concentrations from 0.01 to 0.05% of carbon nanotubes, they preferentially localized at oily sticks (cholesteric topological defects) and suppressed the network of oily streaks. Optical density exhibits a nonlinear behaviour and reach at a minimum at around 0.05–0.08 wt% concentration of carbon nanotubes. This unusual behaviour is due to a structural transition from loose, highly anisotropic aggregates to more compact aggregates with denser packing. The location of this minimum, along with the selective reflection maximum (helical pitch) is influenced by partially reversible UV-induced *trans-cis–trans* isomerization effects. Further, only marginal influence of carbon nanotubes addition was observed on UV exposure controlled helical pitch [42].

Studies of the conductivity, optical density and microstructures in single-walled carbon nanotubes dispersed CLC composites show that as compared to pure CLCs, a shift in the positions of maxima and minima in optical density as a function CNT concentration is there under UV irradiation by means of a mercury lamp emitter. It may be due to the interaction between oily streaks and carbon nanotubes network in addition to other factors [43]. Ordering in cholesteric deteriorates due to *trans–cis* isomerization under UV irradiation and partial restoration of population of *trans*-isomers by restoring irradiation.

Mytykyuk et al. 2021 [44] With marginal concentrations of single, double and multiwalled carbon nanotubes in cholesteric-nematic mixture of LCs results in highly sensitive active medium for nitrous dioxide (NO_2) sensor. This is due to the reason that carbon nanotubes are highly sensitive towards NO_2 gas. These composites are studied for spectral characteristics and results depict that desired maximum coefficient of spectral sensitivity can be achieved by changing concentrations of carbon nanotubes and LC mixture.

5.6 Graphene Nanoplatelets

Graphene nanoplatelets are versatile nanofillers resulting into nanocomposites with modified performance and functions. Due to which they have widespread range of applications in the fields of energy harvesting, structural functional composites, smart sensors, and flexible electronic devices [45].

Dielectric behaviours and microscopic observations of low-frequency texture transitions in graphene nanoplatelets dispersed CLCs using temperature dependent dielectric spectroscopy and voltage-and-frequency dependent electro-optic measurements. The pre-treated graphene nanoplatelets used are of sixteen graphene layers with an average thickness of 10 nm and a diameter of 5 μm. It is found that low concentration (0.5 wt%) of graphene significantly suppresses ionic behaviors as evidenced by reductions in.

(a) Ionic density by approximately 32%
(b) Diffusivity by enhanced LC viscosity
(c) Relaxation frequency due to ion-trapping effects.

Furthermore, the electro-optic parameters of grapheme nanoplatelets dispersed CLC cell reveals a decrease in the operating voltage required to switch from the planar to focal conic state and absence of the low-frequency transition from conic to uniform lying helix texture as well. These results are attributed to the role of graphene nanoplatelets nuclei within the CLC medium that damps both ionic and electro-hydro-dynamic effects [46].

Comparison of the effect of various concentrations of two types of carbon NPs viz. shungite carbon and multiwalled carbon nanotubes on mesomorphic and rheological

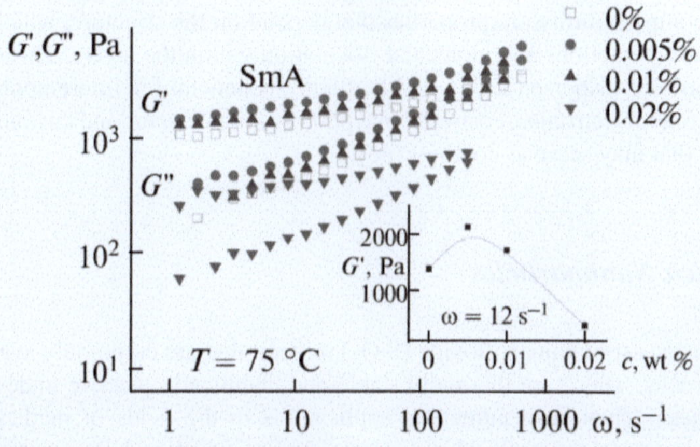

Fig. 5 Frequency and concentration dependence of moduli. (Reproduced with permission from Ref. [47])

properties of CLCs has been done in terms of viscosity, elastic modulus, loss modulus and complex viscosity. Figure 5 shows the variations of storage G′ and loss G″ moduli respectively with the frequency in Smectic A phase at 75 °Cf or different concentrations of carbon NPs. Concentration dependence of storage modulus is shown in the insert at fixed frequency.

Rheological properties play vital role for tribological applications. Shungite carbon is found to be more effective lubricating coolant additive component as compared to multiwalled carbon nanotubes. It is due to the results that show decrease in the value of viscosity in shungite carbon dispersed CLC composites as compared to multiwalled carbon nanotubes dispersed CLC composites. An increase in the storage modulus has been observed in both the composites. Furthermore, the textural studies using optical microscopy reveal that the dispersion of shungite carbon does not disturb the orientation of CLC molecules as textures of pure CLC and composite are identical [47].

5.7 *Ferroelectric Nanopowder/$BaTiO_3$Nanopowder*

Electrical and acoustical properties of CLCs are altered when dispersed with ferroelectric nanopowder of barium titanate ($BaTiO_3$). By using ultrasonic interferometer at room temperature at fixed frequency, acoustical properties like ultrasonic velocity, Rao constant, density, adiabatic compressibility, acoustic impedance and Wada constant could be determined. Barium titanate nano powder enhanced the sensitivity of CLC that in turn enhance LC performance. The ultrasonic velocities in pure CLC solutions are lower and increase upon the addition of ferroelectric nanopowder, indicating a greater association of molecules. Ultrasonic velocity and adiabatic compressibility are properties that depend on the structure, which relates to interlayer interactions and molecular orientation within the layers. This suggests that nanopowder-dispersed LCs could be highly beneficial for future applications, such as in microelectronics, cost-effective photovoltaic devices, and custom-shaped containers that may serve as coatings [48].

5.8 Silica Nanoparticles

Silica NPs consist of silicon dioxide (SiO_2) mineral that are commonly available in sand and rocks. These are thermally stable and chemically inactive under normal circumstances. They have numerous applications in the fields of medical, food, catalysis, drugs. In the molecular dynamic studies of silica NPs organization at the interface of CLCs droplets in aqueous solutions, interfacial NPs assemblies shaped by elastic fields of LCs are achieved. Complex ordered surface patterned NP-decorated emulsions with high density and cross-linkable arrangements are produced by utilizing surfactant-modified NPs that attach to water-liquid interface. These emulsions can be controlled via the underlying elastic field. Such approach relies on surface anchoring i.e. the molecular orientation of LCs at the boundaries rather than bulk defects.

Additionally, it is observed that the NPs aggregation is influenced by the elastic energy of CLC, indicating that their adsorption behavior resembles nucleation and growth kinetics. In nutshell, CLCs can act as templates, allowing silica NPs to be arranged into configurations with adjustable length scales down to hundreds of nanometers by varying both the surfactant and chiral dopant concentrations. Furthermore, making silica NPs more hydrophobic reveals that the dynamics of their assembly at the interface resemble nucleation and growth. These arrangements can be easily cross-linked by altering the surrounding aqueous phase. The properties of these assemblies can be further tailored since NP surface modifications can be achieved through simple, in situ electrostatic adsorption. Many materials can undergo similar surface modifications, and silica NPs with different core compositions can also be synthesized. To ensure NPs localize in regions of favorable anchoring, a balance in their hydrophobicity is necessary, allowing them to be interfacially active yet mutually repulsive. This sensitivity to hydrophobicity represents a new area of exploration in NP-decorated emulsions. The LC-patterned, NP-decorated emulsions are established as a system that merges interfacial phenomena with elasticity, enabling not only the design of structures but also exploring NP behaviors at interfaces [49].

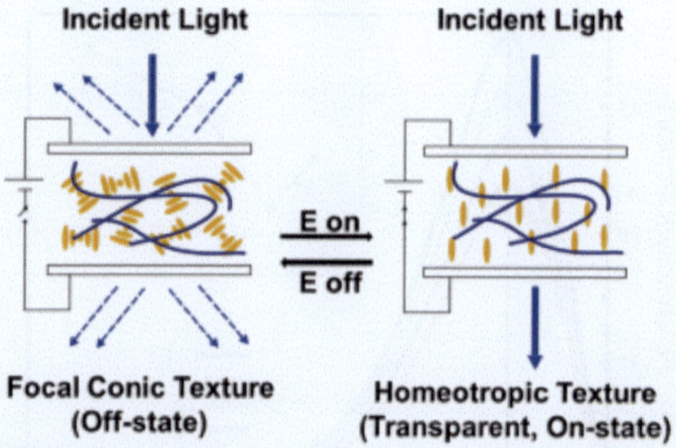

Fig. 6 Schematic for off and on states. (Reproduced with permission from Ref. [50])

5.9 Cellulose Nanoparticles

It is interesting to disperse chiral cellulose NPs in PC05 CLC for LC display cells. The resulting chiral cellulose NPs-doped PC05 CLC cell exhibits a remarkable transparency of 97.4% and a fast response time of 7.6 milli seconds. Similarly, the PDAT-doped PD2T polymer stabilised CLC cell achieved a high transmittance of 93.6% and a rapid response time of 13 milli seconds. A schematic diagram for off and on-states is shown in Fig. 6.

LC display cells on substrates without an alignment layer by incorporating polymeric chiral NPs into CLC mixtures were successfully fabricated. The addition of 2 wt% chiral cellulose NPs increased the transmittance of the CLCs from 3.4 to 97%. This innovative chiral dopant approach offers a straightforward method for LC display fabrication [50].

5.10 Titanium Oxide Nanoparticles and Nanorods

Titanium oxide (TiO_2) NPs are used number of applications including pharmaceutical, biomedical and cosmetic industries due to their exceptional characteristics like capable of absorbing UV radiations transparency, being less toxic, antibacterial and resistant to body fluids. Optical studies of titanium oxide NPs dispersed CLCs suggest an increase in photoluminescence for these composites despite of the reduction in absorbance and band gap as clearly visible in the Fig. 7.

Dielectric studies show a decrease in relaxation frequency with TiO_2 NPs addition as compared to pure CLC. A homogeneous alignment is clearly visible in morphological studies and a blue shift is also depicted with the temperature increase in the

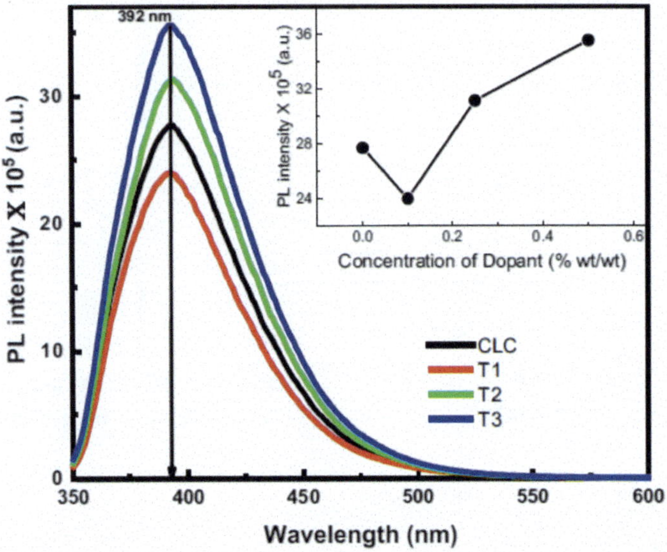

Fig. 7 Variation of photoluminescence intensity with nanoparticles concentration. (Reproduced with permission from Ref. [51])

reflection spectra of the composite material. These composites are potential materials for fast electro-optic applications and luminescent display [51]. A recent study of titanium oxide nanorods dispersed CLCs composites for the detection of ethanol gas suggests that these composites are promising materials for highly sensitive gas sensors in the future technology [52]. A change in the resistance of the CLC matrix due to interaction of the titanium oxide porous nanorods with the molecules of ethanol gas is supposed to be the possible reason behind the fast gas detection in 21 s of response time and modified high sensitivity.

5.11 *Zinc Oxide Nanoparticles*

Zinc oxide (ZnO) NPs are highly catalytic and chemically stable and widely used in biomedical research, sensors, ceramic industry, solar cells, dentistry, cosmetics etc. A latest study of zinc oxide NPs dispersed CLC shows that these composites are ideal for tunable smart window. However, the behavior is quite opposite in focal conic and homeotropic states. There is an increase in the transmittance in homeotropic state from 4 to 90% with a response time of less than one millisecond and 87 to 4% decrease in the focal conic state with the increase in zinc oxide NPs concentration. The device is quite stable for around 170 s when the electric field is repeatedly switched on and off. These composite materials are fairly useful as windows for buildings and automobiles for temperature control and energy saving [52].

6 Conclusions

In nutshell, the incorporation of nanomaterials into CLCs represents a promising improvement in the field of condensed matter physics. The various effects observed not only alter the dynamic properties and stability of the LCs but also open new opportunities for applications in display technologies, sensors, tunable photonic devices and lasers etc. While challenges remain, such as optimizing dispersion methods and understanding the long-term stability of these composites, the potential for innovative solutions in various fields and industries is substantial. Continued research in this area will likely yield exciting developments, paving the way for next-generation materials with unique functionalities.

References

1. Chandrasekhar, S.: Liquid Crystal, 2nd edn. Cambridge University Press, Newyork (1992)
2. De Gennes, P.G., Prost, J.: The Physics of Liquid Crystals, 2nd edn. Clarendon Press, Oxford (1973). https://doi.org/10.1016/j.optmat.2022.112243
3. Atkins, P.: Physical Chemistry, 7th edn. Oxford University Press, Oxford (2004)
4. Kumar, P., Neeraj, K.S., Lee, S.H., Raina, K.K.: Analysis of dichroic dye-doped polymer-dispersed liquid crystal materials for display devices. Thin Solid Films. **520**(1), 457–463 (2011). https://doi.org/10.1016/j.tsf.2011.06.038
5. Neeraj, K.P., Raina, K.K.: Analysis of dielectric and electro-optic responses of nanomaterials doped ferroelectric liquid crystal mixture. J. Mater. Sci. Technol. **27**(912), 1094–1098 (2011). https://doi.org/10.1016/S1005-0302(12)60002-4
6. Mekuye, B., Abera, B.: Nanomaterials: an overview of synthesis, classification, characterization, and applications. Nano Select. **4**(8), 486–501 (2023). https://doi.org/10.1002/nano.202300038
7. Suhag, D., Thakur, A., Thakur, P.: Integrated Nanomaterials and their Applications. Springer, Singapore (2023). https://doi.org/10.1007/978-981-99-6105-4
8. Tahir, M.B., Sagir, M., Asiri, A.M.: Nanomaterials: Synthesis, Characterization, Hazards and Safety. Elsevier Publishers (2021). https://doi.org/10.1016/C2020-0-00287-2
9. Abdullaeva, Z.: Nano- and Biomaterials: Compounds, Properties, Characterization, and Applications. Wiley-VCH Publishers. Weinheim, Germany (2017). https://doi.org/10.1002/9783527807024.fmatter
10. Chavez-Hernandez, J.A., Velarde-Salcedo, A.J., Navarro-Tovar, G., Gonzalez, C.: Safe nanomaterials: from their use, application, and disposal to regulations. Nanoscale Adv. **6**(6), 1583–1610 (2024). https://doi.org/10.1039/D3NA01097J
11. Manhas, N., Adimule, V., Lakshminarayana, P., Keri, R., Kumar, L.S., Bathula, C.: Utilization of CO^{2+} doped Eu_2O_3 nanostructures and their effect on the dielectric and photoisomerization of liquid crystals. Phys. Fluids **35**(10) (2023). https://doi.org/10.1063/5.0171578
12. Adimule, V., Bathula, C., Sharma, K., Keri, R.: Study of structural, optical, dielectric properties poly-3-butyl thiophene reinforced with CO_3O_4 nanoparticles. Nano-Struct. Nano-Objects **38**, 101150 (2024). https://doi.org/10.1016/j.nanoso.2024.101150
13. Adimule, V., Sharma, V., Kumar, P., Keri, R., Sharma, K.: Perylene-3, 4, 9, 10-tetracarboxylic dianhydride dye intercalated liquid crystal molecules: synthesis, spectroscopic characterizations and photoluminescence dynamics. J. Mol. Liq. **413**, 25990 (2024). https://doi.org/10.1016/j.molliq.2024.125990

14. Adimule, V., Yallur, B.C., Gowda, A.H.: Advanced sensors based on carbon nanomaterials. In: Carbon Nanomaterials-Based Sensors, pp. 259–268. Elsevier (2022). https://doi.org/10.1016/B978-0-323-91174-0.00004-4
15. Neeraj, R.K.K.: Nickel nanoparticles doped ferroelectric liquid crystal composites. Opt. Mater. **35**(3), 531–535 (2013). https://doi.org/10.1016/j.optmat.2012.10.014
16. Garbovskiy, Y., Glushchenko, I.: Nano-objects and ions in liquid crystals: ion trapping effect and related phenomena. Crystals **5**(4), 501–533 (2015). https://doi.org/10.3390/cryst5040501
17. Neeraj, R.K.K.: Multiwall carbon nanotubes doped ferroelectric liquid crystal composites: a study of modified electrical behavior. Physica B: Condensed Matter. **434**, 1–6 (2014). https://doi.org/10.1016/j.physb.2013.10.031
18. Mitov, M., Portet, C., Bourgerette, C., Shoeck, E., Verelst, M.: Long-range structuring of nanoparticles by mimicry of a cholesteric liquid crystal. Nat. Mater. **1**, 229–231 (2002). https://doi.org/10.1038/nmat772
19. Mitov, M., Bourgerette, C., de Guerville, F.: Self-assembly of metallic nanoparticles in a cholesteric liquid crystal. In: Proceedings of the SPIE 5220, Nanofabrication Technologies, San Diego, California, United States (2003). https://doi.org/10.1117/12.503558
20. Mitov, M., de Guerville, F., Bourgerette, C.: Evidence of surface segregation in the organization of metallic nanoparticles dispersed in a cholesteric liquid crystal. Mol. Cryst. Liq. Cryst. **435**(1), 13/[673]–19/[679] (2005). https://doi.org/10.1080/15421400590955596
21. Jeng, S., Hwang, S., Hung, Y., Chen, S.: Cholesteric liquid crystal devices with nanoparticle aggregation. Opt. Express **18**(21), 22572–22577 (2010). https://doi.org/10.1364/OE.18.022572
22. Gahrotra, R., Sharma, V., Dogra, A.R., Malik, P., Kumar, P.: Performance augmentation of bistable cholesteric liquid crystal light shutter-effect of dichroic dye on morphological and electro-optical characteristics. Opt. Mater. **127**, 112243 (2022). https://doi.org/10.1016/j.optmat.2022.112243
23. Gahrotra, R., Sharma, V., Dogra, A.R., Malik, P., Kumar, P.: Temperature dependent morphological and electro-optical characteristics of dye doped cholesteric liquid crystal. ECS Trans. **107**(1), 5481 (2022). https://doi.org/10.1149/10701.5481ecst
24. Chen, S.C., Lin, J.D., Lee, C.R., Hwang, S.J.: Multi-wavelength laser tuning based on cholesteric liquid crystals with nanoparticles. J. Phys. D: Appl. Phys. **49**(16), Article 165102 (2016). https://doi.org/10.1088/0022-3727/49/16/165102
25. Bitar, R., Agez, G., Mitov, M.: Cholesteric liquid crystal self-organization of gold nanoparticles. Soft Matter **7**, 8198–8206 (2011). https://doi.org/10.1039/C1SM05628J
26. Ayeb, H., Grand, J., Sellame, H., Truong, S., Aubard, J., Felidj, N., Mlayah, A., Lacaze, E.: Gold nanoparticles in a cholesteric liquid crystal matrix: self-organization and localized surface plasmon properties. J. Mater. Chem. **22**(16), 7856–7862 (2012). https://doi.org/10.1039/C2JM30622K
27. Bitar, R., Agez, G., Mitov, M.: Cholesteric liquid crystals doped with gold nanoparticles. In: Liquid Crystals XVI, vol. 8475, p. 847503. SPIE (2012). https://doi.org/10.1117/12.9283599
28. Pendery, J.S., Merchiers, O., Coursault, D., Grand, J., Ayeb, H., Greget, R., Donnio, B., Gallani, J.L., Rosenblatt, C., Félidj, N., Borensztein, Y., Lacaze, E.: Gold nanoparticle self-assembly moderated by a cholesteric liquid crystal. Soft Matter **9**(39), 9366–9375 (2013). https://doi.org/10.1039/C3SM51736E
29. De Sio, L., Cataldi, U., Guglielmelli, A. Bürgi, T. Tabiryan N., Bunning T.J.: Dynamic optical properties of gold nanoparticles/cholesteric liquid crystal arrays. MRS Commun. **8**, 550–555 (2018). https://doi.org/10.1557/mrc.2018.80
30. Liu, Y., Lin, G., Chen, Y., Wei, X., Du, M., Zhang, Y., Li, Z., Zhao, L.: UV-triggered optical asymmetry modification of gold nanorods helical assemblies in a cholesteric liquid crystal host. Adv. Opt. Mater. **12**(24), 2400862 (2024). https://doi.org/10.1002/adom.202400862
31. Liu, Y., Wang, Y., Wei, X., Liu, H., Zhang, Y.: Enhanced laser emission by incorporating helically twisted silver nanoflakes within polymer-stabilized cholesteric liquid crystals. ACS Photonics (2024). https://doi.org/10.1021/acsphotonics.4c01406
32. Lisetski, L., Bulavin, L., Lebovka, N.: Effects of dispersed carbon nanotubes and emerging supramolecular structures on phase transitions in liquid crystals: physico-chemical aspects. Liquids. **3**(2), 246–277 (2023). https://doi.org/10.3390/liquids3020017

33. Lu, S., Chien, L.: Carbon nanotube doped liquid crystal OCB cells: physical and electro-optical properties. Opt. Express **16**, 12777–12785 (2008). https://doi.org/10.1364/OE.16.012777
34. Lisetski, L.N., Minenko, S.S., Zhukov, A.V., Shtifanyuk, P.P., Lebovka, N.I.: Dispersions of carbon nanotubes in cholesteric liquid crystals. Mol. Cryst. Liq. Cryst. **510**(1), 43/[1177]–50/[1184] (2009). https://doi.org/10.1080/15421400903058056
35. Koysal, O.: Conductivity and dielectric properties of cholesteric liquid crystal doped with single wall carbon nanotube. Synth. Met. **160**(9–10), 1097–1100 (2010). https://doi.org/10.1016/j.synthmet.2010.02.033
36. Chang, C.K., Chiu, S.W., Kuo, H.L., Tang, K.T.: Cholesteric liquid crystal-carbon nanotube hybrid architectures for gas detection. Appl. Phys. Lett. **100**(4), 043501 (2012). https://doi.org/10.1063/1.3679680
37. Lisetski, L.N., Minenko, S.S., Fedoryako, A.P., Lebovka, N.I., Soskin, M.S.: Dispersions of carbon nanotubes in cholesteric liquid crystals: features of aggregate formation. Funct. Mater. **20**(2), 153–157 (2013). https://doi.org/10.15407/fm20.02.153
38. Yaroshchuk, O., Tomylko, S., Gvozdovskyy, I., Yamaguchi, R.: Cholesteric liquid crystal–carbon nanotube composites with photo-settable reversible and memory electro-optic modes. Appl. Opt. **52**(22), E53–E59 (2013). https://doi.org/10.1364/AO.52.000E53
39. Samoilov, A.N., Minenko, S.S., Lisetski, L.N., Lebovka, N.I., Soskin, M.S., Torgova, S.I.: Dispersions of carbon nanotubes in cholesteric liquid crystals with photoactive components. Funct. Mater. **21**(4), 373–378 (2014). https://doi.org/10.15407/fm21.04.373
40. Berezina, E.V., Godlevskiy, V.A., Usoltseva, N.V.: Investigation of cholesteric liquid crystals and carbon nanotubes additives on mineral oil antifrictional and rheological characteristics. Procedia Eng. **150**, 579–583 (2016). https://doi.org/10.1016/j.proeng.2016.07.044
41. Petryshak, V.S., Mikityuk, Z.M., Vistak, M., Gotra, Z., Akhmetova, A., Wójcik, W., Assembay, A.: Highly sensitive active medium of primary converter SO_2 sensors based on cholesteric-nematic mixtures, doped by carbon nanotubes. PrzeglądElektrotechniczny **1**, 119–122 (2017). https://doi.org/10.15199/48.2017.03.27
42. Samoilov, A.N., Minenko, S.S., Lisetski, L.N., Soskin, M.S., Torgova, S.I., Lebovka, N.I.: Anomalous optical properties of photoactive cholesteric liquid crystal doped with single-walled carbon nanotubes. Liq. Cryst. **45**(2), 250–261 (2017). https://doi.org/10.1080/02678292.2017.1314560
43. Lisetski, L.N., Minenko, S.S., Samoilov, A.N., Lebovka, N.I.: Optical density and microstructure-related properties of photoactive nematic and cholesteric liquid crystal colloids with carbon nanotubes. J. Mol. Liq. **235**, 90–97 (2017). https://doi.org/10.1016/j.molliq.2016.11.125
44. Mykytyuk, Z.M., Vistak, M.V., Kogut, I.T., Petryshak, V.S.: Highly sensitive active medium of sensor NO_2, based on cholesteric nematic mixture with impurities of carbon nanotubes. Phys. Chem. Solid State. **22**(3), 426–431 (2021). https://doi.org/10.15330/pcss.22.3.426-431
45. Jimenez-Suarez, A., Prolongo, S.G.: Graphene nanoplatelets. Appl. Sci. **10**(5), 1753/1–1753/3 (2020). https://doi.org/10.3390/app10051753
46. Wu, P., Lisetski, L.N., Lee, W.: Suppressed ionic effect and low-frequency texture transitions in a cholesteric liquid crystal doped with graphene nanoplatelets. Opt. Express **23**, 11195–11204 (2015). https://doi.org/10.1364/OE.23.011195
47. Usoltseva, N.V., Smirnova, M.V., Kazak, A.V., Smirnova, A.I., Bumbina, N.V., Ilyin, S.O., Rozhkova, N.N.: Rheological characteristics of different carbon nanoparticles in cholesteric mesogen dispersions as lubricant coolant additives. J. Friction Wear. **36**, 380–385 (2015). https://doi.org/10.3103/S1068366615050165
48. Mani, S., Pradhan, M., Sharma, A., Chawade, S., Hadkar, S., Mishra, K., Amare, J., Sarawade, P.: Effect of ferroelectric nanopowder on electrical and acoustical properties of cholesteric liquid crystal. Non-Metallic Mater. Sci. **2**(1), 12–16 (2020). https://doi.org/10.30564/omms.v2i1.1821
49. Tran, L., Kim, H.N., Li, N., Yang, S., Stebe, K.J., Kamien, R.D., Haase, M.F.: Shaping nanoparticle fingerprints at the interface of cholesteric droplets. Sci. Adv. **4**(10), eaat8597 (2018). https://doi.org/10.1126/sciadv.aat8597

50. Kuo, C.Y., Emelyanenko, A.V., Hung, S.C., Chen, W.C., Liu, C.Y.: Improving the electro-optical properties of cholesteric liquid crystal devices via cellulose nanoparticle dopants. Polym. J. **56**(5), 541–551 (2024). https://doi.org/10.1038/s41428-023-00879-1
51. Pathinti, R.S., Gollapelli, B., Jakka, S.K., Vallamkondu, J.: Green synthesized TiO_2 nanoparticles dispersed cholesteric liquid crystal systems for enhanced optical and dielectric properties. J. Mol. Liq. **336**, 116877 (2021). https://doi.org/10.1016/j.molliq.2021.116877
52. Pathinti, R.S., Dasari, S.G., Gollapelli, B., Gogula, S., Ramana Reddy, M.V., Vallamkondu, J.: Enhanced ethanol gas detection using TiO_2 nanorods dispersed in cholesteric liquid crystal: synthesis, characterization, and sensing performance. J. Alloy. Compd. **994**, 174731 (2024). https://doi.org/10.1016/j.jallcom.2024.174731

Recent Advances on Polymer-Stabilized Cholesteric Liquid Crystal: Development and Applications

Amid Ranjkesh and Mahendran Vellaichamy

Abstract Cholesteric liquid crystals (CLCs) are of great interest to researchers due to their special self-assembled helical structure and selective light reflection properties. The cholesteric phase is considered a special state of the nematic phase and is often referred to as chiral nematic liquid crystals. CLCs exhibit a spontaneous helical structure with the twist axis perpendicular to the local orientation. The CLC molecules are arranged in layers, with the long axis of the molecules parallel to the layer planes, and each plane rotates at a certain angle to its neighboring planes. Research on CLCs often requires a continuous description of chiral structures and their properties from the nanometer range to the macroscopic scale. The pointing vector of LC molecules in the layer direction returns to the initial orientation state after 360° of rotation, and this periodic layer spacing is called the pitch (p). The pitch can change with the chemical environment, temperature, electric field, etc. Due to this unique helical structure of CLCs, they exhibit special optical properties, such as selective light reflection, circular dichroism, and rotational properties, which make them widely used in various fields. The helicoidal structure of CLC can be stabilized via in situ photopolymerization of liquid crystal monomers in a CLC mixture, resulting in polymer-stabilized CLCs (PSCLCs). PSCLCs exhibit a dynamic optical response that external stimuli, including electric fields, heat, and light can induce. In this chapter, we discuss the electro-optic response of PSCLCs on film, particles, fiber, and droplets containing broadening bandwidth, red and blue tuning, switching the reflection notch, the size of PSCLC droplets, and the thermal response of PSCLC particles.

Keywords Liquid crystals (LCs) · Cholestric liquid crystals (CLCs) · Polymer-stabilized cholesteric liquid crystals (PSCLCs) · Polymer-stabilized nematic liquid crystals (PSNLCs) · CLC bandgap · CLC films · CLC photonics

A. Ranjkesh (✉) · M. Vellaichamy (✉)
Department of Condensed Matter Physics, Jožef Stefan Institute, Ljubljana, Slovenia
e-mail: amidranjkesh@gmail.com

M. Vellaichamy
e-mail: mahendran.vellaichamy@ijs.si; chamy.mahendra@gmail.com

P. Kumar et al. (eds.), *Cholesteric Liquid Crystals*, Engineering Materials,
https://doi.org/10.1007/978-981-96-3821-5_12

1 Introduction

Liquid crystals (LCs) are the materials, well recognized for their spontaneous orientational order as anisotropic fluids, and are widely preferred in varied applications such as displays, smart windows, lasers, photonic crystals, diffusers, lenses, and augmented reality (AR) technology. The nematic LC (NLC) phase, a prevalent liquid crystal (LC) state, exhibits anisotropic birefringence properties defined by $\Delta n = n_e - n_o$, where n_e and n_o represent the extraordinary and ordinary refractive indices, respectively. The NLC phases are converted into the cholesteric LC (CLC) and blue phases by adding chiral dopant with different range of concentrations. Polymers are commonly used as additives in LCs to improve their electro-optical properties, giving rise to polymer-dispersed liquid crystals (PDLCs) and polymer-stabilized liquid crystals (PSLCs). PDLCs are typically created by dispersing LC droplets in a polymer matrix, which produces strong light scattering due to the refractive index contrast between the polymer and LC components [1]. On the other hand, PSLCs are composite materials made from oriented polymer networks and low-molecular-weight LCs, usually formed through photopolymerization of a small proportion (generally < 10 wt.%) of photo-curable LC monomers dissolved in these low-molecular-weight LCs. Upon UV irradiation, the liquid crystal monomers (LCMs) are polymerized to form liquid crystal polymer networks (LCPNs) by blending a minor amount of LCMs into non-reactive LCs in a homogeneous solution. These films, known for their exceptional thermal, mechanical, dielectric, and optical properties, are applied as brightness-enhancing films in displays, contributing to energy efficiency [2]. The performance and durability of PSLCs are influenced by the structures of the polymer networks, which are, in turn, affected by factors such as the structures of mono- and difunctional LCMs, the ratio of LCMs to non-reactive LCs, UV light curing intensity, curing duration, and polymerization temperature. Notably, the monomers used for polymer stabilization are not limited to LCMs [3, 4]. The preservation of the original LC director configuration in PSLCs is achieved as polymer networks form a template for the LC alignment [5], or as the macroscopic alignment of LC directors is maintained by a crosslinked polymer network. PSLCs are mainly applied in smart windows, broadband reflectors, and diffractive optical devices [6].

In general, without an applied electric field, the LC molecules in polymer-stabilized nematic LCs (PSNLCs) align according to the polymer network's orientation. When an electric field is applied, the LC molecules reorient while the polymer networks limit their movement, altering the optical properties of PSNLCs. This transition is mainly due to the shearing force exerted by the applied electric field, which causes LC molecules to change alignment with improved switching times. This enables applications in smart windows, dimming glasses, and microstrip line phase shifters [7–11]. The selective reflection of CLCs can be dynamically controlled by various external stimuli, including thermal, optical, electrical, and magnetic inputs. Polymer-stabilized CLCs (PSCLCs) are engineered to enhance the response of CLCs to external stimuli, allowing for electrically tunable dynamic optical responses by incorporating CLCs with negative dielectric anisotropy. This selective reflection's

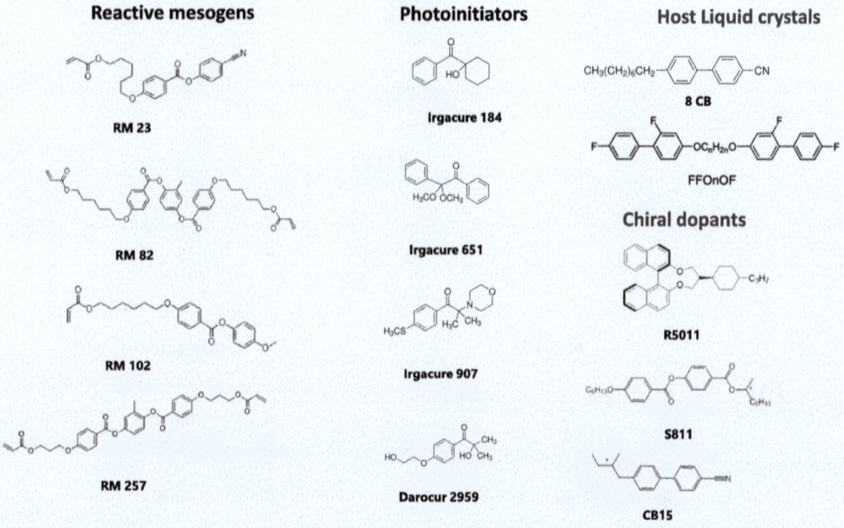

Fig. 1 Molecular structures of different constituent molecules used in PSLC and PSCLCs

position or bandwidth can be adjusted, showing significant potential in sensors, reflective smart windows, bistable devices, and rewritable materials [8]. In general, the performance of PDLCs for a specific application depends on various parameters, such as the morphology of LC and polymers, polymer–LC ratios, LC phase, curing conditions, a surface alignment layer, the size of the device, and the type of materials used, including LC, chiral dopants, monomers, and photoinitiators. Figure 1, shows a representation of different constituent molecules widely used in PDLC/PDCLC preparation.

This chapter summarizes recent research advancements in PSLCs and reviews their innovative applications in enhanced device functionality. Looking forward, the development of PSLCs poses various industrial production challenges, alongside opportunities for versatile applications and emerging technologies.

2 Polymer-Stabilized Nematic Liquid Crystals (PSNLCs)

PSLCs are prepared in a three-step process. The first step involves preparing the mixture of NLC, monomer, and photoinitiator. In the second step, this mixture is introduced into a cell with an alignment layer, which defines the preorientation of NLC. The final step is polymerizing the mixture under suitable curing conditions. Depending on the intended application, the curing process may be carried out with or without the application of voltage. Working principle of a typical PSNLC device is shown in Fig. 2. Figure 2a shows type of materials used to make PSNLC. The operating mechanism of the PSNLCs is illustrated in Fig. 2b. In the OFF state of

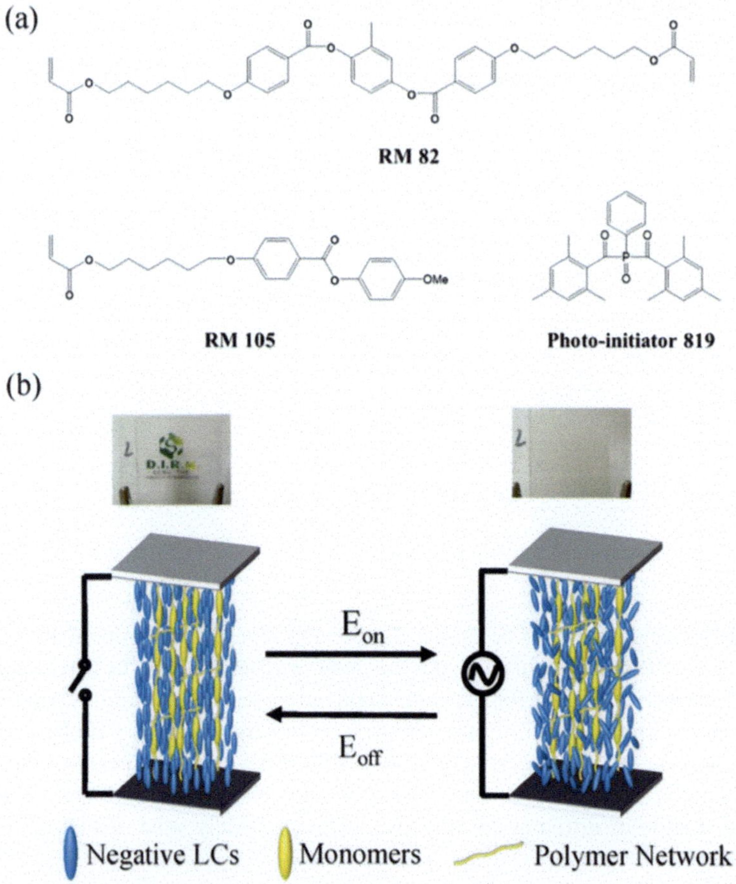

Fig. 2 Working principle of PSLC device. **a** Materials used for PSLC compositions. **b** Schematic and photo of PSLC device in the field-off state (transparent state) and in the field-on state (scattering state). (Reproduced with permission from Ref. [13]; John Wiley & Sons)

applied electric field, pre-alignment of the CLC cell produces a homeotropic LC orientation (Fig. 2b). As a result, the cell appears transparent (Fig. 2b) due to the matching refractive indices of the NLCs with negative dielectric anisotropy and the polymer networks [12]. In the ON state of applied electric field, however, the NLCs adopt a multi-domain structure, causing a mismatch in refractive indices between the LC domains and the polymer networks, which leads to scattering in the PSNLC film (Fig. 2b) [12].

NLCs are commonly used in PSLC fabrication due to their high birefringence and responsiveness to electric, magnetic, optical, and thermal fields. Given the high fluidity of PSLCs, their properties can be adjusted by modifying the cross-link density of the polymer networks. Additionally, the birefringence of NLCs is retained after polymerization when monomers are doped into the LCs. Applying a small electric field can induce a strong photorefractive effect [14]. The applied electric field induced a shearing force between the surface of the substrate used to prepare the cell. This share force alters the director orientation of the LC, thereby affecting the optical properties of the film. The magnitude of reorientation change depends on both the polymer morphology and the intensity of the electric field.

Figure 3 shows an SEM image of the polymer networks formed without LCs, prepared with different monomers and fixed photoinitiator concentrations. The cross-sectional view (Fig. 3a) and the top view (Fig. 3b), clearly show that the polymer network morphology changes drastically with increasing monomer concentration [13]. These morphological changes lead to abrupt changes in the electro-optic property of the PSLC device. When the monomer concentration increases, both the strength and density of the polymer network decrease. This causes a huge reduction in threshold voltage by ~50%, and response time by ~90% along with a decrease in the contrast ratio [13]. This is an example of how the preparation process can significantly impact the overall behaviour of the device (Saadaoui et al. 2024) [15]. The quick switching of transparent and scattering states of PSNLCs, makes them suitable for various applications such as optical shutters and smart windows [13]. In the absence of applied voltage, the LCs and polymer networks are homogeneously aligned, resulting in a transparent, non-scattering state. When the voltage was applied, the LCs rearranged into a poly-domain structure due to the alignment consequence of the polymer network, causing partial scattering of the incident light, which exits on the display's viewing side. The reverse-mode PSNLC films were prepared using polymer walls with photomasks and enhance the stability and electro-optical behaviour [13]. Given that PSNLCs are primarily limited to visible light, the incorporation of antimony-doped tin oxide (ATO) nanoparticles (NPs) enables shielding of over 80% of infrared radiation [16]. In NLCs, exposure to vapor reduces the clearing temperature of the LCs, the POM textures showed the optical appearance shift from bright to dark. Using PSNLCs, a technique has been introduced to obtain a two-dimensional (2D) profile of toluene vapor concentration in a microchannel by controlling the vapor concentration. By tuning the interference colors of the PSNLC under POM, the diffusion coefficient of toluene within the PSNLC is calculated at 1.01×10^{-6} cm^2/s [16].

The component molecules of NLCs have permanent dipole moments; when a DC voltage is applied, reflection symmetry along the LC director is shattered, resulting in net polarization [17]. The inclusion of polymer networks in PSNLCs typically increases the required driving voltage, however adding particles to PSLCs reduces the threshold voltage and switching times. A PSNLCs featuring functionalized single-walled carbon nanotubes (SWCNTs) have been developed, with the addition of CNTs lowering the threshold voltage [18]. Additionally, a polarizing light waveguide plate (LWGP) derived from PSNLCs have been created to convert unpolarized light into linearly polarized light [19]. Other than the display applications, PSNLCs are also potential candidates as tunable optical components, such as electrically tunable lenses [20, 21] switchable Fresnel lenses [22] and phototunable

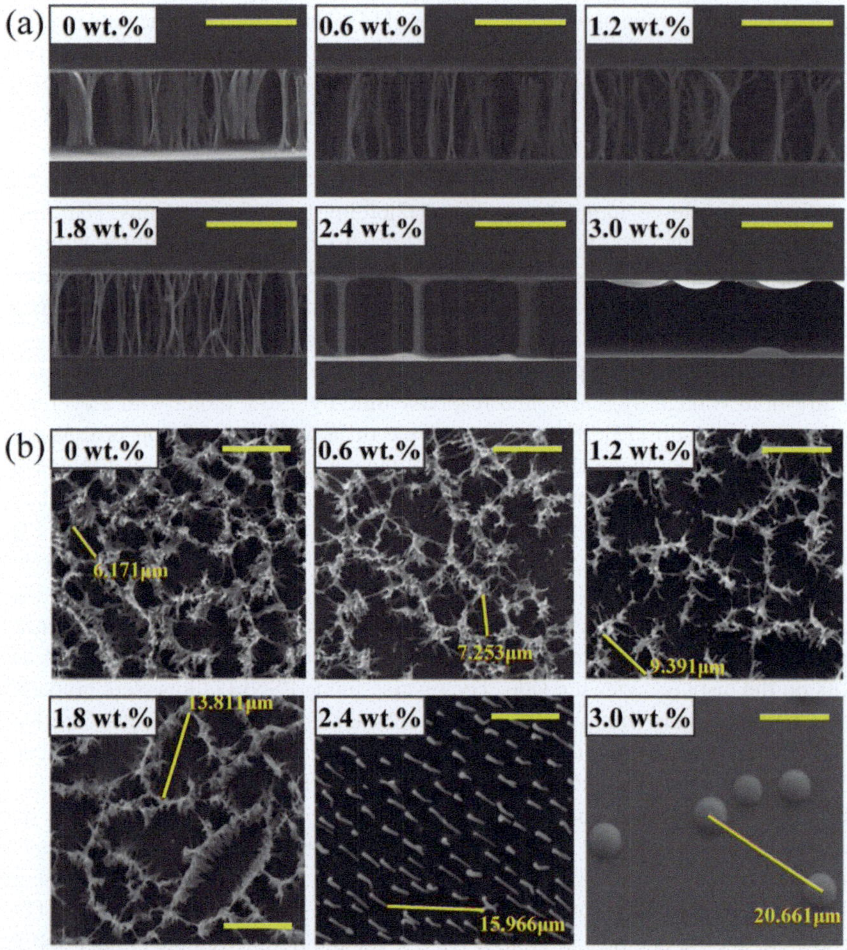

Fig. 3 The effect of monomer concentration on the polymer network morphologies. SEM micrographs of the polymer networks with different mono-acrylate ratios from **a** cross-section view and **b** top view (all scales are 10 μ). (Reproduced with permission from Ref. [13]; John Wiley & Sons)

microlens array [23]. Instead of the conventional curing process with uniform illumination, using a strongly focused light for in situ photopolymerization creates the refractive index distribution to produce lenses [24].

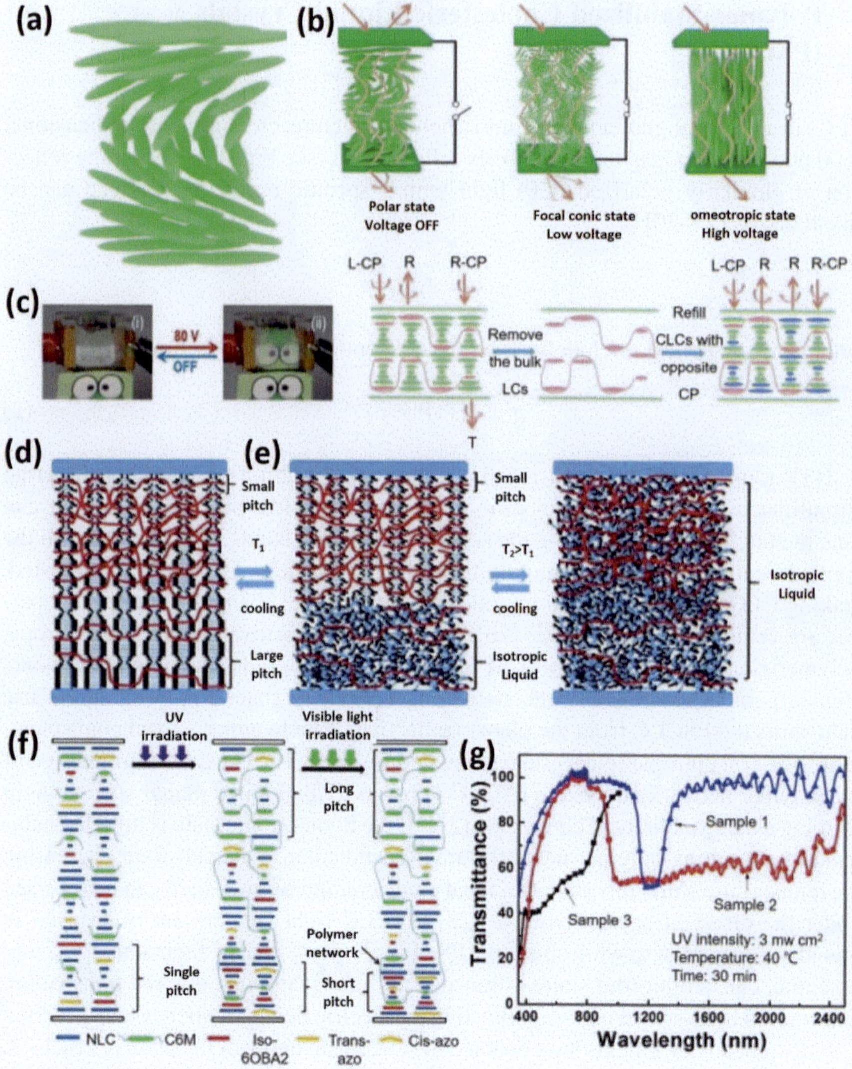

Fig. 4 **a** Illustration of the helical structure in the CLC phase; **b** schematic of the PSCLC light shutter operation; **c** photographs showing the reflection and transmission of the left-handed (LH) and right-handed (RH) PSCLC assembly, both without an electric field and with an applied 80 V electric field, Reproduced with permission from Ref. [31]; **d** schematic of the PSCLC film preparation process, where "R" denotes reflection and "T" denotes transmission; **e** schematic diagram illustrating the mechanism for a thermally switchable broadband reflector. Reproduced with permission from Ref. [34] Royal Society of Chemistry; **f** preparation steps and mechanism for achieving broadband reflection in a PSCLC; **g** transmittance spectra of PSCLCs following UV irradiation. (Reproduced with permission from Ref. [36] Royal Society of Chemistry)

3 Polymer-Stabilized Cholesteric Liquid Crystals (PSCLCs)

CLCs in a planar alignment gather into a helical structure, creating a one-dimensional (1D) photonic bandgap that selectively reflects light (see Fig. 4a). The Bragg reflection of circularly polarized (CP) light with a specific reflection bandgap can be calculated as [25–30]

$$\Delta\lambda = \Delta \mathrm{n} \times p \tag{1}$$

where p is the helical pitch defined by the concentration (c) as following,

$$p = (\mathrm{HTP} \times c)^{-1} \tag{2}$$

HTP represents the helical twisting power of the chiral dopant [25]. External stimuli such as temperature, electric and magnetic fields, pressure, and light can tune the pitch of CLCs, shifting the selective reflection band. Additionally, with the applied electric field through interdigitated electrodes, the p of CLCs can be adjusted, making CLCs highly effective in optical sensors, as the changes in optical properties are visible to the naked eye [26]. In CLCs with positive dielectric anisotropy, polymerizing the LCM or non-LCM in the planar state produces polymer networks aligned parallel to the substrate, stabilizing the planar state. Applying an electric field shifts the PSCLC from the planar (reflective) state to either a focal conic (scattering) or homeotropic (clear) state (shown in Fig. 4b). The elastic restoring force of the polymer network allows the PSCLC to revert to its original planar state without forming a metastable focal conic state [27]. If the homeotropic state is initially stabilized, the resulting polymer networks are perpendicular to the substrate; increasing the temperature shifts this state to a focal conic configuration, which can be reversed under the effect of applied voltage [28]. PSCLC light shutters are often used in greenhouse windows, while bistable PSCLC devices, due to their energy-saving potential, can exhibit four states—transparent, light-scattering, colored transparent, and colored light-scattering—by adjusting the electric field's frequency and direction [29]. Recent advancements have enabled the electrical tuning of the pitch in PSCLCs. Significant, stable control over the spectral reflection bandwidth of PSCLCs has been demonstrated, with reflection bands broadened by up to 300%. Due to electromechanical displacement from trapped ions, PSCLCs tend to have slower switching times compared to LC reorientation-based mechanisms [27]. In PSCLCs with negative dielectric anisotropy, LCs remain in a reflective state under an electric field, allowing reversible blue-shift tuning of the selective reflection [30]. Color-switchable mirrors have also been created from electrically tunable bandwidth broadening of PSCLC CP reflections (Fig. 4c), allowing the selective reflection to switch dynamically from colored to broadband mirror-like reflection. Initially, the word "mirror" is visible; applying a DC bias of 80 V causes the word to disappear, producing a mirror effect [31].

A wash-out/refill method (Fig. 4d) has enabled PSCLCs with double-handed CP reflection bands, reflecting both right-circularly polarized (R-CP) and left-circularly polarized (L-CP) light [32]. In this approach, a cell filled with a mixture of monomers and LCs is UV-cured, washed with cyclohexane and tetrahydrofuran to remove non-polymerized LCs, and refilled with a different LC to form a PSCLC film combining polymer networks with CLCs of opposite helicity, producing both R-CP and L-CP reflections [32]. In CLCs, the LC director forms a helical axis, and in the oblique helicoid (ChOH) state, it tilts at an angle below 90° relative to this axis. This pitch and tilt angle depend on the electric field strength [33]. PSCLC films can serve as broadband reflectors. Polymer network density is higher on the top side (short pitch side) due to polymerization-induced diffusion. Applying a low voltage or temperature T_1 (Fig. 4e) causes CLCs in low-density polymer areas to reorient homeotropically, while CLCs in high-density areas remain stable. As the electric field or temperature T_2 increases, long-wavelength reflection decreases, forming a narrowband reflector [34]. Because of CLCs' thermotropic nature, PSCLC reflection switches thermally when transitioning from the LC to the isotropic state. By cooling below the melting point, dual CP bands in the infrared range are preserved at room temperature [35].

Azobenzene compounds such as QL76 or chiral dopant 2C can adjust LC orientation, extending PSCLC reflection wavelengths to cover the 1000–2400 nm range [36–38]. Figure 4f illustrates the broadband reflection mechanism; after UV irradiation, cells are exposed to visible light. Transmittance spectra for samples with different 2C ratios and chiral dopants (shown in Fig. 4g) reveal that doping PSCLC films with 2C broadens the reflection range to 1000–2400 nm, with sample 1 (without 2C) showing a narrower 1130–1310 nm range. Increasing chiral dopant concentration allows the reflection range to be tuned from 400 to 800 nm. The polymer stabilization of azo-CLCs also shortens the transition time between isotropic and CLC phases [37]. Doping fluorescent molecules into PSCLCs with gradient pitches achieves unique circularly polarized luminescence (CPL) emissions from each cell surface and broadens PSCLC reflection from 150 to 500 nm, modifying CPL signals [39]. Beyond broadband reflection, paraboloid-shaped PSCLC microlens arrays exhibit polarization-dependent optical properties ideal for sensing polarization states of incoming light [40]. Dispersing CLC droplets in polymer solutions and using 3D printing with water evaporation yields oblate CLC droplets, enabling bio-inspired, stimuli-responsive systems capable of color change and shape deformation [41]. Further, CLC droplets dispersed in polymer films can be mechanically stretched from spherical to oblate shapes, increasing the Bragg reflection range and creating structural color, useful for smart sensors [42]. CLC droplets show considerable potential in smart functional materials, photonic devices, sensors, and soft robotics applications [43].

To enhance the transmission levels of PSCLC, Lee et al. reduced the polymer content to 6–8 wt.%, significantly increasing transmission values [44]. In this setup, applying a 160 V DC field tuned the reflection notch from 700 nm to nearly 2500 nm, while maintaining high out-of-band transmission across the tuning range, as shown in Fig. 5a. By adjusting the chiral dopant concentration, tuning in the visible spectrum was also achieved. However, a noticeable drift in the reflection notch position occurred due to the lower polymer content. This drift, or "creep," was more pronounced at around 5 wt.% polymer concentration, whereas samples with 8 wt.% polymer showed greater stability. Figure 5b illustrates the PSCLC's capacity to shift the reflection notch. The electric field was set at 50 V DC and held for 30 min. Upon switching off the field, the reflection band returned to its initial position. This process was repeated for displacements of 200, 300, and 400 nm. The inset of Fig. 5b shows POM images capturing the reflection color as DC voltage increased from 0 to 75 V. Over the 30-min duration under the electric field, the reflection notch continued to drift, with the extent of drift dependent on the field strength—largely due to the polymer network's deformation relative to the length of DC field application.

In 2017, Lee and colleagues demonstrated the blue shifting of the reflection notch [29]. In their experiment, a PSCLC with an initial reflection notch centered at 700 nm was shifted to 430 nm by applying a 35 V DC field. The transmission spectra for this sample are displayed in Fig. 6a, while Fig. 6b shows photographs of the reflection colors along with their corresponding wavelength positions. This blue shifting effect was found to be highly sensitive to both the photo-polymerization conditions and the photoinitiator used. When the PSCLC was cured under lower UV intensity (250 mW cm^{-2} at 365 nm), red-shifting occurred after three minutes, and the reflection bandwidth broadened between 5 and 10 min. Only after 15 min of curing did the blue shift become apparent. Conversely, when the sample was cured at a higher UV intensity of 700 mW cm^{-2} at 365 nm, blue shifting occurred after 10 min of curing. The choice of photoinitiator also played a key role in achieving blue shifting of the reflection notch. Morpholino-keto-type initiators, such as Irgacure 369 and 907,

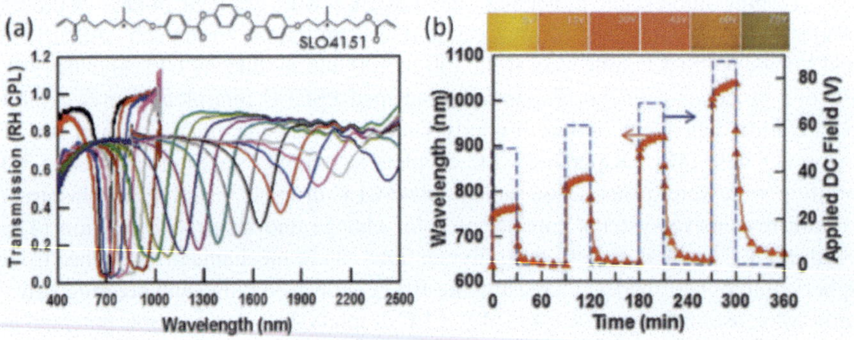

Fig. 5 **a** Transmission spectra of a PSCLC containing 1 wt.% Irgacure 369 and a chiral liquid crystal monomer (SL04151, shown in inset), cured at 80–100 mW cm^{-2} for 3 min, with voltage applied from 0 to 160 V. **b** Reflection wavelength variation in a PSCLC with 5 wt.% polymer concentration, subjected to DC voltages of 50, 60, 69, and 87 V. Voltages were directly applied to the cells and maintained continuously for 30 min. Polarized optical microscope (POM) reflection images display increasing DC voltages from 0 to 75 V. The red triangle marks the notch position of the PSCLC, while the blue dashed line indicates the applied DC field. Cells with a 15 μm thickness were used. Reproduced with permission from Ref. [44]

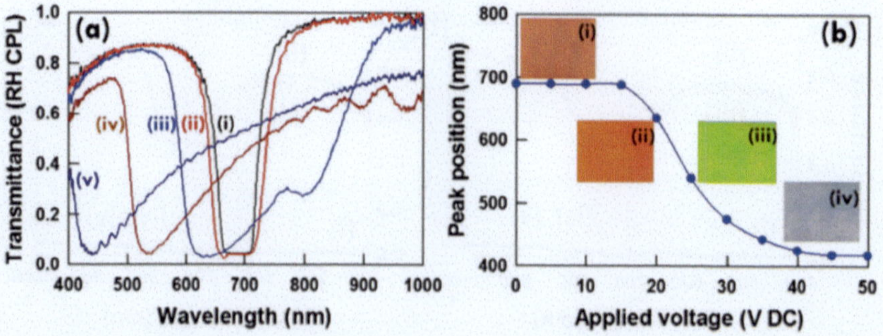

Fig. 6 **a** Blue shifting of the selective reflection of a 15 μm thick PSCLC at (i) 0 V, (ii) 15 V, (iii) 20 V, (iv) 25 V, and (v) 35 V. **b** The wavelength of the selective reflection is plotted as a function of applied DC voltage. Photographs of the reflection at (i) 0 V, (ii) 19 V, (iii) 23 V, and (iv) 36 V are provided as insets. The PSCLC shown was cured at 700 mW cm^{-2} for 30 min. Reproduced with permission from Ref. [29]

enabled blue shifting at concentrations as low as 0.5 wt.% after 30 min of irradiation at 250 mW cm^{-2}. In contrast, when a non-morpholino-keto-type initiator like Irgacure 651 was used, red shifting occurred, even at concentrations between 0.5 and 5 wt.% under similar curing conditions.

Lee and colleagues introduced a switchable PSCLC utilizing negative dielectric anisotropy LCs [44, 45]. In contrast to positive dielectric anisotropy LCs, where the director aligns with the direction of the electric field, the negative dielectric anisotropy LCs in planar geometry remain unaffected by the applied electric potential. The authors demonstrated the switching response of these negative dielectric anisotropy PSCLCs when a DC field was applied. The PSCLC samples were fabricated with a low polymer concentration (less than 1.5 wt.%), resulting in the formation of a weak polymer network. By applying a high electric field, this polymer network was detached from the alignment layer. The application of the DC field caused the polymer network to move toward the negative electrode, which facilitated the switching behavior. Figure 7 illustrates the bistable switching of the PSCLC, which toggles between clear and scattered states by adjusting the applied electric field, reaching a maximum voltage of 200 V DC.

Autonomous, stimulus-responsive PSCLCs are promising materials with various applications. Ranjkesh and colleagues demonstrated thermal and electrical wavelength tuning of Bragg reflection in PSCLCs using UV light absorbers [9]. Their PSCLC device autonomously shifts its selective reflection wavelength with temperature changes or can be adjusted with an electric field. UV absorbers within the PSCLC not only protect the substrate but also stabilize the polymer network by converting harmful UV radiation into heat. By carefully selecting UV absorbers with different structures, absorption spectra, and concentrations, optimal performance can be achieved under various UV intensities. This thermally autonomous and electrically tunable reflector offers potential for polymer-templated LC devices, tunable

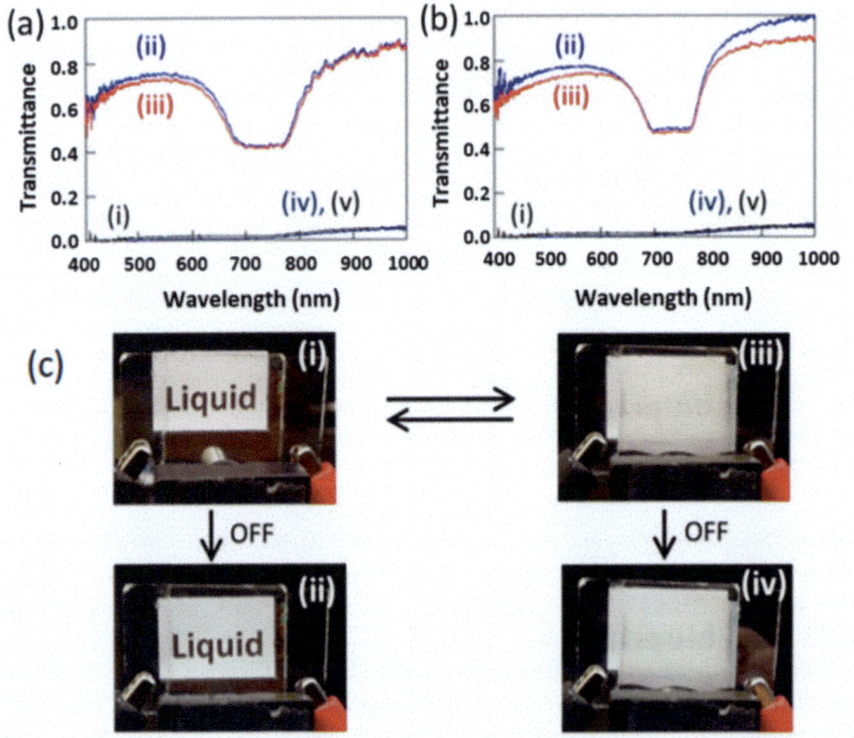

Fig. 7 Bistable switching behavior of PSCLCs with **a** a gap thickness of 30 μm: normal mode showing (i) 0 V, (ii) 210 V DC, (iii) OFF (0 V), (iv) −45 V DC, and (v) OFF (0 V); **b** showing (i) 0 V, (ii) 100 V AC, (iii) OFF (0 V), (iv) 45 V DC, and (v) 0 V; and **c** photographs of samples with a thickness of 15 μm at (i) 50 V AC, (ii) 0 V, (iii) 100 V DC, and (iv) 0 V. Reproduced with permission from Ref. [44]

optical filters, optical and thermal sensors, greenhouse technologies, light signage, and even toys (Fig. 8).

PSCLCs are widely used displays, and also being a promising materials for tunable lenses and laser sources [46]. Yu et al. demonstrated the effect of polymer concentrations on the effective tuning of photonic band gaps in PSCLCs. By incorporating an optical gain medium into PSCLCs, it is possible to precisely control the band-edge lasing, enabling their use as tunable laser sources. Wang et al. utilized the tunable photonic band-edge property to create intriguing multicolor patterns through multiple reflections. By adjusting the chiral dopant concentration, photoinitiator, and curing time, they demonstrated the feasibility of multilayer multicolor patterning [47].

Due to the anisotropic nature of LC and their tunable molecular orientation in response to stimuli, the refractive index can be precisely controlled. This property makes LC a promising material for tuneable focal-length lenses without altering their shape. Perera et al. demonstrated plano-concave nematic LC microlens arrays with constant thickness, which can function as either gradient-index (GRIN) lens elements

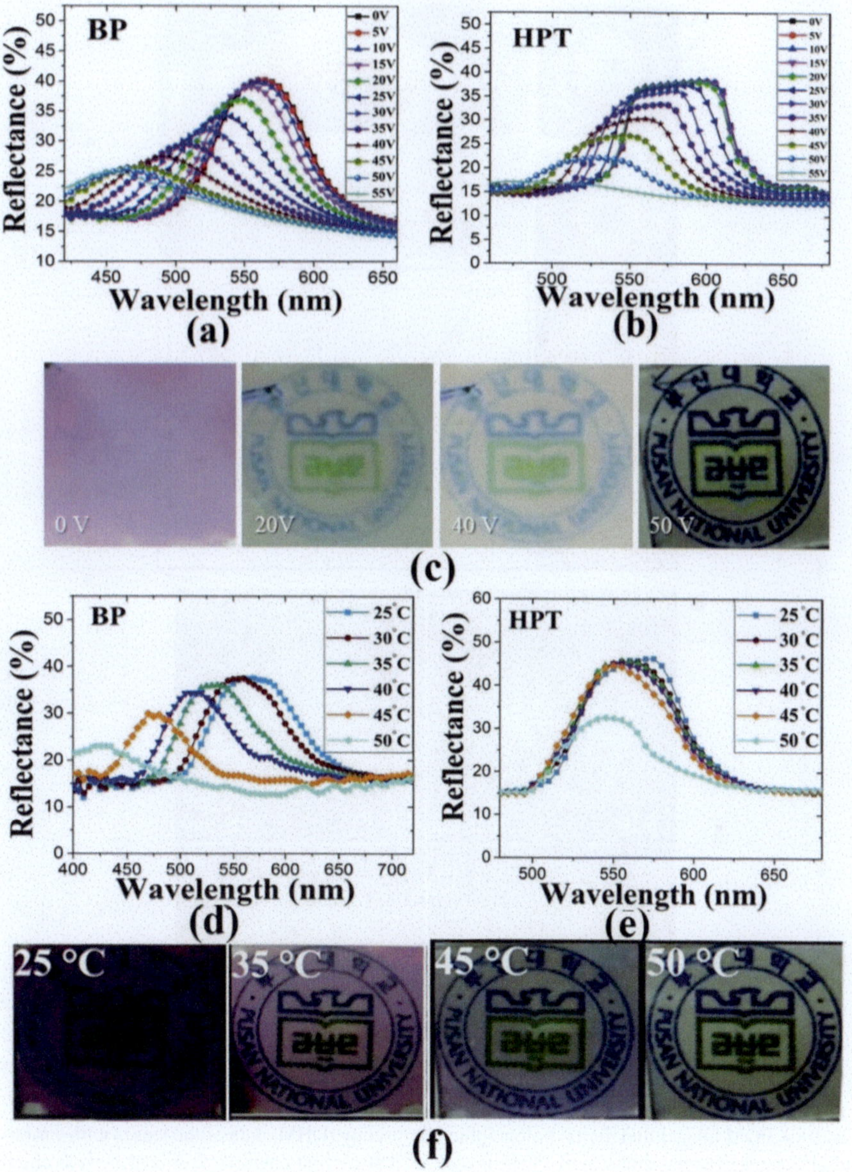

Fig. 8 Impact of various UV light absorbers on the wavelength tuning range of the PSCLC. The reflection spectra of the cell with voltage variations are shown for **a** BP-, and **b** HPT-type UV absorbers, each tested under identical conditions of UV intensity at 1 mW/cm^2 and a concentration of 3 wt%. **c** Fabricated PSCLC device under different applied voltages. **d**–**e** Peak positions corresponding to temperature changes for each type of UV absorber. **f** Temperature-dependent behavior of the cell with the BP-type UV absorber. Adapted from Ref. [9] with permission from the Royal Society of Chemistry

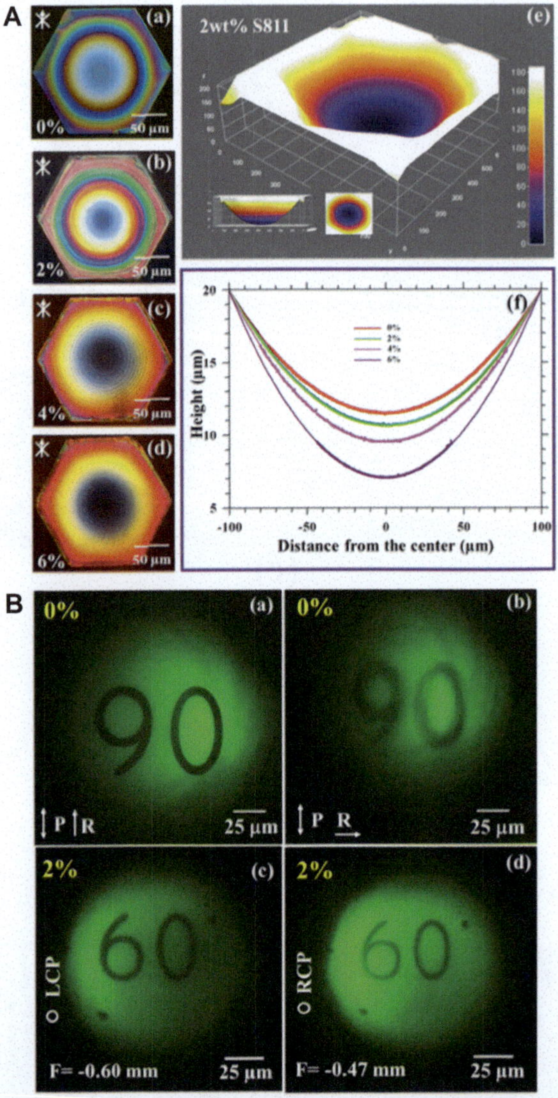

Fig. 9 **A** POM images and digital holographic microscopy (DHM) data of the lenses with various chiral dopant concentrations. **a–d** POM images of lenses containing 0, 2, 4, and 6 wt% chiral dopant S811. **e** 3D recreation of the surface of the 5CB + 2 wt% S811 lens. **f** Height profiles as a function of radial distance from the vertex. Data are shown by different symbols. Dotted lines with colors matching with the color of the data points are fitted by parabolic equations. **B** Mages of numbers "90" and "60" created with 0 and 2 wt% samples are shown in the top and bottom rows, respectively. **a** 0% sample with a linear polarizer in front of the object "90." The objective of the inverted microscope focuses on the image "90" when the polarizer is parallel to the rubbing direction. **b** The image gets defocused when the polarizer is rotated perpendicular to the rubbing direction indicating the dependence of the focal length on the polarization direction. **c** 2 wt% sample between left circular polarizers. **d** 2 wt% sample between right circular polarizers. In both pictures, the objective of the inverted microscope focuses on the image "60," but the focal distances are different, as shown in the images. Reproduced with permission from Ref. [48]

or Pancharatnam–Berry (PB) phase lenses. The plano-convex shape is achieved by drop-casting LC onto spinning TEM grids, causing a fraction of the LCs to flow outward, forming a paraboloid-shaped lens. The metastable lens is then stabilized through polymerization of the LC, either in the nematic or cholesteric phase. The nickel TEM grid, glued onto a polyimide-coated glass plate with planar alignment at ± 45°, is used as the substrate. A controlled step-spinning ramp produces stable nematic and CLC lenses. Figure 9A, a–d shows polarized optical microscopic images of chiral 5CB lenses with varying S811 concentrations. The chiral dopant plays a dual role in lens fabrication: first, it minimizes defects and controls the radius of curvature; second, it stabilizes the lens shape and provides sufficient time for polymerization [48]. As the concentration of the chiral dopant increases, the radius of curvature decreases from 0.59 mm (nematic) to 0.39 mm, while the contact angle increases from 9.6° to 14.5°. The imaging characteristics of the polymer-stabilized lenses are illustrated in Fig. 9B. Additionally, by incorporating an electrode onto the surface, it is possible to tune the focal length using an electric field [21, 49].

4 Conclusions

Polymer stabilization has garnered significant attention in LC research since its inception in 1992. This technique holds substantial promise for applications in tunable lasers, optical sensors, and liquid crystal displays. Devices such as electrically, thermally, and optically switchable PSLC systems have been designed and fabricated to enhance performance, underscoring the crucial role of polymers in advancing LC technologies. In the context of smart windows created with PSLCs, features like multi-modulation and environmental adaptability are essential for achieving comfortable indoor conditions and meeting diverse customer needs. For instance, programmable and colorful patterned PSLC films are increasingly sought after for practical applications like interior decoration. By introducing novel LCMs, researchers can achieve stronger anchoring energy between polymer networks and LC molecules without increasing the operational voltage of PSLC films, facilitating stable and scalable industrial production. Historically, polymer stabilization has raised the threshold voltage of LCs. However, when combined with other systems, such as particle-laden or polymer-dispersed methods, enhanced electro-optical performance in PSLCs can be achieved. The integration of various fluorescent molecules into PSLCs can yield circularly polarized light (CPL) performance alongside selective Bragg reflection in PSCLC films. By carefully designing the structures of PSCLCs, materials with significant dissymmetry factors and high emission quantum yields can be developed, offering potential applications in information storage and processing. PSCLC lasers benefit from a wide tunable range, rapid response times, and long-term stability, making them suitable for various optoelectronic devices. For polymer-stabilized blue phase liquid crystal (PSBPLC) lasers, broad-temperature operation can be attained through the incorporation of dyes or other compounds. However, the dynamic tunability of photonic bandgaps in

PSBPLCs still requires enhancement due to undesirable reflections. With their excellent polarization and wavelength selectivity, PSBPLCs are poised to play a critical role in AR displays. The electro-mechanical properties of polymer networks enable the use of PSLCs in sensors and actuators. Nevertheless, simplifying the fabrication process for large-scale industrial applications remains a significant challenge, primarily due to the weak bonding between LCs and substrates. Furthermore, the durability and environmental stability of the films need to be improved to meet the demands for energy conservation and environmental protection. Additionally, the driving voltages for PSLCs tend to be high, and their response times are relatively slow.

References

1. Mucha, M.: Polymer as an important component of blends and composites with liquid crystals. Prog. Polym. Sci. **28**(5), 837–873 (2003). https://doi.org/10.1016/S0079-6700(02)00117-X
2. Lee, K.M., Ware, T.H., Tondiglia, V.P., McBride, M.K., Zhang, X., Bowman, C.N., White, T.J.: Initiatorless photopolymerization of liquid crystal monomers. ACS Appl. Mater. Interfaces **8**(41), 28040–28046 (2016). https://doi.org/10.1021/acsami.6b09144
3. Kasch, N., Dierking, I., Turner, M.: Stabilization of the liquid crystalline blue phase by the addition of short-chain polystyrene. Soft Matter **9**(19), 4789–4793 (2013). https://doi.org/10.1039/C3SM00065F
4. Radka, B.P., Lee, K.M., Godman, N.P., White, T.J.: Electro-optic characteristics of stabilized cholesteric liquid crystals with non-liquid crystalline polymer networks. Soft Matter **18**(15), 3013–3018 (2022). https://doi.org/10.1039/D2SM00203E
5. Dierking, I.: Fractal and non-fractal structure-property relationships of polymer-stabilized liquid crystals. Adv. Func. Mater. **14**(9), 883–890 (2004). https://doi.org/10.1002/adfm.200305031
6. Li, X., Guo, Y., Zhang, M., Zhang, C., Niu, R., Ma, H., Sun, Y.: Colorable light-scattering device based on polymer-stabilized ion-doped cholesteric liquid crystal and an electrochromatic layer. ACS Appl. Mater. Interfaces **15**(5), 7184–7195. https://doi.org/10.1021/acsami.2c17770
7. Chen, C.-W., Li, C.-C., Jau, H.-C., Lu-Chun, Y., Hong, C.-L., Guo, D.-Y., Wang, C.-T., Lin, T.-H.: Electric field-driven shifting and expansion of photonic band gaps in 3D liquid photonic crystals. ACS Photonics **2**(11), 1524–1531 (2015). https://doi.org/10.1021/acsphotonics.5b00314
8. Lee, K.M., Tondiglia, V.P., White, T.J.: Photosensitivity of reflection notch tuning and broadening in polymer stabilized cholesteric liquid crystals. Soft Matter **12**(4), 1256–1261 (2016). https://doi.org/10.1039/C5SM01985K
9. Ranjkesh, A., Yoon, T.-H.: Thermal and electrical wavelength tuning of Bragg reflection with ultraviolet light absorbers in polymer-stabilized cholesteric liquid crystals. J. Mater. Chem. C **6**(45), 12377–12385 (2018). https://doi.org/10.1039/C8TC04563A
10. Wang, L., He, W., Wang, Q., Meina, Y., Xiao, X., Zhang, Y., Ellahi, M., Zhao, D., Yang, H., Guo, L.: Polymer-stabilized nanoparticle-enriched blue phase liquid crystals. J. Mater. Chem. C **1**(40), 6526–6531 (2013). https://doi.org/10.1039/C3TC31253D
11. Xu, C.-T., Zhang, D.-W., Yuan, R., Chen, Q.-M., Liang, X., Hu, W.: Optical orbital angular momentum processors with electrically tailored working bands. Laser Photonics Rev. **17**(7), 2370031 (2023). https://doi.org/10.1002/lpor.202370031
12. Hu, X., Zhang, X., Yang, W., Jiang, X.-F., Jiang, X., de Haan, L.T., Yuan, D., Zhao, W., Zheng, N., Jin, M., Shui, L., Schenning, A.P.H.J., Zhou, G.: Stable and scalable smart window based on polymer stabilized liquid crystals. J. Appl. Polym. Sci. **137**(30), 48917 (2020). https://doi.org/10.1002/app.48917

13. Zhou, Y., You, Y., Liao, X., Liu, W., Zhou, L., Zhang, B., Zhao, W., Xiaowen, H., Zhang, L., Yang, H., Zhou, G., Yuan, D.: Effect of polymer network topology on the electro-optical performance of polymer stabilized liquid crystal (PSLC) devices. Macromol. Chem. Phys. **221**(18), 2000185 (2020). https://doi.org/10.1002/macp.202000185
14. Wiederrecht, G.P., Wasielewski, M.R.: Photorefractivity in polymer-stabilized nematic liquid crystals. J. Am. Chem. Soc. **120**(13), 3231–3236 (1998). https://doi.org/10.1021/ja974033z
15. Saadaoui, L., Yang, D., Hassan, F., Qiu, Z., Wang, Y., Fan, Y., Drevensek-Olenik, I., Li, Y., Zhang, X., Xu, J.: Electro-optic response of polymer-stabilized cholesteric liquid crystals with different polymer concentrations. Polymers **16**(17) (2024). https://doi.org/10.3390/polym16172436
16. Zhang, Z., Zhang, R., Liangge, X., Li, J., Yang, L., Yang, Y., Bolshakov, A., Zhu, J.: Visible and infrared optical modulation of PSLC smart films doped with ATO nanoparticles. Dalton Trans. **50**(29), 10033–10040 (2021). https://doi.org/10.1039/D1DT01575C
17. Hicks, S.E., Hurley, S.P., Yang, Y.C., Yang, D.K.: Electric polarization frozen by a polymer network in nematic liquid crystals. Soft Matter **9**(14), 3834–3839 (2013). https://doi.org/10.1039/C3SM27594A
18. Krishna Prasad, S., Baral, M., Murali, A., Jaisankar, S.N.: Carbon nanotube reinforced polymer-stabilized liquid crystal device: lowered and thermally invariant threshold with accelerated dynamics. ACS Appl. Mater. Interfaces **9**(31), 26622–26629 (2017). https://doi.org/10.1021/acsami.7b08825
19. Moheghi, A., Nemati, H., Yang, D.-K.: Polarizing light waveguide plate from polymer stabilized liquid crystals. Opt. Mater. Express **5**(5), 1217–1223 (2015). https://doi.org/10.1364/OME.5.001217
20. Ren, H., Wu, S.-T.: Tunable electronic lens using a gradient polymer network liquid crystal. Appl. Phys. Lett. **82**(1), 22–24 (2003). https://doi.org/10.1063/1.1534915
21. Presnyakov, V.V., Galstian, T.V.: Electrically tunable polymer stabilized liquid-crystal lens. J. Appl. Phys. **97**(10), 103101 (2005). https://doi.org/10.1063/1.1896436
22. Fan, Y.-H., Ren, H., Shin-Tson, W.: Switchable Fresnel lens using polymer-stabilized liquid crystals. Opt. Express **11**(23), 3080–3086 (2003). https://doi.org/10.1364/OE.11.003080
23. Xiong, G.-R., Han, G.-Z., Sun, C., Hua, X., Wei, H.-M., Zhong-Ze, G.: Phototunable microlens array based on polymer dispersed liquid crystals. Adv. Func. Mater. **19**(7), 1082–1086 (2009). https://doi.org/10.1002/adfm.200801335
24. Presnyakov, V.V., Asatryan, K.E., Galstian, T.V., Tork, A.: Polymer-stabilized liquid crystal for tunable microlens applications. Opt. Express **10**(17), 865–870 (2002). https://doi.org/10.1364/OE.10.000865
25. Nemati, H., Liu, S., Zola, R.S., Tondiglia, V.P., Lee, K.M., White, T., Bunning, T., Yang, D.-K.: Mechanism of electrically induced photonic band gap broadening in polymer stabilized cholesteric liquid crystals with negative dielectric anisotropies. Soft Matter **11**(6), 1208–1213 (2015). https://doi.org/10.1039/C4SM02283A
26. Mulder, D.J., Schenning, A.P.H.J., Bastiaansen, C.W.M.: Chiral-nematic liquid crystals as one dimensional photonic materials in optical sensors. J. Mater. Chem. C **2**(33), 6695–6705 (2014). https://doi.org/10.1039/C4TC00785A
27. Radka, B.P., Pande, G.K., White, T.J.: The contribution of network elasticity to electro-optic response in polymer stabilized cholesteric liquid crystals. Soft Matter **19**(25), 4634–4641 (2023). https://doi.org/10.1039/D3SM00225J
28. Bao, R., Liu, C.-M., Yang, D.-K.: Smart bistable polymer stabilized cholesteric texture light shutter. Appl. Phys. Express **2**(11), 112401 (2009). https://doi.org/10.1143/APEX.2.112401
29. Gruzdenko, A., Dierking, I.: Electro-optic properties of polystyrene particle-laden polymer-stabilized liquid crystals. J. Mater. Chem. C **11**(16), 5438–5449 (2023). https://doi.org/10.1039/D3TC00437F
30. Lee, K.M, Tondiglia, V.P., Godman, N.P., Middleton, C.M., White, T.J.: Blue-shifting tuning of the selective reflection of polymer stabilized cholesteric liquid crystals. Soft Matter **13**(35), 5842–5848 (2017). https://doi.org/10.1039/C7SM01190C

31. Lee, K.M., Tondiglia, V.P., McConney, M.E., Natarajan, L.V., Bunning, T.J., White, T.J.: Color-tunable mirrors based on electrically regulated bandwidth broadening in polymer-stabilized cholesteric liquid crystals. ACS Photonics **1**(10), 1033–1041 (2014). https://doi.org/10.1021/ph500259h
32. Guo, J., Yang, H., Li, R., Ji, N., Dong, X., Hao, W., Wei, J.: Effect of network concentration on the performance of polymer-stabilized cholesteric liquid crystals with a double-handed circularly polarized light reflection band. J. Phys. Chem. C **113**(37), 16538–16543 (2009). https://doi.org/10.1021/jp903394r
33. Rumi, M., Bunning, T.J., White, T.J.: Polymer stabilization of cholesteric liquid crystals in the oblique helicoidal state. Soft Matter **14**(44), 8883–8894 (2018). https://doi.org/10.1039/C8SM01278D
34. Khandelwal, H., Timmermans, G.H., Debije, M.G., Schenning, A.P.H.J.: Dual electrically and thermally responsive broadband reflectors based on polymer network stabilized chiral nematic liquid crystals: the role of crosslink density. Chem. Commun. **52**(66), 10109–10112 (2016). https://doi.org/10.1039/C6CC04721A
35. Agez, G., Mitov, M.: Cholesteric liquid crystalline materials with a dual circularly polarized light reflection band fixed at room temperature. J. Phys. Chem. B **115**(20), 6421–6426 (2011). https://doi.org/10.1021/jp2014622
36. Chen, X., Wang, L., Chen, Y., Li, C., Hou, G., Liu, X., Zhang, X., He, W., Yang, H.: Broadband reflection of polymer-stabilized chiral nematic liquid crystals induced by a chiral azobenzene compound. Chem. Commun. **50**(6), 691–694 (2014). https://doi.org/10.1039/C3CC47438K
37. White, T.J., Bricker, R.L., Natarajan, L.V., Serak, S.V., Tabiryan, N.V., Bunning, T.J.: Polymer stabilization of phototunable cholesteric liquid crystals. Soft Matter **5**(19), 3623–3628 (2009). https://doi.org/10.1039/B906758B
38. White, T.J., Bricker, R.L., Natarajan, L.V., Tabiryan, N.V., Green, L., Li, Q., Bunning, T.J.: Phototunable azobenzene cholesteric liquid crystals with 2000 nm range. Adv. Func. Mater. **19**(21), 3484–3488 (2009). https://doi.org/10.1002/adfm.200900396
39. Li, Z., Lan, R., Bao, J., Hu, W., Wang, M., Zhang, L., Yang, H.: Tunable circularly polarized luminescence with a high dissymmetry factor emitted from luminogen-bonded and electrically controlled polymer-stabilized cholesteric liquid crystals. ACS Appl. Mater. Interfaces **14**(6), 8490–8498. https://doi.org/10.1021/acsami.1c24264
40. Perera, K., Padmini, H.N., Mann, E., Jákli, A.: Polymer stabilized paraboloid liquid crystal microlenses with integrated pancharatnam-berry phase. Adv. Opt. Mater. **10**(2), 2101510 (2022). https://doi.org/10.1002/adom.202101510
41. Yang, C., Baiheng, W., Ruan, J., Zhao, P., Chen, L., Chen, D., Ye, F.: 3D-printed biomimetic systems with synergetic color and shape responses based on oblate cholesteric liquid crystal droplets. Adv. Mater. **33**(10), 2006361 (2021). https://doi.org/10.1002/adma.202006361
42. Yang, C.-J., Dong, C.: Researches and applications of cholesteric liquid crystal droplets. Chin. J. Liq. Cryst. Displays **37**(8), 1070–1078 (2022). https://doi.org/10.37188/CJLCD.2022-0002
43. Yang, C., Baiheng, W., Ruan, J., Zhao, P., Shan, J., Zhang, R., Yoon, D.K., Chen, D., Liu, K.: Mechanochromic responses of cholesteric liquid crystal droplets with nanoscale periodic helical structures showing reversible and tunable structural color. ACS Appl. Polym. Mater. **4**(1), 463–468 (2022). https://doi.org/10.1021/acsapm.1c01362
44. Lee, K.M., Tondiglia, V.P., White, T.J.: Bistable switching of polymer stabilized cholesteric liquid crystals between transparent and scattering modes. MRS Commun. **5**(2), 223–227 (2015). https://doi.org/10.1557/mrc.2015.40
45. Lee, K.M., Tondiglia, V.P., Lee, T., Smalyukh, I.I., White, T.J.: Large range electrically-induced reflection notch tuning in polymer stabilized cholesteric liquid crystals. J. Mater. Chem. C **3**(34), 8788–8793 (2015). https://doi.org/10.1039/C5TC01320H
46. Perera, K., Nilanthi Haputhantrige, M., Hlinel, E.H., Mostafa, M., Adaka, A., Mann, E.K., Lavrentovich, O.D., Jákli, A.: Polymer stabilized paraboloid liquid crystal microlenses with integrated pancharatnam-berry phase. Adv. Opt. Mater. **10**(2), 2101510 (2021). https://doi.org/10.1002/adom.202101510

47. Yu, R., Cao, Y., Chen, K., Li, Y., Liu, W., Li, B., Li, H., Yang, Y.: Light intensity-selective photopolymerization and photoisomerization for creating colorful polymer-stabilized cholesteric liquid crystal patterns. ACS Appl. Mater. Interfaces. **14**(33), 38228–38234 (2022). https://doi.org/10.1021/acsami.2c10763
48. Perera, K., Padmini, H.N., Mann, E., Jákli, A.: Polymer stabilized paraboloid liquid crystal microlenses with integrated pancharatnam-berry phase. Adv. Opt. Mater. **10**(2), 2101510 (2021). https://doi.org/10.1002/adom.202101510
49. Perera, K., Nilanthi Haputhantrige, M., Himel, S.H., Mostafa, M., Adaka, A., Mann, E.K., Oleg, A.J., Lavrentovich, D.: Electrically tunable polymer stabilized chiral ferroelectric nematic liquid crystal microlenses. Adv. Opt. Mater. **12**(7), 2302500 (2023). https://doi.org/10.1002/adom.202302500

Enhancement in Optical and Dielectric Properties of Quantum Dots Doped Cholesteric Liquid Crystal

Pooja, Vandna Sharma, Pankaj Kumar, and K. K. Raina

Abstract Cholesteric liquid crystals (CLCs) exhibit distinctive optical properties due to their helical structure, making them valuable for various applications, from displays to sensors. The incorporation of quantum dots (QDs) as dopants into CLCs introduces new avenues for enhancing and expanding their functionality and performance. This chapter explores the effects of doping QDs into Cholesteric liquid crystal (CLC), highlighting how these QDs influence the optical, electro-optical, and dielectric properties of CLCs. In this chapter, the studies related to the processes used for the synthesis of QDs, and respective characterizations have also been discussed. The detailed review on the enhancement of CLC applications such as CdSe QDs in display devices for enhanced colour contrast and brightness, PbS QDs with size tunable photonic devices, Mn-doped ZnS QDs for biosensors, and $CuInS_2$ QDs for smart windows etc. has been explored. These advancements in QDs doped CLCs have opened the great scope of CLCs from experimental research to their practical usage in next generation energy efficient devices.

Keywords Cholesteric liquid crystals · Quantum dots · Optical · Electro-optical · And Dielectric properties

Pooja (✉) · V. Sharma · P. Kumar
Centre for Liquid Crystal Research, Chitkara University Institute of Engineering and Technology, Chitkara University, Punjab, India
e-mail: pooja1156as.phd22@chitkara.edu.in

V. Sharma
e-mail: vandnasharma@chitkara.edu.in

P. Kumar
e-mail: pankaj.kumar@chitkara.edu.in

K. K. Raina
M. S. Ramaiah University of Applied Sciences, Bangalore, India

P. Kumar et al. (eds.), *Cholesteric Liquid Crystals*, Engineering Materials,
https://doi.org/10.1007/978-981-96-3821-5_13

1 Introduction

Cholesteric liquid crystals (CLCs) are known for their unique helical structures, which enable them to exhibit remarkable electro-optical and dielectric properties, including selective reflection and tunable optical characteristics. These features make CLCs highly suitable for applications in display technologies, sensors, and optical switches [1].

In 1888, liquid crystalline behavior was first observed by an Austrian botanist named Friedrick Reinitzer who observed that a material known as para cholesteryl benzoate had two distinct melting points (145.5° and 178.5 °C). This observed phase possesses properties of both liquid as well as crystalline solids. Liquid crystal (LC) was basically formed by anisotropic molecules (directional dependent properties) thus shows birefringence when placed under polarized light. Further studies in LC allow researchers to classify LC broadly into two categories (thermotropic and lyotropic). Thermotropic LCs are heat dependent while lyotropic are concentration dependent. Among thermotropic LC phases, the cholesteric or chiral nematic phase has chirality due to the doping of chiral molecules in nematic LC and thus shows pitch dependent behavior. The thermo-optical and electro-optical properties of CLCs fascinate researchers for exploring this field with enhancement in the properties of polymer dispersed pure LC systems in normal and reverse modes and obtained the improved opaque state [2–6]. Further, the effect of nanoparticles in the polymer dispersed nematic LC [7], nanoparticle [8] and dyes [9] to achieve the homeotropically aligned LCs on the conductive surfaces [10].

In continuation, researchers modify the properties of nanoparticles such as structural, optical, photoluminescence, and dielectric properties [11–13] to widen their area for varied applications. Further, more enhancements were attained by various researchers on opting the theoretical estimations and implemented for the experimental investigations using different dopants such as nanoparticles [14], ions [15], dyes [16, 17] and QDs [18, 19] in the pure LC, CLC and polymer stabilized CLC to form composite films with better performance for varied applications.

Different dopants work on different principles like dyes work on the absorption [20], polymer work on the network formation and QDs shows the quantum confinement effect and thus plays a significant role in improving the contrast ratio and lowering the operating voltage. Various research works on nanoparticles dispersed in LCs have been carried out, wherein the shape/size of NPs and the self-assembling nature of the LCs are widely taken into consideration. It has been reported that the dispersions of a variety of NP geometries such as spherical colloids [21], magnetic nanoplates [19] titanium dioxide [22], metal oxide nanowires [23], multiwall carbon nanotubes [24], graphene and its derivatives [25] have produced many fascinating results regarding the interaction between particle shape and alignment in the LC phase. Nevertheless, there are also some disadvantages and shortcomings of these NPs, like intricacy to obtain series of these particles simply differing in size and large size distribution. To overcome these shortcomings QDs, with their size-dependent electronic properties (ability to emit light at varying wavelengths) and high photoluminescence efficiency, offer a compelling avenue for improving the performance of CLCs. QDs such as are Mn-doped ZnS QDs employed as contrast agents in biological imaging to facilitate the visualization of specific targets in cells or tissues. Moreover, their capacity to effectively transform sunlight into energy renders them very viable for enhancing the efficiency of solar cells. QDs provide varied and potent features that have the potential to change several sectors. QDs have the potential to be employed in medicine to specifically target and transport medications to certain cells in the body, hence enhancing the accuracy and efficacy of treatment. QDs have the potential to be incorporated into LED headlights in the automotive sector, resulting in enhanced brightness and energy efficiency for vehicle illumination. Thus, integration of QDs into CLCs, leading to significant enhancements in their electro-optical and dielectric properties [1] for various application such as CdSe QDs in display devices [26], PbS QDs in photonic devices [27], Mn-doped ZnS QDs for biosensors [28], CdSe QDs for electro-optical devices [18] and $CuInS_2$ QDs for smart windows [29].

For instance, a study by Chauhan et al. [18] demonstrated that the incorporation of CdSe QDs into a cholesteric matrix resulted in enhanced dielectric permittivity and reduced response times, facilitating faster switching speeds in electro-optical devices. The authors attributed these improvements to the strong coupling between the QDs and the LC molecules, which modifies the overall dielectric response of the composite material. Furthermore, the optical properties of CLCs can be significantly altered through QD doping. In a recent work by Zhang et al. [3], it was shown that embedding InP QDs in a CLC led to enhanced photonic bandgap properties, resulting in a broader range of tunability for selective reflection. This enhancement is primarily due to the quantum confinement effects exhibited by the QDs, which influence the refractive index of the CLC and broaden its photonic applications. The interplay between QDs and CLCs is not only a fascinating area of research but also holds the potential for developing next-generation optical devices [18]. By tailoring the size and concentration of QDs, researchers can achieve desired electro-optical properties, leading to applications in tunable lasers, displays, and biosensors. Recent advancements by Chemingui et al. [22] have shown that the incorporation of QDs can lead to synergistic effects that improve both the stability and the performance of CLCs under varying environmental conditions. The schematic illustration of the QD doped CLC light shutter in the Planar (P), Focal-Conic (FC), and Homeotropic (H) states is shown in Fig. 1a–c. In P state, the LC molecules twist along a helical axis that is perpendicular to the LC cell surface without external voltage (Fig. 1a). In this instance, part of the light that strikes the surface of the LC cell is transmitted, and some is reflected in accordance with the Bragg reflection requirements. However, because of their infrared reflection characteristics as composites, the Bragg reflection of CLCs in the P state may be disregarded [25]. When low voltage is applied, the CLC is separated into domains, each of which has a helical axis pointing in a distinct direction attaining a FC state (Fig. 1b). As a result, the refractive indices of the various domains are not matching. As a result, the randomly oriented LC molecules scatter incident light on the LC cell surface. Applying high voltage results in the H state (Fig. 1c). Here, the helical axis of the composite is unwound, a high voltage is delivered to the LC cell, and the LC molecules are oriented parallel to the direction of the electric field. Thus, the appropriately oriented LC molecules transfer the incoming light onto the LC cell while minimizing light scattering [25]. A vast array of technologies could be advanced by incorporating metallic frameworks [30–33] into CLCs. Next-generation photonic devices, sensors, and displays have intriguing new possibilities due to the unusual combination of metals improved optical, electrical, and mechanical capabilities with the tunable, self-organizing structure of cholesteric liquid crystals. The creation of more intelligent, effective, and adaptable materials for a range of scientific and industrial applications may result from this hybrid approach. To a certain extent, many research organizations have conducted various studies to improve CLC applications.

Thus, this chapter will explore into the mechanisms behind the enhancement of electro-optical and dielectric properties in QDs doped CLCs. The role of fundamental interactions, including the influence of QD size, concentration, and material composition on the properties of the host LC matrix has been explored. By analyzing recent

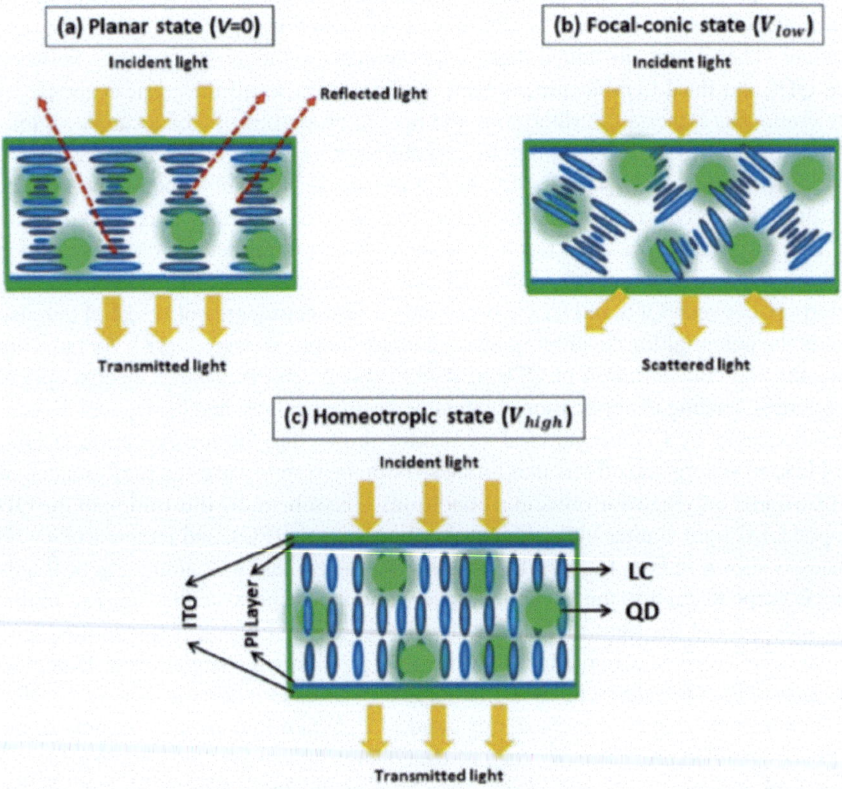

Fig. 1 Schematic illustration of the QD doped CLC light shutter in the **a** P state, **b** FC state and **c** H state. Reproduced with permission from Ref. [25]

experimental findings and theoretical models, a comprehensive overview of how quantum doping leveraged to develop advanced electro-optical devices with superior functionality for next generation energy efficient applications, has been given in this chapter. CLCs with inclusion of QDs could successfully enhance optical and electro-optical properties of these materials but there are some selection guidelines for the inclusion.

2 Selection Criteria for Interaction Between Cholesteric Liquid Crystals (CLCs) and Quantum Dots (QDs)

LCs with inclusion of QDs could successfully enhance optical and electro-optical properties of these materials but there are some selection guidelines for the inclusion of QDs.

2.1 Basic Properties of CLCs

- *Helical Molecular Arrangement*: CLCs are a tremendous class of material, whose properties are defined by an exceptional helical molecular arrangement. The helical structure leads to an amazing range of optical properties such as selective reflection, birefringence, optical activity, and polarized light transmission. Such properties give CLCs superiority over other materials in a wide variety of applications, like displays, optical filters, sensors, and many others [34].
- *Selective Reflection Ability*: One of the primary attributes of CLCs is their selective ability to reflect light. This reflection depends on the pitch of the helix and determines the wavelength of the reflected light. In this way, selective reflection is often termed as a "photonic band gap," which allows CLCs to reflect some wavelengths in their spectrum; for example, in the range from the visible, UV, up to the IR range [35, 36]. According to Bragg's Law, reflection by CLCs is described in terms of the proportionality of the wavelength of the reflected light with the pitch of a helical arrangement and the refractive index of the material [34]. The modulation of reflection accounts for why CLCs have been exploited for applications related to the creation of tunable reflective surfaces. Additionally, while metallic mirrors reverse the polarization of light on reflection, CLCs reflect without changing polarization type, and indeed circular polarization. Such a property is very useful in applications where circular polarized light is being used, for example, in 3D displays and advanced photonic devices [35].
- *Dynamical Tunability*: Another important feature of CLCs is their dynamical tunability. In CLCs, since it is a fluid-like material, its optical properties can change as a response to external stimuli, for example temperature or electric fields, and even light. For example, with an application of electric field, orientation of the

CLC molecules may be altered, thus changing the pitch of the helix and shifting the reflected wavelength of light. Such a tunable nature of CLCs allows for adaptive optical devices, such as the creation of tunable filters and smart windows, in which light transmission can be dynamically controlled [37].

- *Temperature Sensitivity*: The pitch of the helix varies as the temperature varies, and the reflected light wavelength shifts accordingly. By taking advantage of this, researchers have developed "smart" temperature-sensitive materials with a color change in a CLC film with temperature change. Such thermochromic behavior is applied in temperature sensors and indicators [35]. Furthermore, CLCs can be utilized in laser applications especially in laser dyes where with its helical structure, such material acts as a natural resonator for light waves since lasing can be very efficient at certain wavelengths [36].
- *Bistablity*: CLCs can exhibit bistability-that is, it retains two different states optically without the need for continuous energy supply. Bistability is one of the most attractive features in the low-power technology of display, in which the CLC can be switched into a reflective planar state or a light-scattering state under an electric field from its focal conic state. After it has been switched on, the CLCs remain in the state they were switch, till another electrical signal is applied to switch them on and off again. This saves energy, especially in such devices as e-paper displays [35].
- In addition, CLCs also exhibit versatility in mechanical properties. As naturally soft materials, CLCs can easily be integrated into flexible substrates, thus opening potential applications in flexible displays and wearable devices. Even when bent or stretched, their optical properties are maintained and make them superior to more rigid materials used in classical displays. Thus, in general, CLCs are superior because they have optical properties which can be adjusted, dynamic responses to numerous stimuli and high reflective capability with preservation of polarization [36].

2.2 QDs: Synthesis and Properties

2.2.1 Synthesis of QDs

Different synthesis methods of QDs represented in the Fig. 2, offer various advantages in size control, scalability, and environmental friendliness. Major synthesis methods commonly applied for QD production are discussed below:

- *Hydrothermal Synthesis*

Hydrothermal synthesis is a widely used technique for the production of QDs in an aqueous environment at higher temperature and pressure, thus favoring the nucleation process followed by the growth of nanocrystals. This approach has been proven to be the most optimized method for preparing water-dispersible QDs, as demonstrated

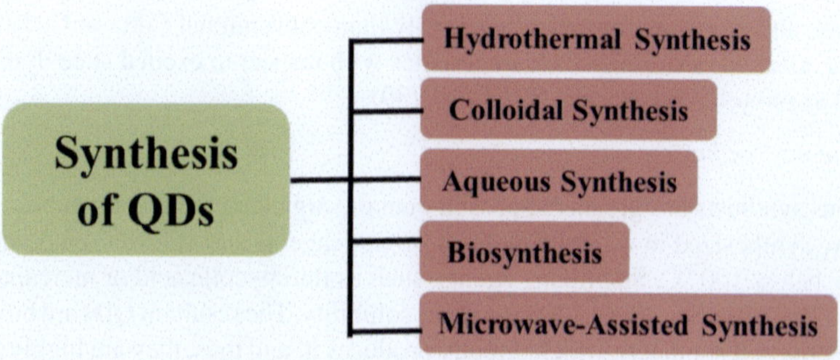

Fig. 2 Different methods for the synthesis of the QDs used as dopants in the CLC mixtures

with cadmium-based CdS QDs. In hydrothermal process, usually, a precursor solution containing cadmium chloride and sulfur-containing compounds like thiourea is mixed with a stabilizing agent that controls the size of the obtained QDs [38]. The reaction mixture is sealed in a Teflon-lined autoclave and heated to high temperatures of about 100–200 °C. Pressure and temperature conditions favor precursor material dissolution. Following nucleation, growth of QDs then occurs. The size and optical properties of the resulting QDs can be tailored by adjusting reaction parameters, such as precursor concentration, pH, and time. The pH of the reaction mixture usually determines the size of the resulting QD. High pH typically leads to a smaller QD, whereas low pH results in larger nanocrystals. The hydrothermal method is preferred due to its relative ease and scalability, thus supporting large-scale production of high-quality, crystalline QDs with strong photoluminescence. Hydrothermal synthesis comes with various advantages: operates in an aqueous solution. It is both eco-friendly as well as biocompatible. This method does away with the use of costly organic solvents. In addition, hydrothermally synthesized QDs are suitable for biological applications as they are soluble in water, and water solubility is a basic prerequisite for bioimaging and drug delivery [29].

- *Colloidal Synthesis*

The popular method for synthesizing semiconductor QDs is colloidal synthesis. This involves the in-situ synthesis of QDs in solution by dissolving metal precursors into organic solvents and then injecting chalcogenides such as sulfur, selenium, or tellurium into this environment. The process often happens at high temperatures of up to 350 °C, using various agents for stabilization, such as trioctylphosphine oxide (TOPO), to control QD growth and prevent aggregation [39]. A strange advantage of the colloidal method is its ability to give high-quality QDs with narrow size distributions and, therefore, highly tunable optical properties. Furthermore, the QD sizes can be varied within very broad limits by controlling the reaction temperature and precursor concentrations. The method can be scaled up; advances in continuous flow reactors make it suitable for industrial-scale production. The colloidal synthesis approach also becomes feasible for the preparation of Cd-free QDs, which, being less toxic and more environmentally friendly than conventional CdSe or CdTe QDs, possess different performance characteristics with respect to excited state dynamics as well as photoluminescence efficiencies [39].

- *Aqueous Synthesis*

Aqueous synthesis is a greener approach than the organometallic approaches. Here, QDs are synthesized in water using fewer toxic reagents and at lower temperatures, usually below 100 °C. Stabilizing agents, such as thioglycolic acid or mercaptosuccinic acid, are used to cap the QDs for water solubility. The resultant QDs are biocompatible aqueous synthesis, which directly produces it, and thus, they are highly fit for biological applications such as bioimaging and drug delivery. In fact, even at such low temperatures and with green solvents, aqueous synthesis can produce QDs with good photoluminescence and relatively narrow size distributions, though the latter may not be as crystalline as some of those produced by colloidal methods. This would place aqueous syntheses of QDs in a position to promise biomedical research with compatibility for water-based biological systems and clear environmental benefits [38, 39].

- *Biosynthesis*

Biosynthesis is a new approach that involves the use of microorganisms such as bacteria or fungi for the synthesis of QDs. Microorganisms can reduce metal precursors in a controlled environment to produce QDs. Biosynthesis shows very good control over the size and shape of QDs and their surface properties. Further, biosynthesis is very environment friendly because it avoids the use of toxic solvents and high temperatures [24]. Despite the initial development, biosynthesis of QDs has shown promising promise for the production of biocompatible QDs with unique surface characteristics. Applications in medicine and environmental monitoring, catalysis, may soon be granted as the size and shape of QDs can be selectively designed by biological processes [38].

- *Microwave-Assisted Synthesis*

This is a bottom-up approach wherein microwave irradiation provides fast heating effects within the reaction mixture, and QDs are prepared uniformly and quickly. Among the advantages, lie shorter reaction times, the use of less energy consumption, and QDs highly crystalline. Thus, rapid heating contributes to the better controllable size, and reaction conditions can be set to give monodisperse QDs with tunable optical properties. It is more useful in producing carbon quantum dots (CQDs) and other biocompatible QDs that find application in bioimaging, sensing, and drug delivery. It is also scalable and cost-effective, thus is appropriate for large-scale production [40, 41].

2.2.2 Properties of QDs

- *Quantum Confinement Effect*

QDs are nanoscale semiconductor particles that have extraordinary optical and electronic properties that mainly arise from the quantum confinement effect [42]. This is a direct result of the fact that QDs are quite small, in most cases consisting of a size in the nanometer range between 2 and 10 nm (nm), which confines charge carrier motion inside the particle, comprising of electrons and holes. That means the energy levels of them are quantized, leading to the light emission through specific wavelengths that could be effectively controlled by the alteration in the dimensions of the QD. More specifically, in decreasing the size of the QD, that energy gap between the electron and hole gets enhanced, which then leads to the emission of light at shorter wavelength, referred to as blue shift, while longer wavelengths might result if you increase the dimensions of the QDs, which is mentioned as red shift. This tunability in emission makes QDs particularly versatile for applications requiring specific wavelengths, such as in display technologies, optoelectronics and bioimaging [42–44].

Perhaps most importantly, QDs exhibit a quantum yield that is substantial essentially, they can absorb photons and re-emit them at significant efficiency. This suggests great promise for QD-based applications-from bright, stable sources of light, analogous to LEDs, solar cells, or lasers. Also, QDs have a narrow-emission spectral line width but with relatively broad excitation wavelengths, so a QD can be excited by a very broad range of wavelengths but emit light at a very specific and controlled wavelength. This property is quite important in multiplexed biological imaging applications where multiple QDs can be excited simultaneously using a single light source with different emission wavelengths to simulate the tracking of multiple markers of biological processes simultaneously. Furthermore, the QDs are not susceptible to photobleaching because they maintain luminescence over extended periods of time. Hence, QDs are suitable for long-time imaging and showing applications [44].

2.2.3 Concentration of QDs

- *Low Concentration*

At lower concentrations of QDs, you might see subtle modifications to optical properties (like photoluminescence), minor structural changes, and possible enhancement in electro-optical properties without significantly disturbing the cholesteric ordering.

- *High Concentration*

At higher concentrations, you could experience more pronounced changes, including scattering effects, defects in the cholesteric structure, potential aggregation, and altered electro-optical and mechanical properties.

It's important to carefully balance the concentration of QDs to avoid aggregation and preserve the desirable properties of both the liquid crystal and the QDs. Authors optimize the concentration of the QDs for best results such as [18] used CdSe QDs with 0.1 and 0.25 wt% in CLC composite and out of these, 0.1 wt% CLC cell showed better performance and [1] used 0.1, 0.3, and 05 wt% of Indium Phosphide/Zinc Sulfide (InP/ZnS) QDs concentration and stated that 0.1 wt% concentration is optimum for their experiment.

3 Effect on the Electro-Optical and Dielectric Properties of CLC

3.1 Electro-Optical Properties

The doping of QDs into CLCs brings about significant alterations of the electro-optical properties of CLCs and improves their performance in many application areas. The high value and size dependences of QDs optical properties and their strong photoluminescence strongly influence the cholesteric matrix, significantly modifying the response of the material to external stimuli, including electric fields, light, and temperature.

3.1.1 Enhanced Fluorescence and Circular Polarization

The circular polarization and fluorescence emission of the system are greatly improved by the QDs doped into CLCs [45, 46]. This is due to the QD photonic stop band of the CLC, which is the range of wavelengths the CLC reflects because of its helical structure. QDs, which have size-dependent photoluminescence, interact with this structure to produce a modulated emission that is very beneficial for optical devices [46].

The ability of QDs to emit light at a specific wavelength becomes size-dependent, and in the case of a CLC matrix, these particles become very effective because the selective reflection within the photonic stop band is caused by the periodic helical structure [45]. In such a system, inclusion of QDs would enhance the emission and make it brighter than an undoped CLC. This has been primarily ascribed to the coupling between the photonic modes of CLC and the excitonic emission of QDs. Hence, the emitted light becomes not only brighter but also enhanced polarization [47].

This interaction is particularly significant when there is a spectral overlap between the QD emission spectrum and the CLC stop band. Several experiments demonstrated that in such configurations, the spontaneous emission from QDs is modulated with a significantly faster decay time and a higher brightness emission. In some geometries, like planar alignments of CLCs, a new decay mechanism appears, making the emission process much faster and brightening it further. This property makes it very useful in applications such as tunable lasers, optical filters, photonic devices, etc., where control over the emission and polarization of light is especially important [48].

However, the singly out-coupled circular polarization of emitted light is another unusual advantage in QD-doped CLCs. In such CLC systems, inherently helical structure-selectively reflects circularly polarized light-isotropic emission of QDs leads to the idea of generation of circularly polarized fluorescence. This property is very useful in sophisticated optical applications, like 3D displays which need a depth perception by circularly polarized light, and communication devices reliant on polarized signals [47, 49]. This means that CLCs doped with QDs exhibit vast enhancement of fluorescence emission and include circular polarization control. Altogether, QD doped CLCs can be a very promising platform for applications requiring high brightness and polarization control since these are mainly most useful for lasers with tunable features, photonic devices, and complex displays. The combination of the QDs' tunable emission and the CLC's selective reflection creates a versatile system with broad applications in modern photonics [47, 49].

3.1.2 Reduction in Threshold Voltage

The addition of QDs into CLCs has brought great improvements with regard to the threshold voltage of material switching between different optical switching states, such as from the planar state to the focal conic state [50]. These enhancements are due to interference between the QDs and LC molecules that can consequently modify the overall electro-optical behavior of the system. Generally, in LCD, it switches between states—also in this context, especially for applications like displays, shutters, or tunable windows—require an electric field to orient LC molecules. The inclusion of QDs in this matrix changes the anchoring energy of the LC molecules. The energy must be used to reorient the LC molecules with or without reorientation [44, 47]. Kocakulah et al. in 2024 observed the Indium Phosphide/Zinc Sulfide QDs doped CLCs and stated that that doping of QDs decreases the threshold voltage from planar to focal conic state but 0.1wt% QD shows maximum reduction as shown in Fig. 3 [1].

The application of QDs such as CdSeS/ZnS in PSCLCs demonstrates an important reduction of the threshold voltage, largely because the anchoring strength holding the molecules there is weakened by the scattering of the LC molecules at the surface of the QDs. It reduces the effect of anchoring, allowing the molecules to flip much more easily in the desired orientation in response to an external electric field applied. For this, this molecule needs low voltage to activate the switching process; hence it is very energy efficient. It is one of the major reasons for this material being suitable for low power electronics application such as e-paper or reflective displays [44]. In addition to that, QD chemical structure could also contribute to threshold voltage tuning [50]. For instance, CdSeS/ZnS QDs' sulfur shell drops the time response of PSCLCs, thus offering improved efficiency in dynamic applications demanding quicker switching.

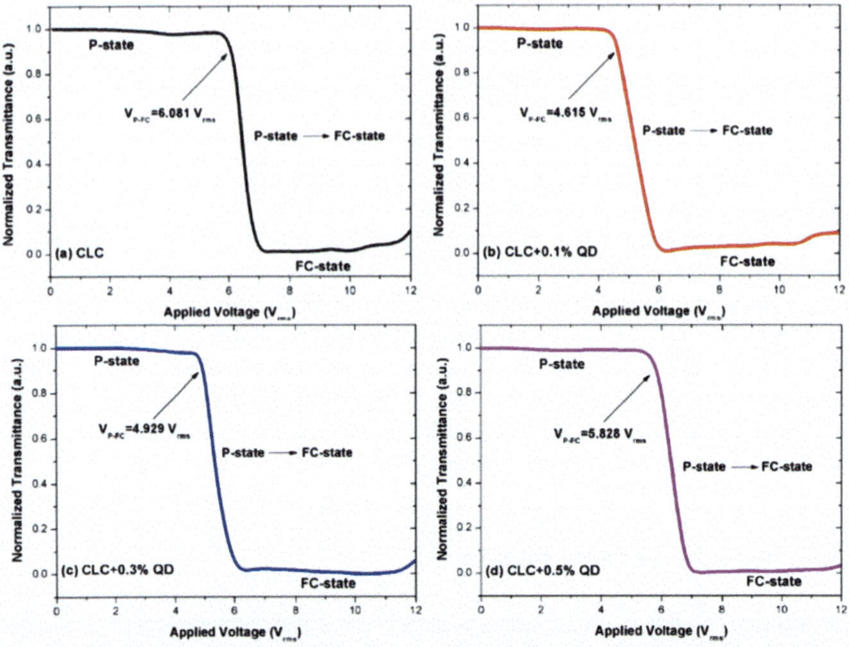

Fig. 3 Normalized transmittance versus applied voltage graphs of **a** CLC, **b** CLC + 0.1% QD, **c** CLC + 0.3% QD and **d** CLC + 0.5% QD samples. Reproduced with permission from Ref. [1]

3.1.3 Faster Response Times

Doping QDs into CLCs significantly improves the response times of such systems to external electric fields. This increased speed is an important feature of the fast modulation of light for applications in displays, optical switches, and smart windows. The interaction of QDs with LC molecules inside the CLC matrix would allow for more efficient realignment of LC molecules when an electric field is applied; this process would be faster when switching between optical states [44, 47].

For example, in the case of QDs doping, the response times are considerably enhanced for, PSCLCs. QDs that have a shell structure especially the type CdSeS/ZnS that considerably reduced the anchoring energy at the surface, which means less constraint on the movement of the molecular ensemble when an electric field is applied. These combined effects result in more rapid reorientation of the LC molecules; the anchoring energy is lowered and the dielectric constant increased with the QDs. This allows for more rapid switching between the planar/focal conic states to be achieved for the LC matrix. It is beneficial for their applications in light shutters and reflective displays [51].

Another reason for the reduction in response time is that QDs can alter the electro-optic environment of CLCs. General Dielectric enhancement by QDs usually decreases the energy barrier for molecular realignment. That, in turn, reduces the

Table 1 Response time for CLC and QDs doped CLC samples

CLC cells	τ_r (ms)	τ_d (ms)	τ_R (ms)
CLC	12.739	1.883	14.622
CLC + 0.1% QD	10.760	2.173	12.933
CLC + 0.3% QD	10.803	2.432	13.235
CLC + 0.5% QD	11.911	2.130	14.041

Reproduced with permission from Ref. [1]

time necessary for the LCs to switch through the various configurations of optics in response to an electric field. The faster switching capability is particularly useful in dynamic display technologies where the high refresh rates are crucial for life-like image transitions, and in optical modulation devices where rapid response to electrical inputs is crucial.

Furthermore, high-speed applications get better response times. It is in such applications like the tunable windows or the optical shutters where switching to a more transparent and opaque state is faster that improve the performance as well as the user experience of the devices. Energy efficiency also has a factor for responsiveness. With quicker switching, the device devours less energy for its improvement and prolongs its lifetime.

In summary, doping QDs into CLCs has a huge impact on speeding up the response time of material to electric fields applied. In light of the applications above, that is, dynamic displays, optical switches, and the like tunable windows where speedy transition between optical states is paramount, thereby improving performance [44, 47]. In 2024, author used Indium Phosphide/Zinc Sulfide (InP/ZnS) QDs in 0.1, 0.3 and 0.5 wt% and calculated τ_r (rise time) and τ_d (decay time) represented in Table 1 provides τ_r, τ_d, and τ_R (response time) values for all values [1]. The table shows that the InP/ZnS QDs contribution caused the τ_r value to decrease and the τ_d value to increase. Furthermore, following the dispersion of InP/ZnS, the τ_R value of QD-doped CLC composites decreased. However, as the concentration of InP/ZnS has increased, so has the value of τ_R. Furthermore, τ_R value of CLC + 0.1% QD has reduced in comparison to pure CLC, while CLC + 0.3% QD and CLC + 0.5% QD have somewhat increased in comparison to CLC + 0.1% QD.

3.1.4 Stabilization and Reduced Phase Separation

Doping of QDs in CLCs enhances the stabilization of the liquid-crystalline phase since phase separation, one of the biggest challenges of the LC technology, is taken care of. Phase separation can occur in undoped LCs due to inherent molecular interactions within the matrix and over time leads to performance degradation. However, QDs act as stabilizers when introduced into the CLC matrix. This skips the phenomenon of phase separation, thereby offering stability [47].

It is found that interaction between the QDs and LC molecules creates a more uniform distribution within the matrix, mainly due to the surface properties of the

QDs, generally capped with stabilizing agents compatible with the LC molecules. These surface ligands help to anchor the QDs within the LC matrix in a way that the aggregation of the LC molecules is avoided, and the structure, in general, remains stable. Hence, the material exhibits better phase stability and maintains its optical and electro-optical properties for a relatively longer period [44].

In addition, the effect of stabilization of QDs reduces the formation of defects within the CLC matrix. These can occur from irregular molecular alignment or mechanical stress from the outside. QDs smear the imperfections, thus creating a more ideal arrangement of molecules. This is critical for usage in devices where defects may otherwise scatter the display of reflective displays or optical filters due to their performance and clarity being compromised. QDs enhance the molecular stability of the CLC, thereby providing higher quality and durability of the material and ensuring consistent performance at a wide range of environmental conditions.

Another important benefit of this stabilization contributes to the enhancement of the thermal stability of the CLC-QD composite. The thermal resistance of the LC phase can be enhanced with QDs, in which material maintains the optical properties in a wider temperature range, hence essential in devices required to function in fluctuating or extreme temperatures-for example, outdoor displays or temperature-sensitive sensors.

Finally, QDs embedded CLCs have significantly improved material stability in the phase with negligible risks associated with phase separation ensuring long-term performance. An improved stability combined with better thermal resistance and defect reduction makes ideal QD-doped CLCs suitable for applications which demand high durability and consistent optical performance such as displays, optical filters, smart windows [47].

3.1.5 Spectral Modulation

These QDs doped CLCs have revealed some interesting properties of modifying the spectral properties of the material, that is, controlling emission and reflection of light. This arises from the interaction of QDs with the CLC helical structure, which acts as a photonic bandgap. The photonic bandgap is the wavelength range that the CLC reflects due to its periodic molecular configuration. The introduction of QDs into this system modulates the emission characteristic(s), depending on the size through the CLC matrix, to achieve a more controlled and tunable output [44, 47].

The most important modulation aspect behind this is that QD-doped CLCs can even modulate the emission wavelength by response to many external stimuli such as electric fields, temperature, or light. These have been demonstrated to influence the CLC's helical pitch, which determines the photonic band gap, and the emission wavelength of QDs could be shifted with it. Author show how chemically produced core–shell CdSe/ZnS QDs implanted in a one-dimensional photonic cavity made of a CLC matrix may be modulated in terms of both spectrum and polarization [52].

This results in a highly tunable system wherein the wavelength of emitted light can be controlled to high precision by adjusting the pitch of the CLC or by controlling the external conditions applied to the system.

Upon overlap between the emission wavelength of QDs and the reflective band of the CLC, it brings about resonance that enhances the emitted light, thus leading to increased intensity. This has applications in optical devices where bright output of light is required, especially through tunable lasers or display technologies. QDs doped CLCs display some unique spectral modulation properties whereby the light output can occur within a given range of wavelengths determined by QD size and concentration along with the CLC matrix tuning.

Finally, it would be useful for the development of very advanced optical filters to be able to control the spectral properties of QD-doped CLCs. This could be done by appropriately choosing the size and type of QDs and modulating the CLC pitch to design such systems to selectively reflect or transmit specific wavelengths of light. This is particularly advantageous for applications in photonics where the reflection and transmission of light need to be well controlled, especially in wavelength-selective filters, sensors, and optical communication devices.

In summary, adding QDs to CLCs is able to enable high spectral modulation. An interaction between the photonic bandgap of the CLC and the tunable emission properties of the QD leads to a system with controllable light emission and reflection. Spectral tunability is particularly useful in a wide range of high-tech applications, including laser technology, optical filters, sensors, and photonic devices, where fine control of wavelength is critical [44].

3.1.6 Photonic Bandgap Manipulation

The introduction of QDs into CLCs significantly modifies the photonic bandgap manipulation. A photonic bandgap is one of the prominent features of CLCs; a helical molecular arrangement results in the selective reflection of light at specific wavelengths. This selective reflection results from the cholesteric structure's periodic nature; a one-dimensional photonic crystal that reflects certain wavelengths while allowing others to pass. Doping QDs into CLCs will alter the photonic bandgap, creating a more material-compliant and more tunable system within which these materials may be suitable for complex photonic applications [53].

Optical properties of QDs depend on their size and can be tuned by altering the size of QDs. However, when doped inside the cholesteric matrix, the QDs interact with the photonic bandgap of CLCs, shifting the reflection wavelength. Due to QD size, concentration, and distribution in CLC, this variation may result in widening the photonic bandgap or a change in the central reflection wavelength. Such tunability becomes critical for particular applications that rely on the manipulation of light, especially in areas concerning tunable lasers, wavelength-selective filters, and optical switches [47].

With this interaction of QDs with the CLC matrix, the wavelength reflecting ability of the system is improved. This is because the addition of QDs brings in local changes to the refractive index of CLC and thereby changes the light-helical structure interaction. Hence, at an increased reflectivity level, photonic control over a wider spectral range is available in QD-doped CLCs. This broadened photonic bandgap allows improved light management in optical devices. This should lead to better performance and further applicability of such devices in photonic communications, display technologies, and optical sensing [44]. Doping CLCs with QDs would lead to the dynamic tunability of their photonic bandgap by using external stimuli such as temperature, electric fields, and mechanical pressure such as Mochalov et al. in 2012 stated that the proposed a Polymer Stabilized QD-doped CLC materials with tunable and wide photonic band gap due to doping of QDs [49]. This kind of dynamic tunability is highly valuable for adaptive optics and smart materials, where the control of light reflection or transmission in real time is important. Such systems can be applied in applications such as tunable optical filters and adaptive lenses, which require environmental adaptability because their ability to tune photonic bandgap corresponds to responding to the external factors.

In summary, the QDs doping into CLCs increases the degree of control and flexibility of photonic band gap. Mutual interaction enables adjustable reflection of transmitted light, which is essential for application in advanced optical and photonic technologies. The ability to tune the photonic band gap through QDs makes QD-doped CLCs highly suitable for next-generation optical technologies such as tunable lasers, adaptive optical devices, and high-performance photonic systems [44, 47].

3.2 Dielectric Properties

Various types of QDs doped in the matrix resulted in considerable change of dielectric properties of CLCs with improved electro-optical and dielectric responses [1, 25, 54]. Such changes can be traced in terms of the real (ε') and imaginary (ε'') parts of the dielectric constant, relaxation time, dielectric anisotropy, and conductivity.

3.2.1 Increase in Dielectric Constant (ε')

Subsequent successful doping of CLCs with QDs led to a considerable increase in the real part of the ε' (dielectric constant), which is associated with the energy stored by the material after the application of an electric field. This increase is most significant at lower frequencies, where the ε' is often maximized. ε' increased as a result of the interaction between LC and QD, thus increasing polarization inside composite material [1, 25].

Here, QDs in CLC-QD systems play the role of polarizable inclusions, increasing the strength of the local electric fields. As a result, the arrangement of molecules to be better aligned by dipoles improves. An interaction between LCs and QDs through their dipoles increases polarization within the system and, thus, ε'. The orientation of LC molecules would increase with an increasing local field, but this is very well pronounced for the homeotropic (H) state, where the molecules are parallel to the electric field. In this state, it would top out at the maximum dielectric constant because of the optimal alignment in the system [25].

Meanwhile, when the system is in the focal conic state, disordered molecules in LCs result in much weaker interaction with QDs. As a consequence, the dielectric constant decrease compared to the homeotropic state. The difference of ε' between the H and FC states also indicates the potential of QDs to enhance molecular orientation and stabilize the electro-optical properties of the material over a change of electric field strengths [1]. The ability to manipulate the dielectric constant in high precision and fine-tuning is achieved with the concentration of QDs, ensuring an added advantage for optimizing device performance under CLC-based devices. Adjustable QD doping concentration controls the degree of polarization in the system of LCs and tailors its material properties to suit specific applications that require higher dielectric constants for better energy storage and electro-optical responses.

In conclusion, doping QDs into CLCs leads to an increase of the dielectric permittivity and exhibits dielectric memory. Thus, allowing better storage of electrical energy in the material. The presence of dielectric memory in CdTe-QD-doped LAHS19 material is represented in Fig. 4. The figure clearly shows that the switched state for CdTe-QDs doped LAHS19 was preserved even after the bias was removed, but in the case of pure LAHS19 (Fig. 4a), it instantly returned to its original state. Similarly, dielectric memory is shown in CdTe-QDs doped LAHS18, FLC6304, and KCFLC 7S materials in Fig. 4b–d. As can be observed from the insets of Fig. 4b–d, it is clear that all FLC materials exhibit dielectric memory by doping CdTe-QDs, while pure material does not. Studies on the memory effect in pure FLC 6304 have revealed that dielectric tests do not reveal any memory effect [55]. This also gives rise to a great potential application of this kind of composites within electro-optical devices. This improvement might be connected with interactions between the QDs and LC molecules, especially when alignments of the molecules are optimized-like in the case of the homeotropic configurations [1, 25].

3.2.2 Reduction in Dielectric Loss (ε'')

The doping of CLCs with QDs also strongly decreases the dielectric loss, which is a measure quantified by the imaginary part of the dielectric constant ε''. Dielectric loss describes the energy loss realized in a material subjected to an alternating electric field. It is also well known that CLCs generally exhibit greater dielectric loss values at lower frequencies because the molecules responding to the applied electric field reorient more slowly. Nevertheless, QD inclusion into the CLC matrix lowers dielectric loss, suggesting more rational energy distribution and less dissipation in the system [1]. The possible reasons for decreased dielectric loss after QD doping are likely as follows: first, added QDs assist in further alignment of LC molecules within the cholesteric phase, which increases the ordered interaction between molecules. This ordering can reduce molecular friction and the energy that would otherwise

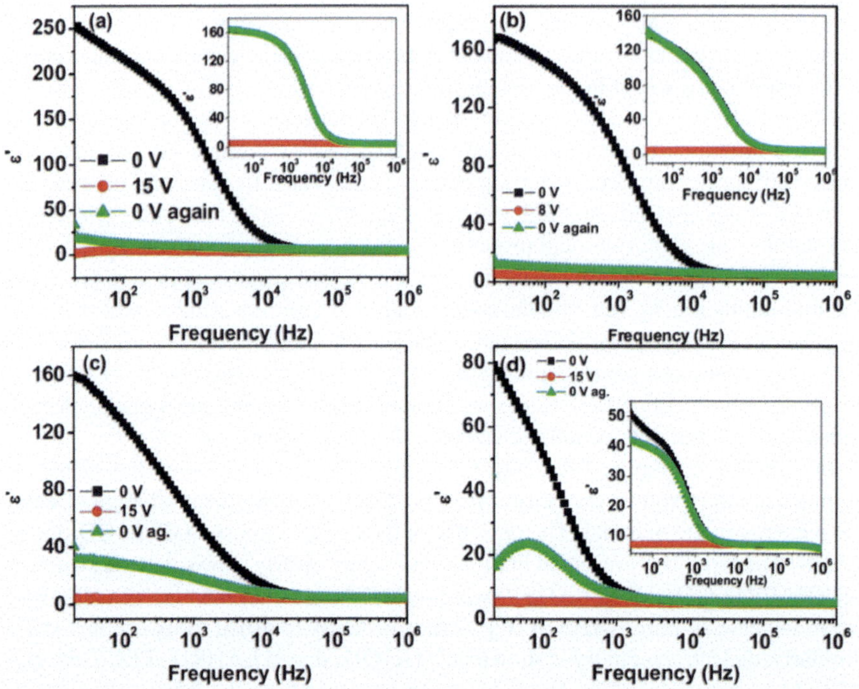

Fig. 4 Dielectric permittivity at different DC biases of CdTe-QD doped FLCS: **a** LAHS19, **b** LAHS18, and **c** FLC 6304 in 4 μm cells; **d** KCFLC 7S in the 10 μm cell. Reproduced with permission from Ref. [55]

be lost in the reorientation process. Consequently, dissipation of energy within the system is decreased, expressed as lower dielectric loss [25].

But the surface properties of QDs, like the existence of stabilizing ligands, will determine the efficiency of dielectric loss reduction. The surface ligands on these QDs interact with LC molecules in their proximity to permit smoother reorientation under an electric field and avoid molecular clumping or misalignment, which might otherwise increase energy dissipation. The QDs also decrease hotspots or localized regions of high energy dissipation through their contribution towards a more uniform electric field distribution within the CLC matrix. The other key factor in this reduction is the concentration of QDs in the CLC matrix. When the QD concentration reaches an optimal value, the nanoparticles can be uniformly dispersed throughout the material, and their stabilizing effects can be maximized, hence minimizing the energy loss. It is worth noting that a too high QD concentration leads to aggregation of the QDs, which complicates the uniform alignment of the LCs and increases the dielectric losses. In order to further facilitate maximum reduction of the dielectric loss in electro-optical performance(s), precise optimization of QD concentration is needed. In conclusion, QD incorporation in CLCs leads to a highly reduced dielectric loss of the system, mainly due to enhanced molecular alignment upon doping with QDs

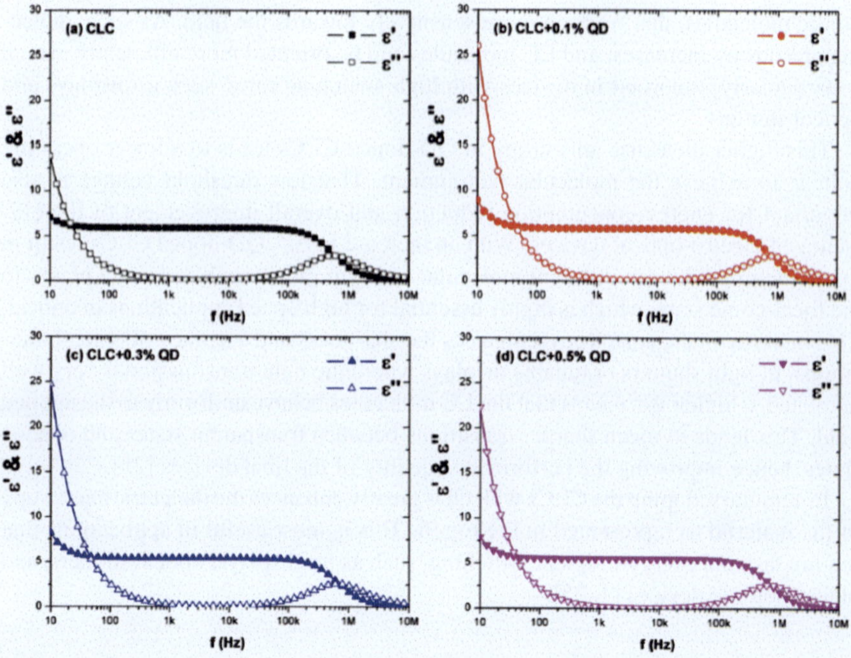

Fig. 5 ε′-f and ε″-f graphs of **a** CLC, **b** CLC + 0.1% QD, **c** CLC + 0.3% QD and **d** CLC + 0.5% QD samples. Reproduced with permission from Ref. [1]

along with greater uniformity of the electric field within the composite material. This is extremely beneficial to electro-optical devices since it brings energy efficiency and enables efficient running [1, 25]. As shown in the Fig. 5, the values of ε′, and ε″, of CLC increased with the contribution of InP/ZnS QD at low frequencies.

3.2.3 Changes in Dielectric Anisotropy (Δε′)

Doping QDs into CLCs increases the value of dielectric anisotropy Δε′. The value of dielectric anisotropy Δε′ is the difference in the values of dielectric constants measured along different axes of the LC molecules. Dielectric anisotropy is an important property in LC systems as it determines the behavior of the material when it has already been submitted to an electric field. An increase in Δε′ in QD-doped CLC systems improves the electro-optical performance of the material, especially concerning switching behavior and alignment under electric fields [1, 25].

Therefore, dielectric anisotropy assumes an important role for the fluidic properties of reorientation within CLCs under the influence of an applied external field. In QDs doped samples, the local electric field within the material gets altered and further, compared with only LCs, QDs enhance the interaction between the fluidic molecules and the applied field, thus promoting better directional alignments of the fluidic molecules, that respond more sensitively towards the field. As such, dielectric anisotropy increases, and LC molecules can be oriented more efficiently, which is particularly important in devices with high switching rates, such as displays and optical shutters.

This higher dielectric anisotropy in QD-doped CLCs leads to a lower operating voltage to achieve the molecular realignment. That low threshold voltage is also important for energy consumption reduction and overall improvement of the efficiency of electro-optical devices. With an increase in $\Delta\varepsilon'$, QD-doped CLCs result in fast switching between different molecular configurations, such as from a planar to the focal-conic state, which is highly essential for high-speed applications in optical. This increase in the anisotropy, enhances the sharpness and accuracy of optical transitions. In light shutters or tunable displays where the light transmission is very well regulated, a higher $\Delta\varepsilon'$ means that the LC molecules behave uniformly to the applied field. This tends to mean sharper transitions between transparent states and opaque states, hence improving the performance quality of the final devices [25].

In summary, doping the CLCs with QDs greatly enhances the dielectric anisotropy of the material as represented in the Fig. 6. This is most useful in applications that require fast and energy-efficient switching, such as for displays, optical shutters, and other photonic devices [1, 25].

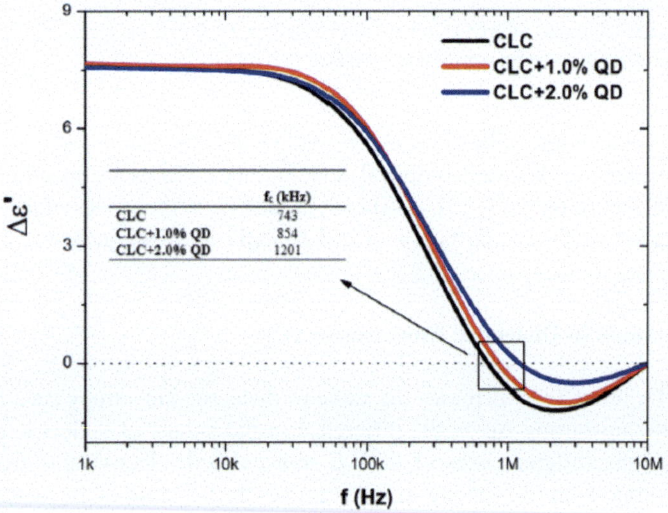

Fig. 6 Dielectric anisotropy of mixtures from 1 kHz to 10 MHz frequency range. Reproduced with permission from Ref. [25]

3.2.4 Improvement in Conductivity (σac)

The introduction of QDs into CLCs leads to a significant increase in the AC conductivity of the material, which directly influences the ability of the material in conduction electric currents upon an applied alternating electric field. The basic reason for this large increment is due to the semiconducting nature of the QDs, which introduces some new charge carriers into the system. These charge carriers, which are injected by QDs, also increase the electrical conductivity of the CLC composite in their entirety at lower frequency as shown in the Fig. 7 [1].

The semiconducting characteristics of QDs like CdSe or CdS allow them to provide free electrons or holes in the system that raise the transportation of charges across the CLC matrix. This is most important at low frequencies because the mobility of charge carriers increases and makes the system highly responsive to the applied electric field. Enhanced charge carrier concentrations improve charge separation and transport, contributing to high conductivity values than in undoped CLCs [18].

Liquid crystal molecule alignments within the cholesteric phase are, therefore, also influenced by the existence of QDs. These act as nucleation sites to help the molecules of LCs align so that conductivity can be improved. The better alignment reduces the resistance of internal charge flow; hence, transportation of currents inside the CLC-QD system is enhanced. It may add some benefits to applications in responsive electro-optical behavior, such as tunable display devices or electrically modulated light shutters [25]. At higher frequencies, the conductivity stabilizes and exhibits less variability between QD-doped and undoped CLCs. This is because, at high frequencies, the LC molecules are less able to rapidly reorient to react to the oscillating electric field and thus tend towards having a more stable conductivity

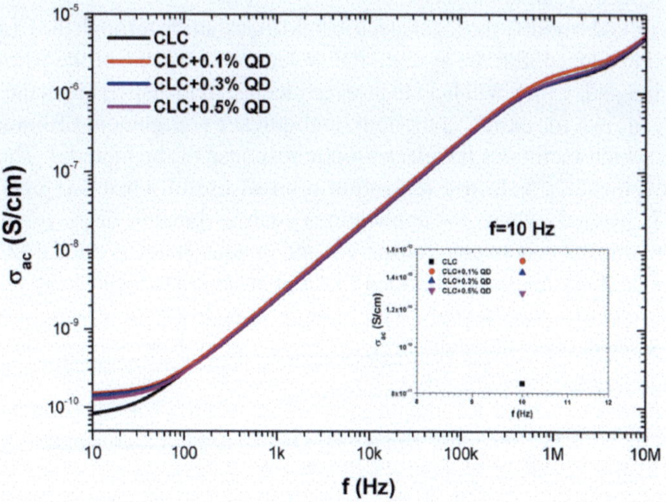

Fig. 7 Conductivity of the CLC materials from 10 Hz to 10 MHz frequency range. Reproduced with permission from Ref. [1]

along the frequency spectrum. However, the enhancements at lower frequencies are notably highly desirable in low-frequency electrical modulation applications, such as in low-power electro-optical devices.

Concluding, the in doped QDs of CLC increases the AC conductivity at relatively lower frequencies. That's due to semiconducting nature of the QDs, which may introduce additional charge carriers into the system. Therefore, this results in enhanced charge transportation and better ordering of the LC molecules in QD doped CLCs, making them highly suitable for advanced electro-optical applications with efficient electrical modulations [1, 25].

3.2.5 Dielectric Relaxation

The CLCs doping highly influences the dielectric relaxation process, making its times of relaxation to become faster. The dielectric relaxation is defined as the time taken by the LC molecules to settle back in their equilibrium orientation after an external electric field has been removed. This would be much slower in usual LC systems, especially when the electric field destroys the molecular alignment. The overall dynamics of molecular realignment, however, are improved with the inclusion of QDs to CLCs, and relaxation times are quicker [1, 25].

The faster dielectric relaxation in doped CLCs is because of the interaction between QDs and LC molecules. QDs act as alignment centers or nucleation sites in the liquid–crystal matrix. They thus stabilize the molecular alignment even in an applied external electric field, so that the friction and resistance due to reorientation are decreased and thus the molecules can quickly return into their equilibrium state after the removal of the electric field. Therefore, the system presents quicker response times and reduced relaxation periods.

Moreover, QDs modify the local electric field distribution through the CLC matrix. If QDs are introduced into the system, the dielectric landscape of the whole system would be changed, which will lead to an even electric field dispersion in the material. Uniformity allows for easier, consistent, and quicker realignment of liquid–crystal molecules, which facilitates the electro-optic response of the material. The remarkable improvement in dielectric relaxation is most useful when fast optical states switching is needed, where for applications such as tunable displays and optical shutters, the transition between transparent and opaque states is critical [25].

Faster relaxation times in QD-doped CLCs also lead to reduced energy consumptions inside electro-optic devices. The system takes lesser time and, hence lesser energy, to reset between the switching cycles as the molecules revert back to their original state sooner. Energy efficiency is a valuable advantage for continuously operating devices such as LC displays or smart windows as they reduce energy consumptions with fast response times [1]. Distribution and concentration of QDs within the matrix are also factors that will affect the relaxation process in QD-doped CLCs. A proper dispersion allows for the highest effect of stabilizing effects from QDs, thus more uniform molecular alignment and faster dielectric relaxation ensues. Higher concentration of QDs, however might cause aggregation that can interfere with the uniform alignment of LC molecules; thus, it can slow down the relaxation process. Optimization of the QD concentration is, therefore a very crucial task to be taken in order to improve the dielectric relaxation in an appropriate manner without any performance drawback Thus, QD into CLCs enhances dielectric relaxation in the presence of fast realignment of the molecules subsequent to the removal of the electric field. This is due to stabilization by QDs acting as alignment centers and also altering the dielectric medium that enables more rapid responses. These alterations make QD doped CLCs highly suitable electro-optical applications in which rate and efficiency in terms of energy are most important [1, 25] (Table 2).

4 Applications of QDs Doped CLCs

As shown in the Fig. 8, there are many applications of the QDs doped CLC mixtures. QDs might be used as active emitters within the CLC matrix, taking benefit from the LC ability to control the light flow while boosting the brightness and purity of the color throughout the whole system [52, 61]. Just in such a scenario, the application of such mixtures can exist in tunable lasers and reflective displays, where definite features of light emission and reflection must be maintained. Of course, besides these features, QDs also could reduce the threshold switching voltage from one optical state to another, hence enhance electro-optic behavior for application purposes of such systems in displays and light modulation devices [43]. The QDs interaction with the helical structure of the CLC causes a development in brightness and also the possibility of tuning the wavelength of emission [44]. Meanwhile, doping by QDs also in LC systems can increase electro-optical properties due to a decrease in the threshold voltage required for a change in the state of the LC between different optical states [62]. This produces higher efficiency in modulation of light, which remains a necessary demand in many devices, including LCDs, optical switches, and light shutters [43]. QDs stabilize the LC structure due to the formation of clusters that it interacts with, and such interactions enhance the structural stability of the material, even though being a responsive nature [44].

- *Displays Devices*

QD-doped CLC systems have also been reported to exhibit excellent optical and thermal stability. Incorporation of QDs improves the general stability of the material, making it possible for the system to be used as a material in long-term applications and ensuring stable optical performance under changing environmental conditions. For instance, experiments in 2023 by Mani et al. demonstrated that the dispersion of CdS QDs within a CLC matrix led to improved structural stability and thermal properties of the material, promising such composites as good materials for use in robust optical devices [51]. This stability, along with superior optical properties, places QD-doped CLCs in crucial positions such as anticounterfeiting labels, tunable reflective

Table 2 Summarized different properties studies by different author in recent year with QDs as dopant in CLC

References	Material used			Result found with doping of QDs
	Liquid crystal	Chiral dopant	Quantum dot	
Kocakulah et al. 2024 [1]	E7	S-811	Indium phosphide/zinc sulfide (InP/ZnS)	Dielectric constants ↑es AC conductivity ↑es Threshold voltage ↓es Response time, ↓es Optical band gap ↓es at low frequencies
Chauhan et al. 2023 [19]	8CB	CB15	CdSe QDs (3.8 nm)	Dielectric permittivity ↑es Threshold Volatge ↓es Saturation Voltage ↓es
Mani et al. 2023 [51]	Cholesteryl Palmitate (cholesteric 97%) and 4′-Pentyl-4-biphenylcarbonitrile (nematic, 98%	Chiral dopant: 4′-Pentyl-4biphenylcarbonitrile	CdS QDs (25–35 nm)	Optical band gap ↓es UV absorption ↑es Phase transition temperatures ↓es mechanical stability ↑es
Kocakulah et al. 2022 [25]	5CB	S-811	CdSeSZnS (6 nm) (1 and 2 wt%)	$V_{P\text{-}FC}$ and $V_{FC\text{-}H}$ switching voltages ↓es Response time ↓es The resonance frequency ↓es The critical frequency ↑es
Gollapelli et al. 2022 [24]	E7	R5011	Carbon QDs (2–4 nm)	Microdroplets display distinct colors in daylight and UV light and Isotropic transition temperature ↓es
Chemingui et al. 2022 [22]	SLC1717	CB15	CuInZnSe-QDs (2 nm)	Optical band gap ↓es PL intensity ↑es UV absorbance ↑es

(continued)

Table 2 (continued)

References	Material used			Result found with doping of QDs
	Liquid crystal	Chiral dopant	Quantum dot	
Pathak et al. 2021 [56]	Cholesteric liquid crystal (CLC)		Octadecylamine-capped CdSe/ZnS QDs (6 nm)	Optical band gap ↓es PL intensity ↑es UV absorbance ↑es
Yao et al. 2021 [57]	The Ch polysiloxane side-chain liquid crystal polymers (ChPSCLCP) P1	R811	CdSe/ZnS QDs	The reflection band shifted about 200 nm with temperature changes The thermal stability observed above 326 °C for fabricated films
Chen et al. 2018 [58]	MDA-03–3970	S811	CsSnI 3 QDs (3~5 nm)	Fabricated Laser with: Low threshold (150 nJ/pulse) Narrow linewidth (0.20 nm) Stable for both humidity and temperature

(continued)

Table 2 (continued)

References	Material used			Result found with doping of QDs
	Liquid crystal	Chiral dopant	Quantum dot	
Cao et al. 2017 [59]	HTG135200	S5011	GQDs (5 nm)	Tunable amplified spontaneous emission (ASE) in laser and display devices. GQD-CLC material is simple, low-cost, and photostable ASE triggered at pump energy greater than 1.25 mJ cm^{-2} Emission peak shifts from 662 to 669 nm with temperature increase temperature changes from 50 to 90 °C PL intensity ↑es
Kumar et al. 2016 [60]	BL036	CB15	CdSe Q Ds (6–7 nm)	Circularly polarised fluorescence intensity ↑es Threshold Voltage ↓es Saturation Voltage ↓es
Mochalov et al. 2012 [49]	E48	CB15	CdSe/ZnS (2.5 or 4.5 nm)	Photonic band gap ↑es Dissymmetry factor ↓es

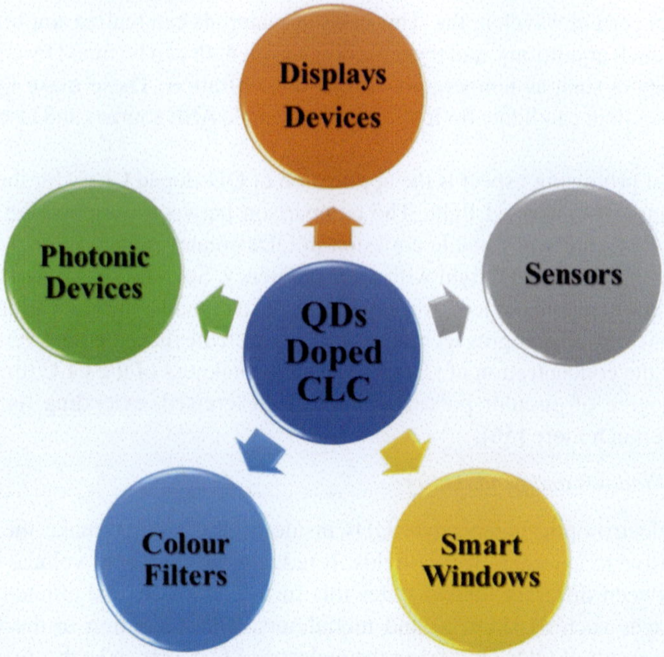

Fig. 8 Different applications of the QDs doped CLC

coatings, and long-lasting displays. Such systems have potential in the future technologies like tunable lasers, sensors, displays, and light modulation devices [44, 51, 60, 63].

- *Photonic Devices*

The major benefit of doping QDs into CLCs is the enhanced photonic bandgap of the CLCs. The ability of CLCs to selectively reflect light at a certain wavelength has been dictated by their helical pitch and can be finely tuned if QDs are introduced into the system. Moreover, due to the size-dependent emission properties of QDs, they have an ability to interact with photonic modes of the CLC and, hence, broaden the photonic bandgap and intensify and make the process much more efficient. This interaction makes QD doped CLCs particularly suitable for tunable laser applications where wavelength control becomes critical [64].

The combination also leads to higher overall luminescence of the system. In reality, QDs are excellent light-emitting particles with significant quantum yields. By their inclusion in CLCs, their photoluminescence increases and, therefore creates a CPL system, especially applicable to three-dimensional displays, optoelectronic devices, and even information storage, where polarization and luminescence must be fully controlled [64]. The interaction between QDs and CLCs is also applied to the development of tunable ASE sources. Doping ASE devices in CLCs makes it possible to fabricate ASE devices that can operate at very low pump energy thresholds with efficiency at certain wavelengths. This class of materials can realize amplified emission under such conditions, and the emission wavelength can be tuned by controlling external factors such as temperature or QD concentration. These make QD-doped CLCs an excellent candidate for low-cost, photostable ASE sources and laser devices [44].

The most promising aspect is the application of QD-doped CLCs for the production of circularly polarized light. The comparison between the photonic bandgap effect in a CLC and size-tunable emission in QDs creates a material that is known to emit circularly polarized light with great efficiency. Such properties are critical to next-generation optical devices, from 3D displays to optical communication systems and perhaps even bio-imaging technologies. However, the recent studies verified that, by varying the concentration of QDs and also the thickness of the CLC film, control over the degree of circular polarization can be exercised, extending its scope of application much more [56].

- *Smart Windows and Modulators*

From an electro-optic perspective, QDs inside a CLC matrix make the material more sensitive to external electric fields. It reduces the threshold voltage to a state change between different optical states this further increases the efficiency of the LC in devices such as shutters, and modulators. This reduction in the threshold voltage is a result of QDs modifying the dielectric LC matrix in such a way that the system becomes more responsive to applied fields. More recently, QD doping has also been reported to improve switching times, which is highly useful for high-data rate applications requiring temporal modulation of optical signals [63]. For instance, addition of CdSe QDs to polymer-stabilized cholesteric liquid crystals (PSCLC) has been reported to significantly enhance electro-optic switching behavior and reduce threshold voltages while improving optical performance [60]. Figure 9a–c represents the macroscopic photographs of the QD doped CLC composite in the P state (light that strikes the surface of the LC cell is transmitted, and some is reflected in accordance with the Bragg reflection requirements), FC state (LC molecules scatter incident light on the LC cell surface), and (c) H state (LC molecules transfer the incoming light onto the LC cell while minimizing light scattering) for the light shutter application [25].

- *Color Filter and Sensor*

Addition of QDs into CLCs, with their helical molecular structures able to selectively reflect light, enables not only a functional improvement in the photonic properties of the system but also modification of the photonic bandgap of CLCs that defines the window of wavelengths reflected [52, 59] for Color Filter and Sensor application.

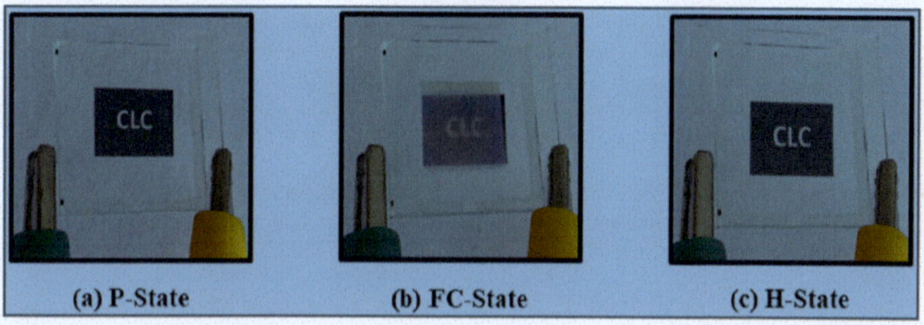

Fig. 9 Macroscopic photographs of the QD doped CLC composite in the **a** P state, **b** FC state, and **c** H state. Reproduced with permission from Ref. [25]

5 Conclusion

QDs interactions with CLCs present significant improvements in many electro-optical and dielectric properties. The alterations in photonic bandgap change, the increase in fluorescence intensity, lower threshold voltages, and inducing circularly polarized light by QD-doped CLC systems make such materials extremely valuable for applications in advanced photonic and optoelectronic devices. Enhancement in the dielectric permittivity and anisotropy of the CLC materials opens new area of applications like in storage devices etc. The decreased response time of the QDs doped CLC mixture increases impact in the smart window applications. Table 2 represents a summarized table with different QDs in different application with CLC. for Thus, QDs in CLC increases the quality of the applications.

References

1. Kocakülah, G., Köysal, O.: Determination of optical band gap, transmittance, and dielectric constant parameters of indium phosphide/zinc sulfide quantum dot doped cholesteric liquid crystals. Opt. Mater. **149**, 115101 (2024). https://doi.org/10.1016/j.optmat.2024.115101
2. He, Z., Zeng, J., Zhu, S., Zhang, D., Ma, C., Zhang, C., Yu, P., Miao, Z.: A bistable light shutter based on polymer stabilized cholesteric liquid crystals. Opt. Mater. **136**, 113426 (2023). https://doi.org/10.1016/j.optmat.2022.113426
3. Zhang, W., Guan, H., Hu, Y., Wang, W., Liu, F., Yang, X., Hao, L.: Advances in two-dimensional (2D) inorganic chiral materials and 2d organic-inorganic hybrid chiral materials. Curr. Chin. Sci. **3**(4), 293–308 (2023). https://doi.org/10.2174/2210298103666230406095730
4. Jain, A.K., Deshmukh, R.R.: An overview of polymer-dispersed liquid crystals composite films and their applications. Liq. Cryst. Display Technol. 1–68. https://doi.org/10.5772/intechopen.91889
5. Sharma, V., Kumar, P., Chinky, C., Gahrotra, R., Raina, K.K., Malik, P.: Effect of nano particles on electro optic properties of polymer dispersed liquid crystal in normal mode. In: AIP Conference Proceedings, vol. 2142, no. 1. AIP Publishing (2019). https://doi.org/10.1063/1.5122510

6. Sharma, V., Kumar, P., Chinky, C., Malik, P., Raina, K.K.: Morphological and electro optic studies of polymer dispersed liquid crystal in reverse mode. In: AIP Conference Proceedings, vol. 1953, no. 1. AIP Publishing (2018). https://doi.org/10.1063/1.5032969
7. Sharma, V., Kumar, P.: Electro-optically oriented Kerr and orientational phase study of normal mode polymer dispersed liquid crystals–effect of dispersion of nanoparticles. J. Mol. Liq. **348**, 118030 (2022). https://doi.org/10.1063/1.5032969
8. Dogra, A.R., Kumar, P., Sharma, V., Malik, P., Khanra, P., Kapoor, M.: Deposition of nanoparticles layer on ITO substrate and its application for alignment control of liquid crystal. In: AIP Conference Proceedings, vol. 2357, no. 1. AIP Publishing (2022). https://doi.org/10.1063/5.0080664
9. Dogra, A.R., Sharma, V., Malik, P., Kumar, P.: In-situ homeotropic alignment of dye doped liquid crystal molecules on multilayered self-assembled nanoparticles in confined cells for next generation display devices. Opt. Mater. **154**, 115731 (2024). https://doi.org/10.1016/j.optmat.2024.115731
10. Dogra, A.R., Sharma, V., Gahrotra, R., Khanra, P., Dogra, A.: Thickness dependent electro-optical behaviour of homeotropically aligned nematic liquid crystal. ECS Trans. **107**(1), 5021 (2022). https://doi.org/10.1149/10701.5021ecst
11. Adimule, V., Bathula, C., Sharma, K., Keri, R.: Study of structural, optical, dielectric properties poly-3-butyl thiophene reinforced with CO_3O_4 nanoparticles. Nano-Struct. Nano-Obj. **38**, 101150 (2024). https://doi.org/10.1016/j.nanoso.2024.101150
12. Adimule, V., Yallur, B.C., Keri, R., Bathula, C., Batakurki, S.: Microstructure, photoluminescence and electrical properties of SmxGd (1–x): SrO hybrid nanomaterials synthesized via facile coprecipitation method. Electron. Mater. Lett. **19**(3), 278–297 (2023). https://doi.org/10.1007/s13391-022-00394-0
13. Adimule, V., Keri, R., Rajendrachari, S., Kendrekar, P., Kulkarni, C.V.: Highly photoluminescence and wide band gap insulating metal hybrid nanoparticles array of samarium-doped SrO: CoO: synthesis, characterizations and sensor characteristics. J. Mater. Sci.: Mater. Electron. **34**(5), 442 (2023). https://doi.org/10.1007/s13391-022-00394-0
14. Pathinti, R.S., Tatipamula, A.K., Vallamkondu, J.: ZnO nanoparticles dispersed cholesteric liquid crystal based smart window for energy saving application. J. Alloy. Compd. **963**, 171198 (2023). https://doi.org/10.1016/j.jallcom.2023.171198
15. Li, X., Guo, Y., Zhang, M., Zhang, C., Niu, R., Ma, H., Sun, Y.: Colorable light-scattering device based on polymer-stabilized ion-doped cholesteric liquid crystal and an electrochromatic layer. ACS Appl. Mater. Interfaces. **15**(5), 7184–7195 (2023). https://doi.org/10.1021/acsami.2c17770
16. Gahrotra, R., Sharma, V., Dogra, A.R., Malik, P., Kumar, P.: Performance augmentation of bistable cholesteric liquid crystal light shutter-effect of dichroic dye on morphological and electro-optical characteristics. Opt. Mater. **127**, 112243 (2022). https://doi.org/10.1016/j.optmat.2022.112243
17. Sharma, V., Kumar, P., Sharma, A., Raina, K.K.: Droplet configuration control with orange azo dichroic dye in polymer dispersed liquid crystal for advanced electro-optic characteristics. J. Mol. Liq. **233**, 122–130 (2017). https://doi.org/10.1016/j.molliq.2017.03.012
18. Chauhan, G., Malik, P., Malik, P., Deep, A.: Improved performance of cadmium selenide quantum dots-doped polymer stabilized cholesteric liquid crystals for light shutter. Liq. Cryst. **50**(15), 2540–2551 (2023). https://doi.org/10.1080/02678292.2023.2264249
19. Köysal, O., Gleeson, H.F., Kocakülah, G.: A double-layer light shutter consisting of polymer dispersed liquid crystal and azo dye/quantum dot. Opt. Mater. (2024). https://doi.org/10.1016/j.optmat.2024.115645
20. Adimule, V., Sharma, V., Kumar, P., Keri, R., Sharma, K.: Perylene-3, 4, 9, 10-tetracarboxylic dianhydride dye intercalated liquid crystal molecules: Synthesis, spectroscopic characterizations and photoluminescence dynamics. J. Mol. Liq. (2024). https://doi.org/10.1016/j.molliq.2024.125990
21. Mathew, C., Pathinti, R.S., Pradhan, S.R., Gollapelli, B., Chithari, K., Ghosh, M., Ashok, N., Vallamkondu, J.: Enhanced security through fluorescent dye-doped cholesteric liquid

crystal shells for anti-counterfeiting. Opt. Mater. (2024). https://doi.org/10.1016/j.optmat.2024.115614
22. Chemingui, M., Yang, D., Wang, Y., Ayeb, H., Cai, W., Tang, A., Drevensek-Olenik, I., Zhang, X., Xu, J.: The influence of quantum dots on the optical properties of a room temperature cholesteric liquid crystal. Liq. Cryst. **49**(15), 2095–2107 (2022). https://doi.org/10.1080/02678292.2022.2103850
23. Lu, X., Zhou, Z., Ni, B., Li, H., Li, Y., Li, B., Liu, W., Wang, Y., Yang, Y.: Tuning the circularly polarized luminescence of polymer-stabilized cholesteric liquid crystal films using chiral dopants. J. Mater. Chem. C **10**(21), 8246–8253 (2022). https://doi.org/10.1039/D2TC00326K
24. Gollapelli, B., Suguru Pathinti, R., Vallamkondu, J.: Carbon quantum dots doped cholesteric liquid crystal films and microdroplets for anti-counterfeiting. ACS Appl. Nano Mater. **5**(8), 11912–11922 (2022). https://doi.org/10.1021/acsanm.2c02933
25. Kocakülah, G., Köysal, O.: Electro-optical and dielectric response of quantum dot-doped cholesteric liquid crystal composites. J. Mater. Sci.: Mater. Electron. **33**(8), 5489–5500 (2022). https://doi.org/10.1007/s10854-022-07739-3
26. Shu, Y., Lin, X., Qin, H., Hu, Z., Jin, Y., Peng, X.: Quantum dots for display applications. Angew. Chem. **132**(50), 22496–22507 (2020). https://doi.org/10.1002/ange.202004857
27. Pang, C., Deng, Y.H., Kheradmand, E., Poonkottil, N., Petit, R., Elsinger, L., Detavernier, C., Geiregat, P., Hens, Z., Van Thourhout, D.: Integrated PbS colloidal quantum dot photodiodes on silicon nitride waveguides. ACS Photonics **10**(12), 4215–4224 (2023). https://doi.org/10.1021/acsphotonics.3c00945
28. Patel, J., Jain, B., Singh, A.K., Susan, M.A.B.H., Jean-Paul, L.: Mn-doped ZnS quantum dots–an effective nanoscale sensor. Microchem. J. **155**, 104755 (2020). https://doi.org/10.1016/j.microc.2020.104755
29. Singh, G.: Recent advances on cadmium free quantum dots-liquid crystal nanocomposites. Appl. Mater. Today **21**, 100840 (2020). https://doi.org/10.1016/j.apmt.2020.100840
30. Challa, M., Yallur, B.C., Ambika, M.R., Adimule, V.: Influence of nano particles on optical properties of Cu-MOFs. Adv. Mater. Res. **1173**, 23–33 (2022). https://doi.org/10.4028/p-vn4hd4
31. Nandi, S.S., Adimule, V., Kadapure, S.A., Kerur, S.S.: Rare earth based nanocomposite materials for prominent performance supercapacitor: a review. Appl. Mech. Mater. **908**, 3–18 (2022). https://doi.org/10.4028/p-rff302
32. Nagalakshmi, G., Nandeesh, I.M., Yallur, B.C., Adimule, V., Batakurki, S.: Synthesis and optical properties of copper terephthalate metal organic frame works. Eng. Chem. **2**, 3–11 (2023). https://doi.org/10.4028/p-fdqs03
33. Adimule, V., Nandi, S., Sharma, K., Keri, R.: Microstructure, zeta potential and enhanced photocatalytic dye degradation performances of mesoporous flower like Nb_2O_5@ CeO_2 bimetallic oxide nanomaterials decorated with reduced graphene oxide. Inorg. Chem. Commun. (2024). https://doi.org/10.1016/j.inoche.2024.112592
34. Kats, E.I.: Optical properties of cholesteric liquid crystals. Soviet J. Exp. Theor. Phys. **32**(5), 1004–1007 (1971)
35. Coates, D.: Development and applications of cholesteric liquid crystals. Liq. Cryst. **42**(5–6), 653–665 (2015). https://doi.org/10.1080/02678292.2015.1020454
36. Bunning, T.J.: Cholesteric liquid crystals: properties and applications (2014). https://doi.org/10.1080/1358314X.2014.887505
37. Lukishova, S.G., Schmid, A.W., Supranowitz, C.M., Lippa, N., McNamara, A.J., Boyd, R.W., Troud Jr, C.R.: Dye-doped cholesteric-liquid-crystal room-temperature single-photon source. J Modern Opt. **51**(9–10), 1535–1547 (2004). https://doi.org/10.1080/09500340408235291
38. Mussa Farkhani, S., Valizadeh, A.: Three synthesis methods of CdX (X= Se, S or Te) quantum dots. IET Nanobiotechnol. **8**(2), 59–76 (2014). https://doi.org/10.1049/iet-nbt.2012.0028
39. Pu, Y., Cai, F., Wang, D., Wang, J.X., Chen, J.F.: Colloidal synthesis of semiconductor quantum dots toward large-scale production: a review. Ind. Eng. Chem. Res. **57**(6), 1790–1802 (2018). https://doi.org/10.1021/acs.iecr.7b04836

40. Xu, Z., Jiang, Y., Li, Z., Chen, C., Kong, X., Chen, Y., Zhou, G., Liu, J.M., Kempa, K., Gao, J.: Rapid microwave-assisted synthesis of SnO_2 quantum dots for efficient planar perovskite solar cells. ACS Appl. Energy Mater. **4**(2), 1887–1893 (2021). https://doi.org/10.1021/acsaem.0c02992
41. Adimule, V., Yallur, B.C., Gowda, A.H.: Advanced sensors based on carbon nanomaterials. In: Carbon Nanomaterials-Based Sensors, pp. 259–268. Elsevier (2022). https://doi.org/10.1016/B978-0-323-91174-0.00004-4
42. Kabasares, K.: Quantum Dot Confinement in Cholesteric Liquid Crystals. Doctoral dissertation, University of California, Merced (2017)
43. Gan, P., Zhang, X., Zhao, L., Shi, W., Cao, H., Wang, H., Yang, Z., Wang, D., He, W.: Broadband reflection in polymer-stabilized cholesteric liquid crystal film with zinc oxide nanoparticles film thermal diffusion method. Liq. Cryst. **48**(14), 1959–1968 (2021). https://doi.org/10.1080/02678292.2021.1909765
44. Kuo, C.Y., Emelyanenko, A.V., Hung, S.C., Chen, W.C., Liu, C.Y.: Improving the electro-optical properties of cholesteric liquid crystal devices via cellulose nanoparticle dopants. Polym. J. **56**(5), 541–551 (2024). https://doi.org/10.1038/s41428-023-00879-1
45. Bugakov, M.A., Samokhvalov, P.S., Shibaev, V.P., Boiko, N.I.: Hybrid fluorescent cholesteric materials with controllable light emission containing CdSe/ZnS quantum dots stabilized by liquid crystalline block copolymer. Opt. Mater. Express **11**(7), 1842–1851 (2021). https://doi.org/10.1039/D0RA02442B
46. Bobrovsky, A., Mochalov, K., Oleinikov, V., Sukhanova, A., Prudnikau, A., Artemyev, M., Shibaev, V., Nabiev, I.: Optically and electrically controlled circularly polarized emission from cholesteric liquid crystal materials doped with semiconductor quantum dots. Adv. Mater. **24**(46), 6216 (2012). https://doi.org/10.1002/adma.201202227
47. Aboulaich, A., Billaud, D., Abyan, M., Balan, L., Gaumet, J.J., Medjadhi, G., Ghanbaja, J., Schneider, R.: One-pot noninjection route to CdS quantum dots via hydrothermal synthesis. ACS Appl. Mater. Interfaces **4**(5), 2561–2569 (2012). https://doi.org/10.1021/am300232z
48. Cho, M.J., Park, H.G., Jeong, H.C., Lee, J.W., Jung, Y.H., Kim, D.H., Kim, J.H., Lee, J.W., Seo, D.S.: Superior fast switching of liquid crystal devices using graphene quantum dots. Liq. Cryst. **41**(6), 761–767 (2014). https://doi.org/10.1080/02678292.2014.889233
49. Mochalov, K.E., Bobrovsky, A.Y., Oleinikov, V.A., Sukhanova, A.V., Efimov, A.E., Shibaev, V., Nabiev, I.: Novel cholesteric materials doped with CdSe/ZnS quantum dots with photo-and electrotunable circularly polarized emission. In: Liquid Crystals XVI, vol. 8475, pp. 169–177 (2012). https://doi.org/10.1117/12.929729
50. Zhao, D., Xu, L., Shang, Y., Li, X., Guo, L.: Facet-dependent electro-optical properties of cholesteric liquid crystals doped with Cu_2O nanocrystals. Nano Res. **11**, 4836–4845 (2018). https://doi.org/10.1007/s12274-018-2072-2
51. Mani, S., Khosla, S., Sarawade, P.: Effect of quantum dots dispersion on the structural, optical, and thermal properties of liquid crystal system. Adv. Mater. Res. **1176**, 33–42 (2023). https://doi.org/10.4028/p-82i41e
52. Rodarte, A.L., Shcherbatyuk, G.V., Shcherbatyuk, L., Hirst, L.S., Ghosh, S.: Dynamics of spontaneous emission of quantum dots in a one-dimensional cholesteric liquid crystal photonic cavity. RSC Adv. **2**(33), 12759–12763 (2012). https://doi.org/10.1039/C2RA21167J
53. Lukishova, S.G., Winkler, J.M., Bissell, L.J., Stroud Jr, C.R.: Resonance in quantum dot fluorescence on a band-edge of a 1-D photonic bandgap cholesteric structure under cw-laser excitation. In: Emerging Liquid Crystal Technologies VIII, vol. 8642, pp. 173–179 (2013). https://doi.org/10.1117/12.2021405
54. Manhas, N., Adimule, V., Lakshminarayana, P., Keri, R., Kumar, L.S., Bathula, C.: Utilization of CO^{2+} doped Eu_2O_3 nanostructures and their effect on the dielectric and photoisomerization of liquid crystals. Phys. Fluids **35**(10) (2023). https://doi.org/10.1063/5.0171578
55. Kumar, A., Prakash, J., Khan, M.T., Dhawan, S.K., Biradar, A.M.: Memory effect in cadmium telluride quantum dots doped ferroelectric liquid crystals. Appl. Phys. Lett. **97**(16) (2010). https://doi.org/10.1063/1.3495780

56. Pathak, G., Hegde, G., Prasad, V.: Octadecylamine-capped CdSe/ZnS quantum dot dispersed cholesteric liquid crystal for potential display application: Investigation on photoluminescence and UV absorbance. Liq. Cryst. **48**(4), 579–587 (2021). https://doi.org/10.1080/02678292.2020.1799085
57. Yao, W., Lan, R., Li, K., Zhang, L.: Multiple anti-counterfeiting composite film based on cholesteric liquid crystal and QD materials. ACS Appl. Mater. Interfaces. **13**(1), 1424–1430 (2021). https://doi.org/10.1021/acsami.0c18132
58. Chen, L.J., Dai, J.H., Lin, J.D., Mo, T.S., Lin, H.P., Yeh, H.C., Chuang, Y.C., Jiang, S.A., Lee, C.R.: Wavelength-tunable and highly stable perovskite-quantum-dot-doped lasers with liquid crystal lasing cavities. ACS Appl. Mater. Interfaces **10**(39), 33307–33315 (2018). https://doi.org/10.1021/acsami.8b08474
59. Cao, M., Yang, S., Zhang, Y., Song, X., Che, Y., Zhang, H., Yu, Y., Ding, G., Zhang, G., Yao, J.: Tunable amplified spontaneous emission in graphene quantum dots doped cholesteric liquid crystals. Nanotechnology **28**(24), 245202 (2017). https://doi.org/10.1088/1361-6528/aa67ca
60. Kumar, R., Raina, K.K.: Optical and electrical control of circularly polarised fluorescence in CdSe quantum dots dispersed polymer stabilised cholesteric liquid crystal shutter. Liq. Cryst. **43**(7), 994–1001 (2016). https://doi.org/10.1080/02678292.2016.1155771
61. Mirzaei, J., Reznikov, M., Hegmann, T.: Quantum dots as liquid crystal dopants. J. Mater. Chem. **22**(42), 22350–22365 (2012). https://doi.org/10.1039/C2JM33274D
62. Bobrovsky, A., Samokhvalov, P., Shibaev, V.: An effective method for the preparation of stable LC composites with high concentration of quantum dots. Adv. Opt. Mater. **2**(12), 1167–1172 (2014). https://doi.org/10.1002/adom.201400215
63. Singh, G., Fisch, M.R., Kumar, S.: Tunable polarised fluorescence of quantum dot doped nematic liquid crystals. Liq. Cryst. **44**(3), 444–452 (2017). https://doi.org/10.1080/02678292.2016.1217357
64. Wang, Z., Cao, M., Shao, G., Zhang, Z., Yu, H., Chen, Y., Zhang, Y., Li, Y., Xu, B., Wang, Y., Yao, J.: Coherent random lasing in colloidal quantum dot-doped polymer-dispersed liquid crystal with low threshold and high stability. J. Phys. Chem. Lett. **11**(3), 767–774 (2020). https://doi.org/10.1021/acs.jpclett.9b03409